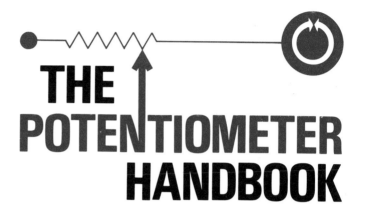

THE POTENTIOMETER HANDBOOK

ACKNOWLEDGEMENTS

PHOTOGRAPHS CONTRIBUTED BY:

AMI MEDICAL ELECTRONICS

BELL AEROSYSTEMS

BIDDLE CO., JAMES G.

CALSPAN CORP.

CENTRAL SCIENTIFIC CO.

CETEC, INC.

DUNCAN ELECTRONICS, INC.

GENERAL RADIO CO.

HEWLETT-PACKARD

INTERSTATE ELECTRONICS CORP.

KRAFT SYSTEMS, INC.

LEEDS AND NORTHRUP

POWER DESIGNS, INC.

SPACE-AGE CONTROL, INC.

TEKTRONIX, INC.

WEYERHAUSER CO.

WILLI STUDER, SWITZERLAND

INDUSTRY STANDARDS

REPRINTED BY PERMISSION OF THE VARIABLE RESISTIVE COMPONENTS INSTITUTE

THE POTENTIOMETER HANDBOOK

USERS' GUIDE TO COST-EFFECTIVE APPLICATIONS

Written for

BOURNS, INC.

By

CARL DAVID TODD, P.E.
CONSULTING ENGINEER

ASSOCIATE EDITORS

W. T. HARDISON
Product Marketing Specialist
Trimpot Products Division
Bourns, Inc.

W. E. GALVAN
Applications Engineer
Trimpot Products Division
Bourns, Inc.

McGRAW-HILL BOOK COMPANY

New York   St. Louis   San Francisco   Auckland   Düsseldorf
Johannesburg   Kuala Lumpur   London   Mexico   Montreal
New Delhi   Panama   Paris   São Paulo   Singapore
Sydney   Tokyo   Toronto

Library of Congress Cataloging in Publication Data

Todd, Carl David.
   The potentiometer handbook.

   1. Potentiometer—Handbooks, manuals, etc.
2. Electric resistors—Handbooks, manuals, etc.
I. Bourns, inc.   II. Title.
TK7872.P6T63   621.37'43   75-20010
ISBN 0-07-006690-6

234567890 MUBP 78432109876

PREFACE

In the decades following the advent of the transistor, electronic technology experienced explosive growth. Thousands of new circuits were generated annually. The demand for variable resistive components to adjust, regulate or control these circuits shared in the expansion.

The number of applications for variable resistive components has increased significantly. This is contrary to predictions of a few soothsayers of the '60's who interpreted the miniaturizing effects of integrated circuit technology as a threat to these components. Potentiometers will continue to enjoy strong growth into the foreseeable future. This optimistic forecast is particularly true in consumer and industrial applications where potentiometers provide the cost-effective solution in trimming applications and the ever-present necessity of control for man-machine interface.

Many articles, booklets, and standards have been published on potentiometers; yet, there is no single, comprehensive source of practical information on these widely used electronic components. It is this void that The Potentiometer Handbook is intended to fill.

One objective of this handbook is to improve communications between potentiometer manufacturers and users. To this end, explanations of performance specifications and test methods, are included. Common understanding of terminology is the key to communication. For this reason, lesser known as well as preferred terminology are included, with emphasis on the latter. Hopefully, this will create the base for easy, accurate dialogue. Over 230 photos, graphs and drawings illustrate and clarify important concepts.

This book assumes the reader has a knowledge of electronic and mathematical fundmentals. However, basic definitions and concepts can be understood by nontechnical personnel. The major portion of this text is written for systems and circuit designers, component engineers, and technicians as a practical aid in design and selection. It is an important reference and working handbook oriented towards practical application ideas and problem solving. For the student, it introduces the basic component, its most common uses, and basic terminology.

Enough objective product design and manufacturing process information is in the text to allow the user to understand basic differences in materials, designs, and processes that are available. This will sharpen his judgment on 'cost-versus-performance' decisions. Thus, he can avoid over-specifying product requirements and take advantage of the cost-effectiveness of variable resistive devices.

Also included are hints and design ideas compiled over the years. As with any discipline, these guidelines are often discovered or developed through unfortunate experience or misapplication. Most chapters conclude with a summary of key points for quick review and reference.

Speaking of misapplication, Chapter 9, To Kill a Potentiometer, is a tongue-in-check *pot*pourri of devious methods to wipe out a potentiometer. This is a lighthearted approach to occasional serious problems caused by human frailties. Not much more need be said except *before* all else fails, *read the book,* or at least this chapter!

Suggestions from readers on improving this volume are encouraged and welcomed. Subsequent editions will include the results of these critiques together with advanced material relating to the state-of-art in potentiometer design and application.

W. T. Hardison

CONTENTS

115 Chapter SIX — APPLICATION AS A PRECISION DEVICE

The potentiometer in systems requiring *high* accuracy. Importance of electro-mechanical parameters in precision applications. Applications in systems and circuits where accuracy is the most important consideration.

LIST OF ILLUSTRATIONS

INTRODUCTION TO POTENTIOMETERS

Chapter 1

CONTROLS FOR FLOW OF ELECTRONS

. . . Long before the mad search for the philosopher's stone or the formula for transmutation of base metals into gold by medieval alchemists, the speculative Greek philosophers had contemplated upon the structure of matter. One, Empedocles brought forth the theory of the structure of matter from the four elements of earth, air, fire and water, but these he subordinated, as complex products composed of primordial indestructible atoms, which were animated by love and hatred. Strangely, our present understanding of the structure of matter could be described in much the same words as these, except that the four elements are now 92 and the indestructible atoms are unit charges of electricity — protons and electrons. Instead of being animated by love and hatred, as Empedocles thought, they are motivated by the repulsion or attraction between like and unlike electrical charges. . . .

Electrons move readily through some substances, called conductors, and scarcely at all through others, called resistors. This happy property of substances, therefore, provides a means by which electronic pressures (voltage) may be controlled by the introduction of resistors of proper dimensions and characteristics into the electrically conducting circuit.

Central Scientific Co., Chicago, Ill.

HISTORICAL BACKGROUND

The italicized quote above is taken from an early 20th century catalog. This particular manufacturer used this bit of technical history as an introduction to variable resistive devices of the type shown in Fig. 1-1, but the history of variable resistive devices is known to predate the turn of the century by more than thirty years.

When Galvani and Volta discovered that elec-

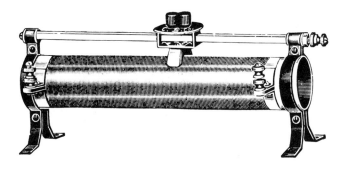

Fig. 1-1 Early 20th century slide-wire rheostat
(Central Scientific Co.)

tricity could be produced by chemical means (c. 1800) they probably gave little thought to in-circuit variability of parameters. However, by the time Ohm presented his famous law in 1827, the first crude variable resistive devices were no doubt being constructed by physicists in all parts of the world. Though its origin can be debated, one certainty is that early forms of variable resistance devices bore very slight resemblance to those available and accepted as commonplace by today's engineer. In the late 19th century, they were found only in laboratories and were large bulky instruments.

One of the earliest devices was a carbon pile shown in Fig. 1-2. Each carbon block was about two inches square and a quarter of an inch thick. An insulated tray held the blocks. Metal blocks, placed anywhere in the stack or pile, provided terminals for connection to external circuitry. Minor adjustment of resistance was accomplished by varying the mechanical pressure exerted by the clamping action of a screw going through one end of the tray and pressing on the metal block at the end of the stack. As the pressure was increased, the carbon blocks were forced closer and closer together, thus reducing the contact resistance from one block to the next, causing the overall resistance from end to end to be decreased. Major changes of resistance could be accomplished by removing some of the carbon blocks and substituting more conductive metal blocks in their place. It was also possible to place terminal-type metal blocks at intermediate points between the ends of the stack to achieve tapping and potential divider applications. This early form, in slightly different configurations, was used for many years.

A later model (c. 1929) is shown in Fig. 1-3. This model offered many improvements over its predecessors. Improvements such as higher wattage dissipation (note cooling fins), wider adjustment range and stability of resistance at high resistance values where blocks are relatively loose.

Many sewing machine motor speed controls in the 1940's used carbon piles of half-inch discs which were only about a sixteenth of an inch thick. In this form, a mechanical linkage from a foot pedal to the pile allowed the operator to vary the pressure on the pile and hence the speed of the motor. The carbon pile is still in use today in such places as telephone circuits and experimental laboratories.

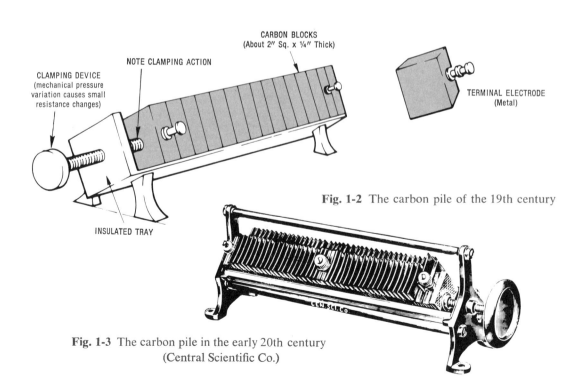

CARBON BLOCKS
(About 2″ Sq. x ¼″ Thick)

NOTE CLAMPING ACTION

CLAMPING DEVICE
(mechanical pressure variation causes small resistance changes)

TERMINAL ELECTRODE
(Metal)

INSULATED TRAY

Fig. 1-2 The carbon pile of the 19th century

Fig. 1-3 The carbon pile in the early 20th century
(Central Scientific Co.)

Another early form of variable resistance device consisted of a length of resistance wire and a sliding contact as shown in Fig. 1-4. The total resistance between A and B could be varied by choosing different types of materials for the wire or by varying the geometrical properties of the wire. It was probably in this simple configuration that early devices originally found their way into measuring instruments of the type shown in Fig. 1-5.

The purpose of this instrument was to measure unknown potentials such as E_x in Fig. 1-5. Two variable resistive devices, R1 and R2, were employed in this circuit. Note that a meter stick was placed adjacent to R2 and served as a scale to determine relative settings of R2's sliding contact. For proper operation, E1 had to be greater than E2 and E2 had to be greater than E_x. The instrument was initially calibrated by placing R2's sliding contact to the full scale (B) position and, with S1 in the calibrate position, was adjusted for a zero on M1 while S2 was being depressed. What was taking place during the calibration procedure was that the voltage across R2 imposed by E1 was being made equal to the voltage across R2 imposed by E2. When this condition was achieved no current flowed in the

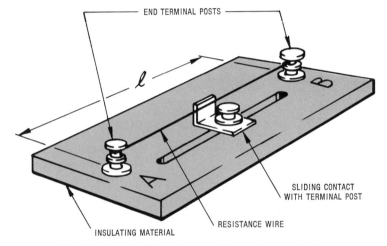

Fig. 1-4 Simple slide-wire variable resistance device

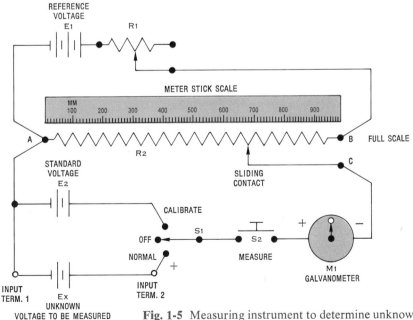

Fig. 1-5 Measuring instrument to determine unknown voltage

section of the circuit containing M_1 and its reading was therefore zero. After the calibration sequence, S_1 was placed in the normal position and the circuit was then ready to measure unknown voltages of magnitudes less than E_2. If an unknown voltage was present at the input terminals 1 and 2, then M_1 would deflect either plus or minus with respect to the calibrated zero.

If the deflection was in the positive direction, then the sliding contact of R_2 could be moved from terminal B toward A until M_1 returned to zero. The value of E_X was then calculated from:

$$E_X = \frac{E_2 R_{AC}}{R_{T2}}$$

where E_2 was the standard voltage (volts), R_{T2}' was the total resistance of R_2 (ohms) and R_{AC} was that portion of R_2's resistance between terminals A and C (ohms).

The unknown voltage could have been determined using the meter stick. If the sliding contact was at 700 mm after the circuit was nulled with the unknown voltage in the circuit, the ratio of R_{AC} to R_{T2}' is:

$$\frac{R_{AC}}{R_{T2}} = \frac{7}{10} = .7 \text{ and}$$

$$E_X = .7\ E_2$$

An even simpler method would have been to calibrate the meter stick in volts and read the unknown voltages directly.

If the galvanometer deflection was in the negative direction, this indicated that E_X was larger than E_2 and therefore was beyond the measuring capability of the instrument. This circuit has been greatly simplified, but there is little doubt that due to this type of application in a *potential* measuring *meter,* the variable resistive device became universally known as the potentiometer.

In the electronics industry today, the term potentiometer has come to mean a component which provides a variable tap along a resistance by some mechanical movement rather than an entire measurement system. However, the basic potentiometer configuration described by Fig. 1-5 is still in use today but utilizes a spiral or helix of linear resistance wire in order to increase its practical length and thus its range and accuracy. Fig. 1-6 is a photograph of a modern commercial instrument using this approach.

Problems of getting enough resistance in a practical amount of space led an inventor named George Little to develop and patent what he called an "Improvement in Rheostats or Resistance Coils" in 1871. This was a structure in which insulated resistance wire was wound around an insulated tube or mandrel in a tight helix as shown by the copy of his patent drawing in Fig. 1-7. The moving slider made contact with the resistance wire along a path where the insulation had been buffed off. It was probably this patent which eventually lead to the style of devices previously shown in Fig. 1-1.

In 1907, H. P. MacLagan was awarded a

Fig. 1-6 Modern instrument for precision ratio measurement (Leeds & Northrup)

4

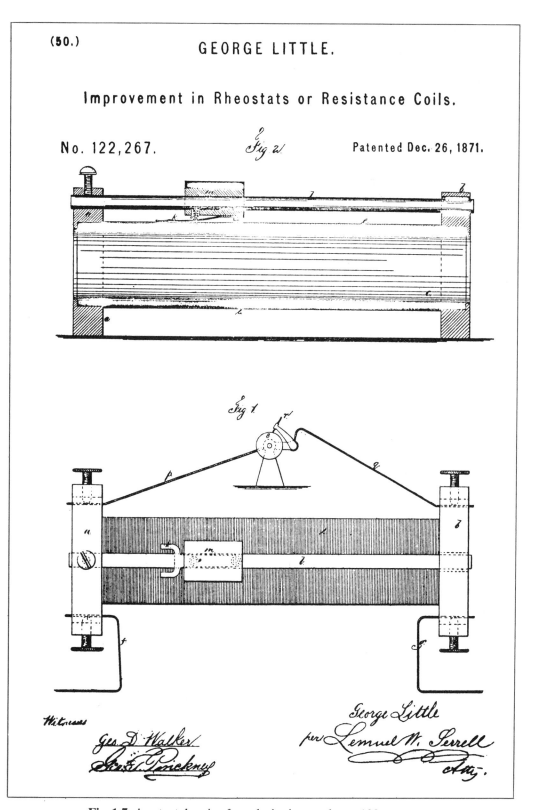

Fig. 1-7 A patent drawing for a device invented over 100 years ago

patent for a rotary rheostat. Fig. 1-8 is a copy of his patent drawing. He had wound the resistance wire around a thin fibreboard card and then formed the assembly into a circle. A wiper, attached to a center post, made contact with the resistance wire on the edge of the card.

The radio era (1920-1940) created a demand for smaller components. Of course, the potentiometer was no exception and the need grew for smaller potentiometers to be used in applications such as volume controls. Resistance materials of wire and carbon were used with the carbon devices proving to be more easily produced in large quantities. The general requirements for the radios of that period were not at all stringent and the carbon volume control became common.

Electronic applications grew by leaps and bounds during World War II, and so did the need for more and better variable resistance devices to permit control, adjustment, and calibration. Components manufacturers strived to improve their products and lower their cost. Of significant note was the development of the first commercially successful 10-turn precision potentiometer by Arnold O. Beckman. He filed patent applications for improvements over earlier efforts in October of 1945. A drawing from the resulting patent is shown in Fig. 1-9.

The post-war years saw the commercializing of television and growth in the commercial aircraft industry. Airborne electronics applications, as well as other critical weight-space needs made size a critical factor.

In May, 1952, Marlan E. Bourns developed a highly practical miniature adjustment potentiometer for applications where infrequent control adjustment was needed. He had combined the advancing technologies of plastic molding and precision potentiometer fabrication and provided the designer with a small adjustment potentiometer with outstanding electrical performance. A copy of his patent drawing is shown in Fig. 1-10.

As the demand for small adjustment devices increased, other manufacturers began to produce similar units. Since the introduction of the miniature adjustment potentiometer, many improvements have been made, yielding better and better performance at lower and lower costs. Fig. 1-11 is a condensed portrayal of adjustment potentiometers available today.

Many of the improvements in the precision potentiometer development came about as a result of their increasing use in analog computers as well as in more complex and precise servo systems. Again and again, potentiometer manufacturers have improved their products to meet the needs of the designers in a continuing process of development.

PRACTICAL DEVELOPMENT OF THE POTENTIOMETER

Let's consider some of the practical factors in building potentiometers. Assume, for a moment, that the potentiometer as you know it does not exist. Then you will proceed to develop it, guided by a high degree of prior knowledge. Initially, you recognize that you need some form of component resistor which has a variable tap whose position can be changed by mechanical motion.

As a start, stretch a piece of uninsulated resistance wire between two terminals. You can now fashion some type of clamp to make contact with the wire at any point between the terminals. The result might look very similar to the device in Figure 1-4 shown previously.

A fundamental equation describing the total resistance, R_T, from A to B is:

$$R_T = \frac{\rho l}{S}$$

Where ρ is the resistivity, given in ohms - centi - meters . The length, l, of the wire is measured in centimeters and S is the cross sectional area of the wire expressed in square centimeters. The calculated R_T will then be given in ohms.

Thus, in order to get a larger value of resistance, either the resistivity or length must be increased, or you might choose to decrease the area. The choices of resistivity are somewhat limited, and increasing the length very quickly produces a bulky and quite impractical component. Using a smaller wire likewise has its problems of increased fragility and difficulty in making proper terminations and contact with the sliding tap.

One way to increase the length of the wire in a practical manner is to wind it around some form of insulating material or mandrel. This could take the form of a fibreboard tube as shown in Fig. 1-12 or a flatter strip of material as shown in Fig. 1-13. A study of either of these potentiometer configurations reveals several possible problems.

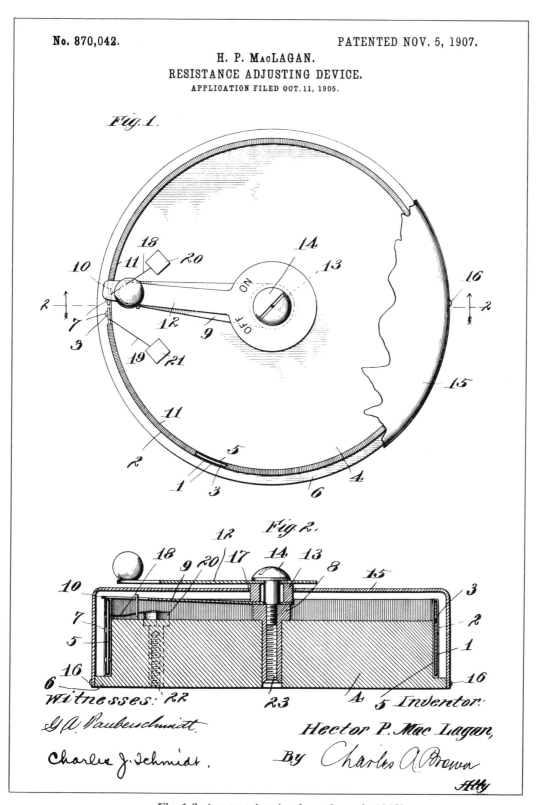

Fig. 1-8 A patent drawing from the early 1900's

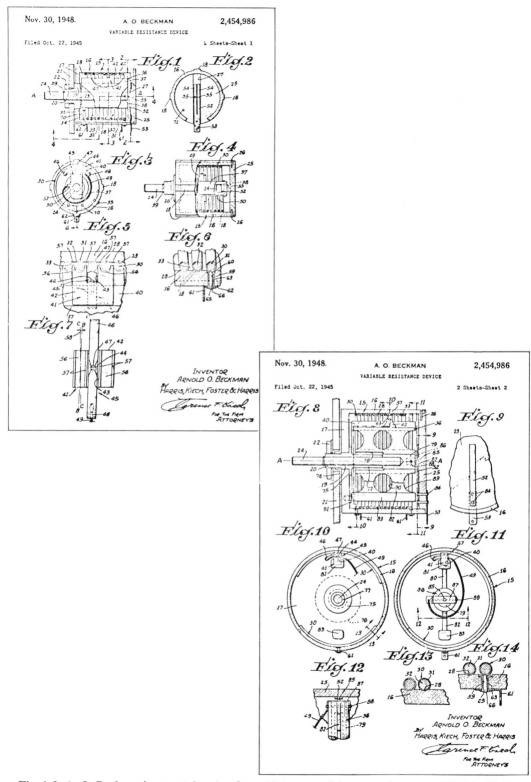

Fig. 1-9 A. O. Beckman's patent drawing for a 10-turn precision potentiometer. Filed in 1945.

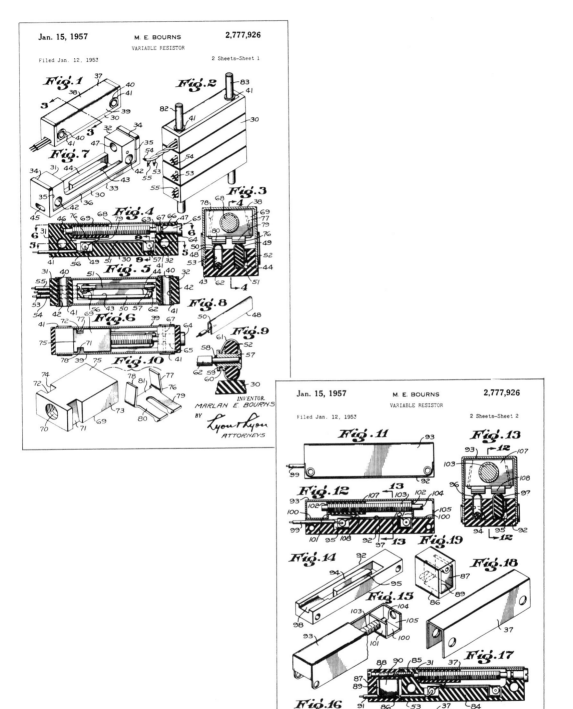

Fig. 1-10 Marlan E. Bourns' patent drawing for a practical miniature adjustment potentiometer. Filed in 1953.

Fig. 1-11 Adjustment potentiometers of today

First of all, the turns of wire need to be close together to prevent any discontinuities with the sliding contact. This presents another problem of possible shorting from one turn to the next. You can use a very light insulation on the wire such that adjacent turns will not short together but which may be easily removed in the path of the sliding contact.

Secondly, unlike our previous straight wire potentiometer, this new version will not permit a smooth and continuous change in the tap position. Now, the tap will electrically jump from one turn to the next with no positions allowed in between, The larger the cross section of the man-

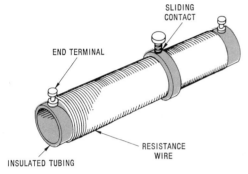

Fig. 1-12 Winding resistance wire on insulated tube allows longer wire in a practical package

drel the greater the resistance, but the greater the jumps will be.

In addition, if you want the relative position of the sliding contact to produce an equivalent change in the effective electrical position of the sliding tap, then you must be very careful to wind the coil of resistance wire uniformly in both tension and spacing throughout the entire length. End terminations must be made and positioned very carefully. You normally would want the extreme mechanical positions to correspond to the electrical ends of the total resistance.

Fig.1-13 A flat mandrel could be used

If the flat mandrel of Fig. 1-13 is curved as shown in Fig. 1-14, two benefits result. First, you can have a longer effective mandrel with less bulk. Then you can easily pivot the sliding contact from a post in the center. Attaching the slider arm to a shaft will allow convenient rotary motion to control the position of the arm.

You may curve the round mandrel potentiometer of Fig. 1-12, if the mandrel's diameter is kept relatively small. A round mandrel is more easily wound and a small size also means that the jumps or steps in resistance as a sliding contact moves from one turn to the next will be less. Furthermore, the length of the mandrel may be curved in the form of a helix as shown in Fig. 1-15. This will allow a long mandrel to be con-

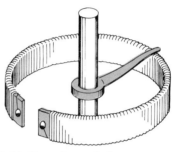

Fig. 1-14 Curved mandrel saves space and allows rotary control

fined to a relatively small space. The helical configuration requires more complicated mechanics to control the position of the slider arm, but the overall performance makes it worth the trouble.

So far, in this imaginary development of potentiometers, only wire has been considered for the resistance element. Other materials are usable that offer advantages but not without introducing some new problems.

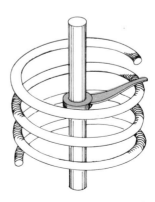

Fig. 1-15 Shaping mandrel into helix puts long length in small space

A resistive element made from a carbon composition material as illustrated in Fig. 1-16A could have a much higher resistance than is possible with wire. In addition, since the element is not coiled, you no longer have to tolerate jumps in the output as you did with wirewound potentiometers. A third benefit comes from the greater ease (less friction) with which the slider can move over the composition element and the corresponding reduced wear which results. A catastrophic failure can occur in the wirewound potentiometer when a single turn is worn through or otherwise broken, but a composition element can continue to function in reduced performance even though extremely worn. Other types of composition elements are shown in Fig. 1-16B and 1-16C.

If you carefully test the composition potentiometer and compare its performance with that of the wirewound versions, you will find that there are several new problems. All of these relate to the properties and nature of carbon compositions. The overall resistance will not be as stable with time and temperature. You may notice that it is even more difficult to get a perfectly uniform change in electrical output of the sliding tap with variation in mechanical position. Terminations are more difficult to make with the composition element. Although potentiometer manufacturers do form single turn units and even helical structures using mandrels coated with a composition material, it is a more complex and critical process than in the case of wirewound devices.

A special problem occurs in developing a variable resistance device for use in a particular application where one of the prime considerations is its setability or adjustability (ease and precision with which output can be set on desired value). In the simplest design configurations, you may find it somewhat difficult to set the potentiometer slider at some exact spot. If you have a unit with linear travel of the sliding contact as in Fig. 1-13, consider adding a lead screw arrangement such as shown in Fig. 1-17. Now, many turns of the lead screw will be required to cause the sliding contact to go from one end to the other. This mechanical advantage means that it will be easier to set the movable contact to any point along the resistive element. Be careful that no excessive play or mechanical backlash exists in the mechanism. This would make it impossible to instantly back the slider up, for a very small increment, if you turn the lead screw past the intended location.

A similar mechanical improvement to the rotary configuration of Fig. 1-14 would be the addition of some form of worm gear. The adjusting screw would be the driving gear and produce a smaller rotation of the main driven gear which would be attached to the shaft controlling the

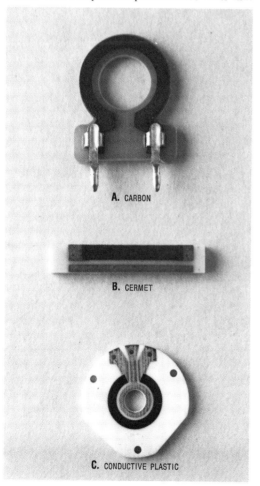

Fig. 1-16 Resistive elements of composition materials

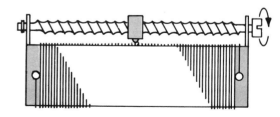

Fig. 1-17 A simple lead screw aids setability

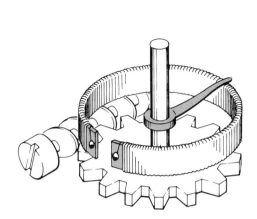

Fig. 1-18 A worm gear may be added to the rotary pot

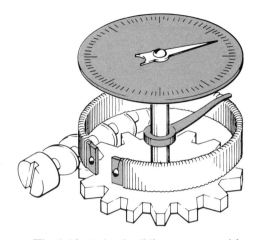

Fig. 1-19 A simple sliding contact position indicating device

sliding contact arm. The end result might look something like that shown in Fig. 1-18.

Further applications of variable resistive devices might require that the relative position of the sliding contact be known to a degree of accuracy better than a simple direct visual estimation. For example, a potentiometer may be used to control the speed of a motor at a location remote from the control center. Some form of indicator is required on the potentiometer so that motor speeds are predictable and accurately repeatable.

The meter stick served as an indicator in the potentiometer circuit arrangement of Fig. 1-5. The scale could have been calibrated in any units desired depending on the particular application involved. A simple indicator for the rotary unit of Fig. 1-18 can be constructed by attaching an appropriately divided scale to the

unit and connecting a pointer to the shaft driving the sliding contact. The result is shown in Fig. 1-19.

Simple dials will not provide adequate accuracy of setability for all applications. More complex mechanisms, such as shown in Fig. 1-20, have been developed by potentiometer manufacturers to meet the constantly increasing demands of the electronics industry.

Thus, in something over 100 years, resistance adjusting devices have evolved from bulky crude rheostats to a whole family of diverse products. Their use has spread from experimental laboratory to sophisticated electronics and critical servomechanisms and even inexpensive consumer items. In fact, most segments of the economy are served by variable resistor devices. Information applied from the following pages will help them serve even more effectively.

Fig. 1-20 Accurate devices for sliding contact position indication

GENERIC NAMES
AND TRADEMARKS

Many common terms used to name variable resistive devices have evolved over the years. Some of them relate to certain applications and will be used in that context later. The more common generic names are listed in Fig. 1-21.

Commercialization of potentiometers has resulted in a proliferation of trademarks in the United States and foreign countries. Manufacturers frequently register their trademarks in the United States Patent Office and identify them with a ® or a statement that they are registered.

Trademarks serve to assure the buyer that certain quality characteristics inherent with a specific manufacturer have been built into the product. It is the reputation behind the trademark that makes it meaningful to the buyer and the user.

Well-known trademarks are usually policed with zeal by their owners. This helps assure that they are not misused and do not fall into common or generic usage which would weaken their value to the public and the manufacturer. The general rule is that a manufacturer's trademark should be used as a *modifier* of the generic name for a product of the manufacturer. For example: TRIMPOT® potentiometers, not trimpots.

adjustable resistors	precisions
adjustment potentiometers	rheostats
adjustments	servo-potentiometers
attenuators	servo-pots
controls	transducer
feedback resistors	trimming potentiometers
gain controls	trimmers
impedance compensators	tweakers
level controls	variable resistive devices
potentiometers	variable resistors
pots	volume controls
precision potentiometers	

Fig. 1-21 Generic names

Notes

Notes

ELECTRICAL PARAMETERS

Chapter 2

"I often say that when you can measure what you are speaking about, and express it in numbers, you know something about it; but when you cannot express it in numbers, your knowledge is of a meagre and unsatisfactory kind; it may be the beginning of knowledge, but you have scarcely, in your thoughts, advanced to the stage of Science, whatever the matter may be."

Lord Kelvin

INTRODUCTION

Electrical parameters are those characteristics used to describe the function and performance of the variable resistive device as a component. These parameters can be demonstrated using simple electronic measurement methods.

Understanding these terms is fundamental to effective communication of application needs and cost-effective product selection. A thorough understanding of this material will aid in interpreting potentiometer manufacturer's data sheets and thus accomplish one of the aims of this book. A summary of electrical parameters is shown in Figure 2-41 for handy reference. Mechanical and environmental specifications can be found in the application chapters.

This chapter is organized for each parameter as follows:

— Definition
— Examples of typical values
— Detailed explanation of factors contributing to the parameter
— Simple electronic circuit to demonstrate the parameter (not for inspection or quality control)

After reading this chapter, further insight into these parameters can be gained by reading the industry standards reproduced in Appendix

I. The Variable Resistive Components Institute (VRCI) has published these standards for precision and trimming potentiometers. Their purpose is to establish improved communication between manufacturer and user. VRCI test circuits are regarded as the industry's standard, while the ones in this chapter are only study aids.

Figure 2-1 is the basic schematic of the potentiometer. This is usually used to show the device in a circuit or system.

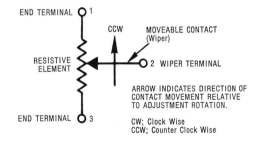

Fig. 2-1 Fundamental schematic representation of variable resistive device

17

TOTAL RESISTANCE, TR

Total resistance, TR, is a simple parameter defined as the resistance between the end terminals of a potentiometer. The end terminals are shown as 1 and 3 in Fig. 2-1.

Total resistance is always specified as a nominal value in units of ohms. A plus and minus percent tolerance from the nominal value is also specified. For example. $10\Omega\pm5\%$, $10K\Omega\pm10\%$, and $100\Omega\pm20\%$.

TR is always specified when defining any potentiometer. It is known by several names including: *value* of the potentiometer, *maximum resistance* or simply the *resistance*.

The major contributor to total resistance is the potentiometer's resistive element. The material and methods used to construct the element determine its resistance. The resistance of the terminals or leads of the potentiometer and the resistance of the termination junctions contribute to total resistance.

A digital ohmmeter is a convenient and accurate device for measuring TR. It is connected to the end terminals of the potentiometer as shown in Fig. 2-2. Total resistance is read directly from the display.

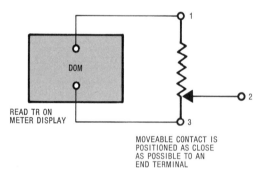

READ TR ON
METER DISPLAY

MOVEABLE CONTACT IS
POSITIONED AS CLOSE
AS POSSIBLE TO AN
END TERMINAL

Fig. 2-2 Fundamental measurement of total resistance

Note in Fig. 2-2 that the potentiometers moveable contact (wiper) is positioned as close as mechanically possible to one of the units end terminals. If the potentiometer were a continuous rotation device, i.e. no mechanical end-stops provided, the wiper would be adjusted to a point completely off of the resistive element. These wiper positions are industry standard test conditions. They are chosen not only to minimize the wiper effect on the TR measurement but also to improve data correlation. For example, when comparing TR measurements taken at different times or from different units, it is known that the wiper was in exactly the same position during each measurement.

Industry standard test conditions specify a maximum voltage for TR measurement. This voltage restriction is necessary to limit the power dissipation in the resistive element. The heating effects of power dissipation will affect the TR measurement. By restricting the test voltage, this heating effect is minimized.

ABSOLUTE MINIMUM RESISTANCE, MR

Absolute minimum resistance, MR, or simply minimum resistance, is the lowest value of resistance obtainable between the wiper and either end terminal.

Minimum resistance is always specified as a maximum. This seems contradictory but the specification is a level of resistance at or *below* which the wiper can be set. For example, 0.5 ohm maximum, or 1.0% maximum, (of total resistance). The design and construction of the potentiometer determines the magnitude of MR. Contact resistance, materials, and termination junctions all may contribute to MR.

For many potentiometers, MR is found when the moveable contact is set at the mechanical end stop near an end terminal. Other designs will exhibit minimum resistance when the wiper is slightly remote from the end stop. Fig. 2-3A shows an example of the latter. Depending on potentiometer design, a termination tab is clipped on or welded to the end of the resistive element. Many turns of resistance wire are bridged by this tab so that resistance within this area is low. The chance of potentiometer failure, due to one wire breaking or loosening, is minimized resulting in higher reliability and longer life. Note that some of the turns between the end-stop and the termination point are not bridged by the termination tab.

As the moveable contact is positioned along the resistive element, the minimum resistance will be achieved when the contact is closest to the termination tab, position A in Fig. 2-3A. If the contact is moved away from position A, in either direction, the resistance between the moveable contact terminal and the reference end terminal will increase. The curve in Fig. 2-3B together with the schematic of Fig. 2-3C serve to further clarify this important parameter.

A wirewound resistive element was chosen in the previous paragraph to demonstrate minimum resistance. Wirewound units often use the construction technique described. However, the occurrence of absolute minimum resistance at a point remote from the end stop is not exclusive to wirewound construction. Some potentiometers utilizing non-wirewound elements will have

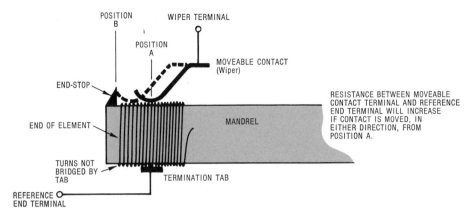

A. WIREWOUND ELEMENT AND TERMINATION

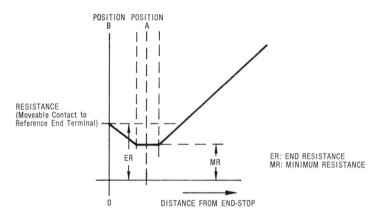

B. DISTANCE FROM END OF ELEMENT vs. RESISTANCE BETWEEN MOVEABLE
CONTACT AND REFERENCE END TERMINAL.

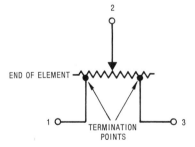

C. SCHEMATIC DRAWN TO ILLUSTRATE TERMINATION POSITION
RELATIVE TO ELEMENT END.

Fig. 2-3 Illustration of minimum resistance and end resistance

their minimum resistance at a point different from the end stop position.

Fig. 2-4 is one possible MR demonstration circuit. With a hookup as shown, the wiper is positioned to a point that gives the minimum resistance reading on a digital ohmmeter.

When measuring MR, the test current must be no greater than the maximum wiper current rating of the potentiometer. High current can cause errors and will damage the potentiometer.

Caution: Never use a conventional volt-ohm-milliammeter, VOM, to measure resistance parameters of a potentiometer.

For the minimum resistance condition the wiper is near one end terminal. Little or no resistance is in the test circuit. In this case, overheating and burn-out can occur even at a *low* voltage.

Since MR is specified as a maximum, production testing can use pass-fail instrumentation. Industry standard test conditions require a special wiper positioning device for fast and accurate adjustment. See Fig. 2-5.

END RESISTANCE, ER

End resistance, ER, is the resistance measured between the wiper and a reference end terminal when the contact is positioned against the adjacent end stop. See position **B** in Fig. 2-3A and 2-3B.

End resistance and minimum resistance are sometimes confused. This is because in many potentiometers the two parameters are, in fact, identical values obtained with the moveable contact in the same position. The only reason for having two parameters relates to the construction technique, which may cause an absolute minimum resistance separate and distinct from the end resistance. Continuous rotation devices have no end stops and therefore, ER is not specified.

End resistance is expressed in terms of a maximum ohmic value or a maximum percentage of the unit's TR. It is common practice for potentiometer manufacturers to specify MR rather than ER.

The test circuit of Fig. 2-4 is perfectly suited to end resistance measurement. All of the cautions outlined for MR measurement in the previous section apply to the measurement of ER.

MINIMUM AND END VOLTAGE RATIOS

Because a potentiometer is sometimes used as a voltage divider, explained in Chapter 3, manufacturers' catalog sheets and components engineers will often specify a minimum voltage and/or an end voltage ratio. End voltage ratio is sometimes referred to as *end setting*. Typical values range from 0.1% to 3.0%.

Fig. 2-6 is a circuit that can be used to demonstrate a potentiometer's minimum and end voltage ratios. Current and voltage levels should only be sufficient to facilitate measurement. In no case should the devices' maximum ratings

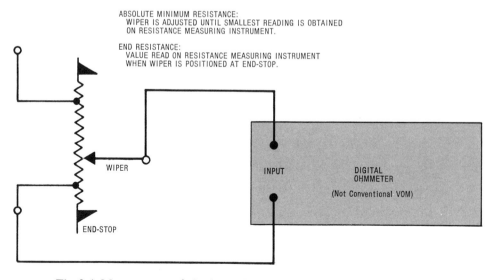

Fig. 2-4 Measurement of absolute minimum resistance and end resistance

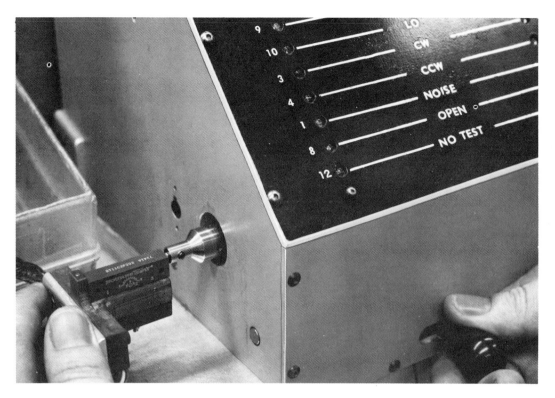

Fig. 2-5 Production testing of absolute minimum resistance

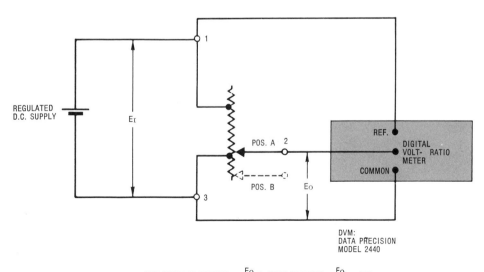

$$\text{DVM DISPLAYS READING} = \frac{E_O}{E_I}\% \quad \text{VOLTAGE RATIO} = \frac{E_O}{E_I} \times 100$$

Fig. 2-6 Measurement of minimum voltage ratio and end voltage ratio

be exceeded. The digital voltmeter shown in Fig. 2-6 displays the ratio of the two voltages present.

To read minimum voltage ratio, the wiper is positioned to give the smallest ratio indication on the DVM. Note that this is position A in Fig. 2-6 and it exactly corresponds to the minimum resistance wiper position. Similarly, if the wiper is positioned against the end stop of terminal 3, position B in Fig. 2-6, the DVM will display the end voltage ratio.

Some potentiometers are constructed using two parallel electrical paths. One path, the resistive element, is connected to the potentiometer's end terminals. The other path, a low resistance collector, is connected to the wiper terminal. When the moveable contact is actuated, it moves along the two paths, making contact with both. This construction and schematic are shown in Fig. 2-7.

For most potentiometer designs, the total resistance of the collector is less than one-half ohm, but it may be as high as two ohms. Assume a unit of the type shown in Fig. 2-7 is tested for its minimum or end-setting characteristics. The reading using end terminal 3 will be greater than the one using end terminal 1. This higher resistance is due to the collector's resistance in series with wiper terminal 2. This small resistance can be very significant in potentiometers of low total resistance.

Chapter 7 provides a detailed discussion of various potentiometer constructions.

CONTACT RESISTANCE, CR

A potentiometer's contact resistance, CR, is the resistance that exists in the electrical path from the wiper terminal to its ultimate contact with the resistive element. Contact resistance can be demonstrated by a simple experiment. Make two very accurate resistance measurements, using a different end terminal as a reference for each measurement. Add the two ohmic values. Compared with a resistance measurement of the device's total resistance, it will be found that the sum of the two *parts* is greater than the *whole*. This is due to the contact resistance which imposes an additional resistance between the moveable contact and the resistive element. This experiment is accomplished in steps 1, 2, and 3 of Fig. 2-8. The equivalent schematic of CR is illustrated by Fig. 2-9.

There are two separate sources of contact resistance. The first contributor to CR is completely analogous to the contact resistance of a switch or connector. It results from the non-perfect junction of the moveable contact with the resistive element.

Surface films of metal oxides, chlorides, and sulfides along with various organic molecules, absorbed gases, and other contaminants can form on either the contact or the surface of the element. These films act as insulators and contribute to contact resistance. Just as with other forms of dry circuit contacts, this portion of CR is voltage and current sensitive. Since distribution of these contaminants is not uniform, some degree of variation in this part of contact resistance will occur. Immediate past history; that is, whether or not the wiper has been moved recently, or cycled repeatedly over the element, can cause a variation in this parameter.

The second contributor to CR results from the non-homogenous molecular structure of all matter and the well known fact that a d.c. current flowing through a material will always follow the path of least resistance. Study the exaggerated drawing of a resistive element and moveable contact in Fig. 2-10. Because of the variation in resistance of the conductive particles, the path of least resistance is irregular through the element from end terminal to end terminal.

The schematic analogy of Fig. 2-10 shows a d.c. measurement made at the wiper terminal 2 with respect to either end terminal. The measurement current will flow from the end terminal along the path of least resistance to a point opposite the moveable contact; across a relatively high resistance path to the element surface; then through the wiper circuit to wiper terminal 2.

In this simplified analysis some liberty has been taken with the physics involved but the cause-effect relationship has been maintained.

As with any resistance, the contact resistance will vary with the magnitude of the measurement current. The variation of CR with current may be different for each element material, contact material, and physical structure, particularly with regard to the force with which the contact is pressed against the resistive element. An example of current versus CR curve for a cermet resistive element is shown in Fig. 2-11. No values are assigned to the curve axis since many combinations of resistance versus current exist. The curve is typical in form, however, dropping very rapidly then flattening to a stable value within a milliamp.

Fig. 2-12 is a simple circuit for observing contact resistance. A constant current source,

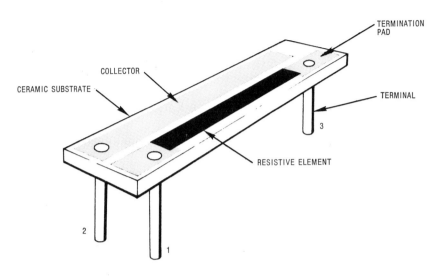

A. A COMMON CERMET POTENTIOMETER CONSTRUCTION

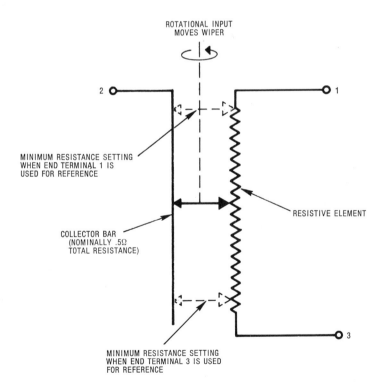

B. SCHEMATIC OF FIG. 2-7A

Fig. 2-7 Construction affects minimum and end-set parameters

IMPORTANT: ADJUST WIPER TO APPROXIMATE CENTER OF RESISTIVE ELEMENT
BEFORE BEGINNING, THEN DO NOT CHANGE POSITION OF THE WIPER
DURING THIS EXPERIMENT

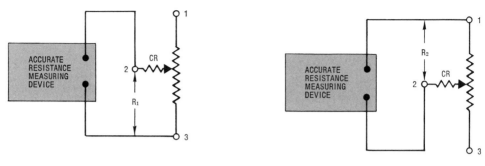

STEP 1: MEASURE AND RECORD R_1 STEP 2: MEASURE AND RECORD R_2

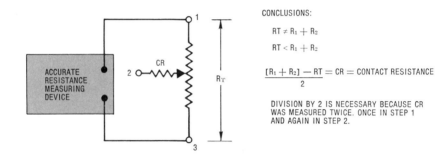

CONCLUSIONS:

$$RT \neq R_1 + R_2$$

$$RT < R_1 + R_2$$

$$\frac{[R_1 + R_2] - RT}{2} = CR = \text{CONTACT RESISTANCE}$$

DIVISION BY 2 IS NECESSARY BECAUSE CR
WAS MEASURED TWICE. ONCE IN STEP 1
AND AGAIN IN STEP 2.

STEP 3: MEASURE THE POTENTIOMETERS RT AND COMPARE WITH $R_1 + R_2$

Fig. 2-8 Experiment to demonstrate contact resistance

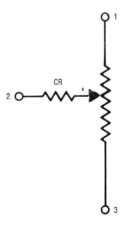

Fig. 2-9 Potentiometer schematic illustrating contact resistance

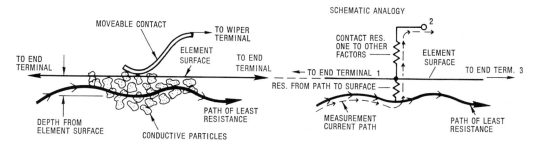

Fig. 2-10 Path of least resistance through element varies in depth from the element surface

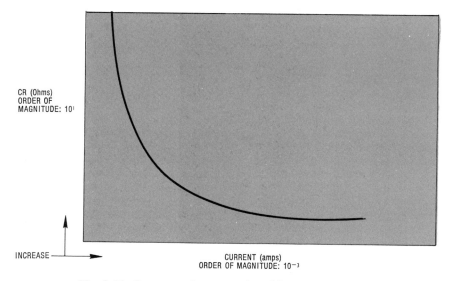

Fig. 2-11 Contact resistance varies with measurement current

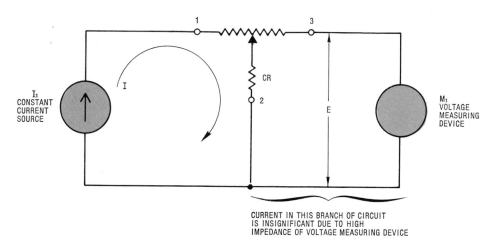

Fig. 2-12 Test configuration for measurement of contact resistance

I₁, provides a test current, I, which is applied through the potentiometer. The current path is in one end terminal; through a portion of the element; through the contact resistance; and then out the wiper terminal 2. The open circuit voltage indicated on M₁ will be proportional to the value of contact resistance.

It is not common procedure to specify or production inspect contact resistance. This is because for any given resistive element, there exists an infinite number of points along its surface where contact resistance could be measured. A very common specification, Contact Resistance Variation, is tested at the manufacturing stage and reflects the variable range of contact resistance as the wiper traverses the element.

CONTACT RESISTANCE VARIATION, CRV

Contact Resistance Variation, CRV, is the maximum, instantaneous change in CR that will be encountered as the result of moving the wiper from one position to another. The limit of CRV is expressed as a percentage of the unit's total resistance or ohms. When the wiper is actuated, the resistance at the wiper terminal, with respect to either end terminal, is apt to increase or decrease by a value within the CRV specification. *1% of TR maximum* and *3 ohms maximum* are typical CRV specifications.

A basic circuit for demonstrating CRV on an oscilloscope is shown in Fig. 2-13. A constant current source, I₁ provides the current I. The path taken by I is indicated by a circular arrow. An oscilloscope with capacitor filter provides a detector that monitors the effective changes in

voltage drop across the contact resistance. The capacitor merely restricts the d.c. voltage component from the oscilloscope display. Only the variation in voltage due to CRV appears on the display.

The current sensitivity of CR, as previously mentioned, imposes restrictions on I. These restrictions are required for accuracy and meaningful data correlation. Fig. 2-14 is a table of typical current values for CRV measurement.

NOMINAL TOTAL RESISTANCE, TR (ohms)	CURRENT I (mA)
< 50	30
50 TO < 500	10
500 TO < 100K	1
100K TO < 2 MEG.	0.1
≥ 2 MEG.	0.05

Fig. 2-14 Current values for CRV measurement of cermet elements resistance variation

Industry standard test conditions require the use of a 100 Hz - 50 kHz bandpass filter in lieu of the capacitor of Fig. 2-13. This filter accomplishes the restriction purpose of the capacitor and, in addition, limits the CRV response to those values within the bandpass spectrum. This limitation is justified because the frequency response of most systems utilizing potentiometers is within the filter bandpass.

The oscilloscope photograph of Fig. 2-15 illustrates a CRV display using the circuit of Fig. 2-13 with a mechanical device to uniformly cycle the wiper. This equipment is pictured in Fig. 2-16. The oscilloscope photograph shows

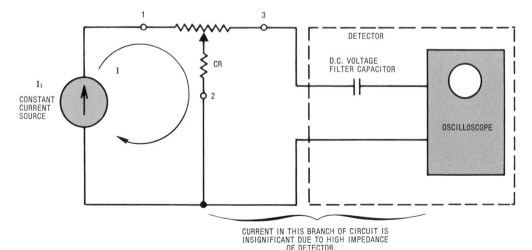

CURRENT IN THIS BRANCH OF CIRCUIT IS INSIGNIFICANT DUE TO HIGH IMPEDANCE OF DETECTOR.

Fig. 2-13 Circuit for demonstration of contact variation

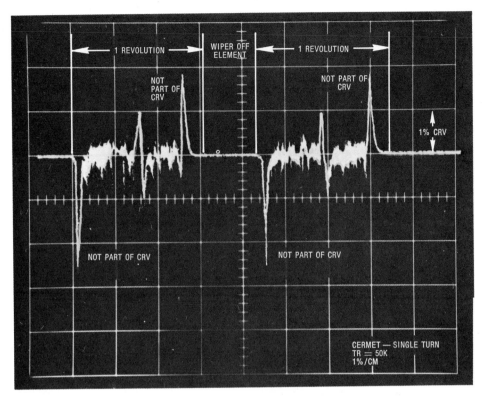

Fig. 2-15 Oscilloscope Display of CRV

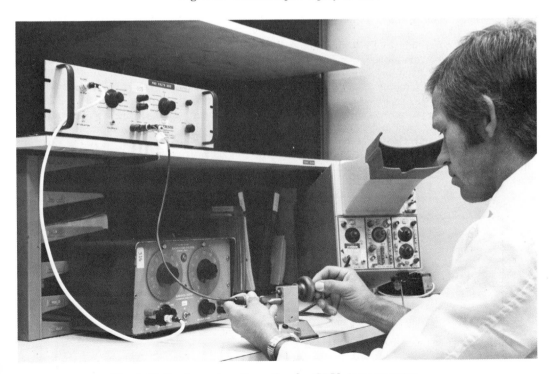

Fig. 2-16 Equipment configuration for CRV demonstration

two complete revolutions of a single turn potentiometer. The extreme variations at the beginning and end of the oscilloscope trace are due to the wiper movement off or onto the termination areas. They are not considered contact resistance variation.

EQUIVALENT NOISE RESISTANCE, ENR

Potentiometers with wirewound elements use the parameter of Equivalent Noise Resistance, ENR, to specify variations in CR.

Before defining ENR, it is necessary to introduce some new terminology. Fig. 2-17 depicts the potentiometer in a voltage-divider mode. Refer to Chapter 3. In this configuration, it is common to refer to the electrical signal present at the unit's end terminals, 1 and 3, as the *input* and the signal present at the wiper terminal 2 as its *output*. If the voltage division performed by the potentiometer was ideal, a graph of the output function as the contact moved from end terminal 3 to end terminal 1 would be a straight line from zero to E_I. It would have a slope equal to the ratio of total input voltage to total resistance. However, when the output is precisely monitored with an oscilloscope, it is observed that the potentiometer not only deviates from the ideal concept, but some degree of electrical noise or distortion is also present on the output waveform. This distortion is imposed by the device itself.

Many factors contribute to ENR, including all of those previously mentioned as contributing to CR and CRV. Oxide film buildup on the surface of the resistive element will act as an insulator until rubbed away by the friction of the wiper. Minute foreign particles resulting from a harsh operating environment may find their way between the wiper and element creating the same effect. Even microscopic bits of metal resulting from friction wear of the parts can lodge

between the resistive element turns affecting ENR.

When these foreign substances interfere with wiper contact they give wirewound potentiometers a dynamic output characteristic which is sporadic and nonrepeatable.

Potentiometer manufacturers specify ENR, a theoretical (lumped parameter) resistance, in series with output terminal 2. This resistance will produce the equivalent loss in an ideal potentiometer. The most common specification of Equivalent Noise Resistance is *100 ohms maximum*.

The earlier discussion on CRV applied only to potentiometers having non-wirewound (film type) resistive elements. These elements present a continuous, smooth path for the wiper. With wirewound elements, the path provided is relatively less smooth and continuous. The wiper effectively *jumps* and *bridges* from one turn of resistance wire to the next. The simplified drawing of Fig. 2-18 emphasizes this bridging action. The wiper usually does not make connection with only one turn of wire but actually touches several at once. This depends on the relative width of the contact to the wire size and spacing.

In Fig. 2-18 the wiper is assumed, solely for illustrative purposes, to be wide enough to touch only two turns when in position A or touch only one turn when at position B. When two turns are simultaneously contacted, that portion of the resistive element bridged by the wiper, is bypassed (i.e., shorted electrically). As a result, the resistance of the *shorted* turn will decrease and change the devices output voltage.

Some aspects of ENR are circuit application dependent, such as the load current I_2, in Fig. 2-17. Most causes are traceable to the variation of contact resistance as the wiper moves across the element. For simple physical demonstration,

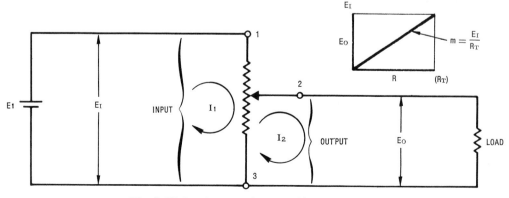

Fig. 2-17 Load current is a contributor to ENR

the circuit in Fig. 2-13 can be used as shown by Fig. 2-19. A one milliamp constant current is passed through the wiper circuit. The resulting voltage drop from wiper to element is monitored by a detector circuit while the wiper is cycled back and forth across the element. The detector circuit consists of an oscilloscope and voltage regulating diode, D. The diode protects the potentiometer from excessive voltage by providing a conductive path for the 1ma current if the wiper circuit becomes open. The display presented on the oscilloscope screen could be

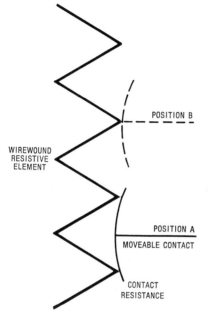

Fig. 2-18 A varying number of resistance wire turns make contact with the wiper

like one of those shown in Fig. 2-20. The waveforms were measured on two different potentiometers. Unit number 1 shows more noise than unit number 2. The calibration of the oscilloscope is only $5\Omega/\text{cm}$. Note that the greatest deviation of unit number 1 is about 4Ω while unit number two remains well below 0.5Ω for its entire travel.

If the oscilloscope were not calibrated in in Ω/cm, the ENR for a particular unit could be determined by first measuring the *maximum* peak voltage drop, E_P with the test current fixed at one milliamp. The Equivalent Noise Resistance could then be calculated by:

$$ENR = \frac{E_P}{I} = \frac{E_P}{.001} = E_P \cdot 10^{-3}$$

ENR will be given in ohms if the value of E_P is in volts. The ENR of a particular unit may be so low that accurate determination of the exact value is difficult, e.g. unit number 2 in Fig. 2-20. However, ENR is specified as a maximum and it is a simple task to determine that any particular unit remains below it's specified maximum.

The industry standard test circuit for ENR uses a low pass filter in place of the capacitor in Fig. 2-19. While this filter limits the amount of noise seen by the oscilloscope, its bandwidth (1 kHz) is in excess of the bandwidth of most systems in which the potentiometer will be utilized. This means it does not filter out any significant distortion.

ENR AND CRV

The analogy of ENR and CRV should be quite obvious. Both specifications are dynamic parameters and are highly dependent upon the fundamental electronic concept of contact resistance. Equivalent Noise Resistance, for wire-

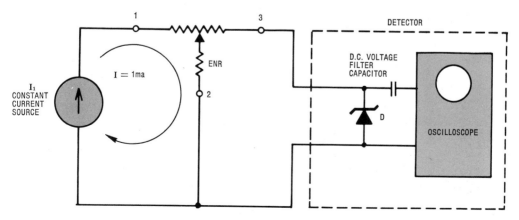

Fig. 2-19 Demonstration of ENR

wound potentiometers, was adopted by manufacturers and users, as a quality indicator a number of years in advance of Contact Resistance Variation.

To better understand today's need for both ENR (wirewound) and CRV (nonwirewound) specifications, some details of element construction must be understood. (For full details, see Chapter 7.)

A major portion of total resistance ranges attainable with wirewounds are made with the same resistance wire alloy. To manufacture a variety of resistance values, the resistance wire size is simply changed and more or less wire is wound on the element. This means the metal-to-metal interface between element and wiper is identical for most resistances.

Total resistance of nonwirewound (carbon and cermet) elements cannot be changed this simply. Instead, slightly different compositions and/or processes are used to change resistance. This means the element to wiper interface (contact resistance) varies with total resistance. This contact resistance is by nature less conductive than the wirewound counterpart.

The result is a more dynamic contact resistance parameter for nonwirewound. From a practical standpoint, although the circuit used

resembles the one for ENR, a different calibration is needed to adequately observe CRV.

Since nonwirewound potentiometers are ideal for many applications, the CRV specification is commonly specified. Wirewounds continue to use the ENR specification.

OUTPUT SMOOTHNESS, OS

Output smoothness, OS, applies to potentiometers with non-wirewound elements used for precision applications. This parameter is the maximum instantaneous variation in output voltage, from the ideal output voltage. It is measured while the wiper is in motion and an output load current is present. Output smoothness is always expressed as a percentage of the total input voltage. A typical specification is *0.1% maximum.*

The factors contributing to contact resistance and contact resistance variation are all causes of output voltage variations. Because these parameters are current sensitive, the presence of an output load current is a significant contributor to output smoothness.

The circuit shown in Fig. 2-21 is the industry standard test circuit but is shown here for dem-

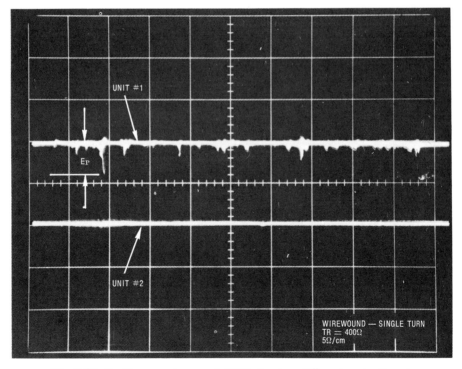

Fig. 2-20 Oscilloscope traces of ENR for two different potentiometers

onstration only. A stable, low noise voltage source E1, is connected as an input to the potentiometer. The output of the device is applied to a load resistor, R1, and to the input of a smoothness filter. The output of the filter is then monitored with an oscilloscope or strip-chart recorder. The ohmic value chosen for R1 is not arbitrary but should be two orders of magnitude (10^2) greater than the potentiometer's total resistance.

The bandpass filter of Fig. 2-21 accomplishes the same major task as the filters used for CRV and ENR demonstrations. These filters remove the d.c. component and restrict noise transistions to frequencies encountered in applications. The choice of a filter is not critical for purposes of demonstrating CRV, ENR or OS. Remember, when interpreting manufacturer's data sheets, that these specifications are based on industry standard test conditions which include the use of a *specific* filter.

All electronic specifications, to be meaningful, assume a set of test conditions. Output

smoothness is no exception. When the circuit of Fig. 2-21 is used as an academic aid, the industry standard test conditions can be simulated by using a mechanical fixture to actuate the moveable contact. The fixture should be capable of driving the potentiometer adjustment mechanism at a rate of 4 revolutions per minute. The resulting strip-chart trace could look something like Fig. 2-22. To determine the device's output smoothness, select the greatest recorded change in output voltage (within a 1% travel increment) and express it as a percentage of the total input voltage, or:

$$OS = \frac{e_{max}}{E_I} \times 100$$

ADJUSTABILITY, A

Although many manufacturer's data sheets specify adjustability, A, this characteristic is the newest potentiometer parameter. It is the result of industry's efforts to further clarify the impor-

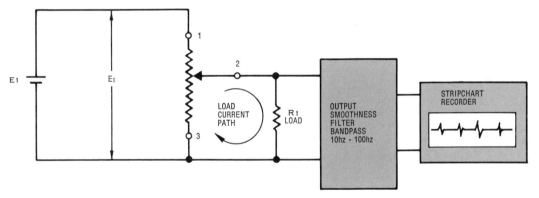

Fig. 2-21 Configuration for output smoothness demonstration

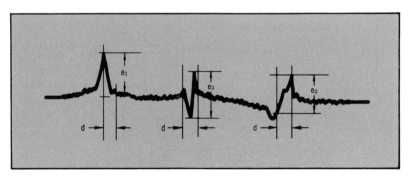

d = 1% OF TOTAL ELECTRICAL TRAVEL
e = PEAK TO PEAK VARIATION WITHIN THE HORIZONTAL MOVEMENT INCREMENT, d

Fig. 2-22 Evaluation of the output smoothness recording

tant effect of wiper-element interaction related to circuit applications.

The specification is new but anyone who has tuned an electronic circuit, has tested a potentiometer's adjustability. This includes common household appliances and complex electronic systems. In some circuits only coarse adjustment is required to produce desired response. In this case, the adjustability of the potentiometer is not critical. In other applications, time consuming fine adjustment is required to achieve desired circuit function. Here, the adjustability is very critical.

Adjustability, as inferred by the previous paragraph, is the accuracy and ease with which the wiper can be positioned to any arbitrarily selected point along the resistive element.

Because the potentiometer is most often applied in one of two modes, (Chapter 3) adjustability parameters are specified for each: Adjustability of in-circuit resistance, (variable current rheostat mode), Adjustability of output

voltage ratio, (voltage divider mode). Adjustability of in-circuit resistance is sometimes referred to as adjustability of output resistance.

Fig. 2-23 illustrates the simplest method of demonstrating adjustability of in-circuit resistance. After setting the wiper as close as possible to 50% of the device's TR, the adjustability of resistance as a percent of TR can be calculated from:

$$A_R\% = \frac{(\text{Achieved reading} - (0.5 \text{ TR})}{\text{TR}} \times 100$$

Fig. 2-24 shows a circuit for measurement of adjustability of output voltage ratio. After attempting to adjust the wiper to achieve a reading of .50 on the DVM, the adjustability of the output voltage ratio as a percent of the attempted setting, is easily calculated from:

$$A_V\% = \left\{ (\text{achieved ratio}) - (.50) \right\} \times 100$$

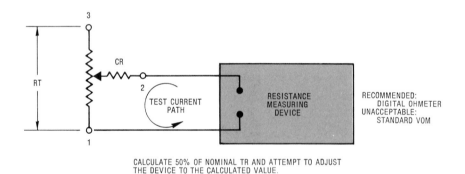

CALCULATE 50% OF NOMINAL TR AND ATTEMPT TO ADJUST THE DEVICE TO THE CALCULATED VALUE.

Fig. 2-23 Adjustability of in-circuit resistance

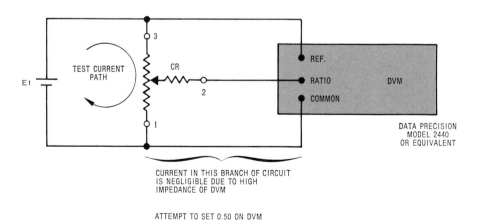

CURRENT IN THIS BRANCH OF CIRCUIT IS NEGLIGIBLE DUE TO HIGH IMPEDANCE OF DVM

ATTEMPT TO SET 0.50 ON DVM

Fig. 2-24 Adjustability of voltage division

Since voltage drop and resistance are directly proportional, A_R and A_V might be erroneously considered equivalent. A comparison of Figures 2-23 and 2-24 will show that the test current path is through the wiper, and hence through the contact resistance, for measurement of A_R. The test current is excluded from the wiper circuit for A_V measurement. This fact causes A_R specifications to be higher than A_V specifications. A typical value for A_R is $\pm\ 0.1\%$. While A_V for the same device could be as low as $\pm 0.05\%$.

The choice of a 50% setting point in the previous examples for adjustability is arbitrary. For consistency and meaningful data correlation, in-dustry standard test procedures specify 30%, 50% and 75% as test settings. These settings must be made within a 20 second time limit.

TEMPERATURE COEFFICIENT OF RESISTANCE, TC

The temperature coefficient of resistance, TC, is an indication of the maximum change in total resistance that may occur due to a change in ambient operating temperature. This parameter is usually specified in parts per million per degree Celsius (Centigrade) or PPM/°C. Temperature coefficient is, to a great extent, dependent upon the type of material used to construct the resistive element and the physical structure

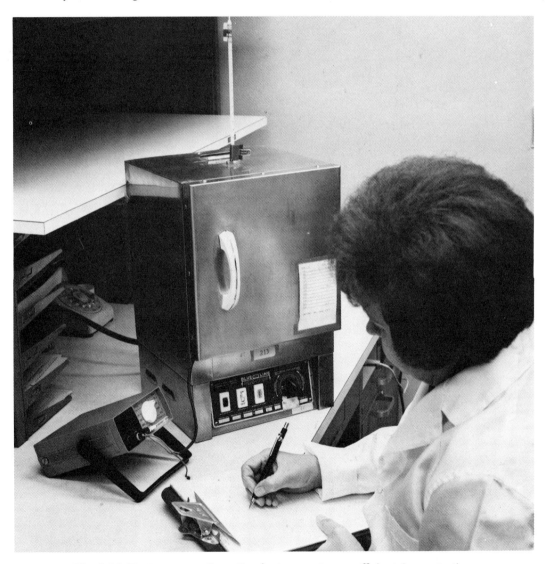

Fig. 2-25 Equipment configuration for temperature coefficient demonstration

of the unit. For example, potentiometers utilizing cermet elements typically have a temperature coefficient of $\pm 100 PPM/°C$. Wirewound element devices typically have $\pm 50\ PPM/°C\ maximum$. It is important to note that total resistance can vary directly or inversely with temperature.

Proper demonstration of a potentiometer's TC requires the use of a temperature chamber and a means for monitoring the chamber's temperature. Also needed is a resistance measuring instrument wired to the potentiometer so that total resistance can be accurately measured in the closed chamber. Fig. 2-25 shows this equipment. To determine the TC for a particular unit, measure and record the devices TR for two ambient temperatures. The two temperatures should be separated by at least 25°C. Allow sufficient time for temperature stabilization at each temperature.

Then:

$$TC = \frac{TR_2 - TR_1}{TR_1(T_2 - T_1)} \times 10^6$$

TC = Temperature coefficient in PPM/°C
TR_1 = TR at ambient temperature T_1
TR_2 = TR at ambient temperture T_2
10^6 is conversion factor to PPM
The resistance must be expressed in ohms and temperatures in °C.

Industry standard test conditions require TR readings to be taken at several temperatures (as many as seven). Using the above formula, the resistance shift for each ambient temperature is evaluated to determine conformance to the TC specification.

Occasionally, in existing literature, the sub-ject of temperature coefficient will include reference to *resistance temperature characteristic,* RTC. This parameter is nothing more than the total resistance change that may occur over a specified ambient temperature range. It is expressed as a percentage of the TR value at a given reference temperature. Mathematically:

$$RTC\% = \frac{TR_2 - TR_1}{TR_1} \times 100$$

RTC = Resistance temperature characteristic from T_1 to T_2
TR_1 = TR at ambient Temperature, T_1 (Reference Temperature)
TR_2 = TR at ambient temperature, T_2
The resistances must be expressed in ohms.

Comparison of the formulas for TC and RTC will show that RTC is simply a percentage change (parts per hundred) in total resistance. For the same measurement conditions, TC is RTC expressed in parts per million — per degree Celsius.

RESOLUTION

There are three types of resolution. Each is a measure of the incremental changes in output with wiper travel characteristic of wirewound potentiometers. For non-wirewound units, output smoothness reflects resolution effects.

Theoretical Resolution. Theoretical resolution, sometimes called *nominal resolution* applies only to linear wirewound potentiometers and *assumes* that the moveable contact can be set to any given turn of resistance wire. Fig. 2-26 shows an example of this type of wirewound element. If N represents the number of active turns in the element, then the theoretical resolu-

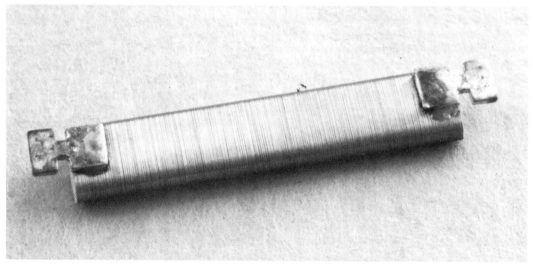

Fig. 2-26 Linear wirewound potentiometer element

tion is given in percent by the following formula:

$$\text{Theoretical Resolution \%} = \frac{1}{N} \times 100$$

The active turns are those turns between the termination tabs which contribute to the potentiometers total resistance. The larger the number of turns, the better or *lower* the theoretical resolution. This also means that, for a given potentiometer construction, the higher resistance values will have a better theoretical resolution because more turns of a smaller diameter wire are used in the element.

As examples, consider a typical wirewound adjustment potentiometer. For a TR value of 1000 ohms, there are approximately 172 turns of wire in the element so the theoretical resolution is 0.58%. If a unit of the same style were constructed for a TR of 20,000 ohms, the element would require about 400 turns, yielding a theoretical resolution of 0.25%.

Travel Resolution. Travel resolution, applicable to wirewound potentiometers only, is the maximum movement of the mechanical input in one direction required to produce an incremental step in the output voltage. For a rotating input it will be specified in degrees but in the case of a linear actuating shaft it will be in thousandths of an inch. This parameter is specified without regard to wiper location on the element.

The typical output of a wirewound potentiometer is a staircase pattern in which the output voltage remains relatively constant for a small amount of wiper travel, then it suddenly changes. Fig. 2-27 is an expanded portion of this output voltage vs. wiper travel pattern. Travel resolution, unlike theoretical resolution, is a measurable output response.

As shown in Fig. 2-27, travel resolution and voltage resolution are related. Since the output

voltage increment is of major concern in most applications, voltage resolution, rather than travel resolution, is specified.

Voltage Resolution. Voltage resolution is defined as the greatest incremental change in output voltage in any portion of the resistance element with movement of the mechanical input in one direction. This parameter is applied only to wirewound units.

Voltage resolution is easily seen from the expanded graph in Fig. 2-27. It is the greatest step height in output voltage resulting from a corresponding change in wiper position.

A circuit suitable for voltage resolution demonstration is shown in Fig. 2-28. A stable voltage source, E_1, supplies 10V as an input to the potentiometer. The output voltage is fed to a load resistor, R_L, and through a high pass filter to a strip-chart recorder. R_L need not be included in the circuit unless it is specified by the end user of the potentiometer. Since R_L is rarely used, the following discussion assumes it is *not* present in the demonstration circuit.

The characteristics of the filter must be such that the charge on the capacitor, C, is allowed to reach a near-steady-state value within the time required for the wiper to move from one turn of resistance wire to the next. The output signal fed to the recorder will be a series of pulses indicating each time a new turn and, hence, a new voltage level is encountered. Fig. 2-29 illustrates the input and output waveforms.

In order to demonstrate these electrical parameters, the time interval, t, between voltage steps, e, may be calculated from the theoretical resolution and the travel time required to traverse the entire electrical length.

$$t = \frac{\text{Travel Time} \times \text{Theoretical Resolution}}{100}$$

$t = $ time interval

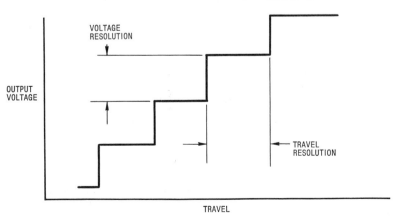

Fig. 2-27 Output voltage vs. travel — illustrating theoretical voltage and travel resolution

Division by 100 is necessary because the theoretical resolution is given in percent. The time interval, t, will be in the same units chosen to express the travel time.

The input resistance of the recorder, R_R, (Fig. 2-28) must be at least 10 times the total resistance of the potentiometer in order to prevent loading errors. It may be that the actual input resistance of the recorder is very large and R_R represents an external shunting resistance.

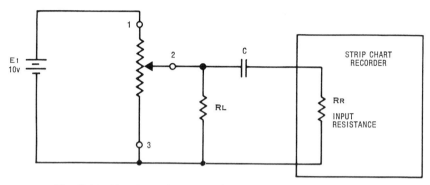

Fig. 2-28 Circuit configuration for demonstrating voltage resolution

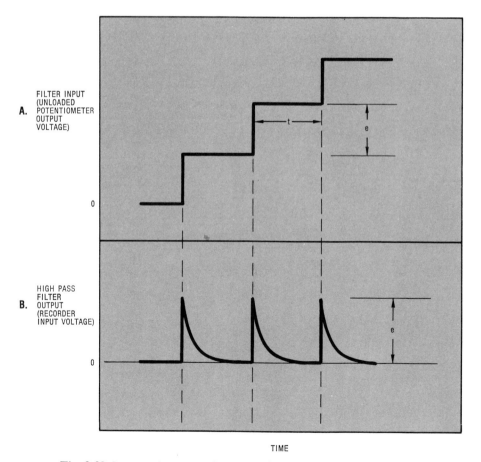

Fig. 2-29 Input and output voltage waveforms for the high pass filter, $R_R C$, in Fig. 2-28

The time constant of the filter approximated by $R_R C$, should be made much less than the time interval, t. Industry standards recommend that $R_R C$ be one tenth the value of t, but a ratio as small as 1 to 5 will contribute negligible error. Thus, the value of the filter capacitance may be calculated from the formula:

$$C = \frac{t}{5R_R}$$

If t is expressed in seconds and R_R in megohms, C will be given in microfarads.

The response of the recorder must be faster than the time constant of the filter, or the true peak value of the output pulse will not be displayed. If a slow response recorder is the only instrument available for demonstration purposes, it may be necessary to move the potentiometer wiper very slowly.

The magnitude of voltage resolution is the ratio of the maximum voltage pulse seen by the recorder to the total input voltage. It is usually expressed in percent. In general:

Voltage Resolution % =

$$\frac{\text{Max. Voltage Pulse}}{\text{Input Voltage}} \times 100$$

The maximum voltage pulse and input voltage must be expressed in like terms. For the circuit of Fig. 2-28 and the waveform of Fig. 2-29B:

$$\text{Voltage Resolution \%} = \frac{e_{max}}{10} \times 100$$

$$= 10e_{max}$$

CONFORMITY

Many precision and special applications of potentiometers require that the output voltage be some well defined nonlinear function of the wiper position and input voltage. Expressing this mathematically:

$$E_0 = E_I f(\theta) \text{ or}$$

$$\frac{E_0}{E_I} = f(\theta)$$

where E_0 represents the output voltage, E_I is the total input voltage, and $f(\theta)$ represents the theoretical output function of the potentiometer.

It is impractical for a manufacturer to meet a given mathematical function specification (ideal output curve) exactly. The function is normally specified with a tolerance or deviation from the theoretical function. This *allowable* deviation of the output curve from a fully defined theoretical function is *conformity*. In other words, it is the tolerance or error band specified about the theoretical (ideal) output curve. This is shown in Fig. 2-30. Before discussing the parameters which are used to characterize conformity it is necessary to define several terms. The illustration of Fig. 2-31 presents a simple method of demonstrating the factors which affect conformity. In addition, Fig. 2-31 provides graphical representation of the following definitions:

Total mechanical travel, Fig. 2-31A, is the amount of angular input rotation θ_M necessary

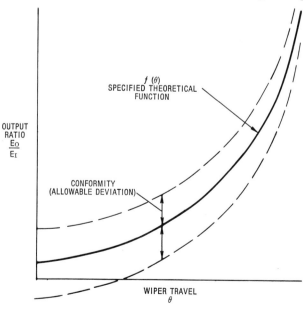

Fig. 2-30 Conformity

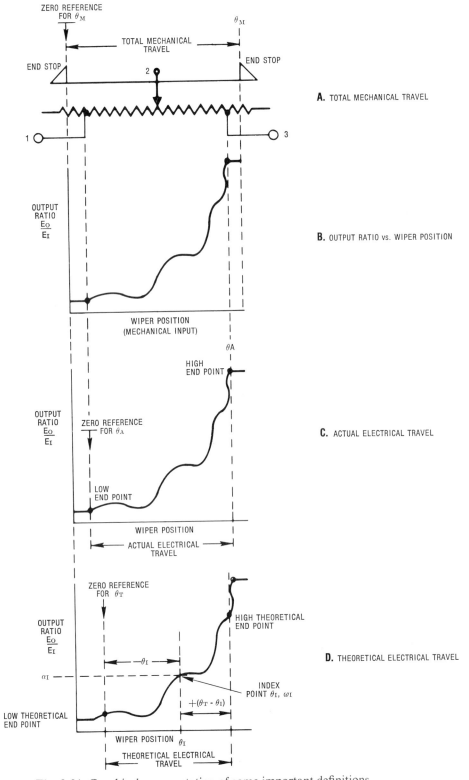

Fig. 2-31 Graphical representation of some important definitions

to move the wiper from one end stop to the other end stop. It is not necessary to use this term when referring to continuous rotation units since end stops are not provided. The output ratio at various wiper positions along the total mechanical travel can be measured and plotted as in Fig. 2-31B. A digital voltmeter with ratio capability is an excellent instrument for this purpose.

Actual electrical travel, is the total amount of angular input rotation, θ_A, over which the output ratio actually varies. This travel range may be easily located by noting the high and low *end points* on the curve of Fig. 2-31B where the output ratio begins or ceases to vary. For the particular potentiometer being considered, these points are shown in Fig. 2-31C.

Theoretical electrical travel, is the amount of angular input rotation, θ_T, defined as the operational range of the potentiometer. This travel range extends between the high and low *theoretical end points* as shown on the curve of Fig. 2-31D. The location of the theoretical electrical travel range is defined by an *index point.* This point is *always* on the output ratio curve. To locate the theoretical electrical travel range, it is only necessary to locate the theoretical end points. To do this, the potentiometer is adjusted until an output ratio of α_I is obtained. The index point specification, clearly defines this output ratio to be θ_I degrees of angular input. Therefore, the theoretical end points can be located by adjusting the potentiometer through angles of $-\theta_I$ and $+(\theta_T-\theta_I)$ from the end points.

In most cases when an index point is required it is specified by the potentiometer manufacturer. The angle and/or ratio of the index point will vary from unit to unit but the manufacturer indicates the index point coordinates on the potentiometer exterior. In some instances, system design may require the angle *or* ratio of the index point to be the same throughout a given quantity of potentiometers. In these cases, the end user specifies one of the index point coordinates and the manufacturer specifies the other coordinate.

Since the index point is always on the output ratio curve, the index point coordinates cannot be guaranteed identical for a given group of units. At least one coordinate, either input angle or output ratio, must vary from device to device.

To apply the above terms to an example, assume that a particular potentiometer has a *total mechanical travel* of 352° and an *index point* at 50% output ratio, 170° rotational input. This potentiometer might have an *actual electrical travel* of 348°, with *end points* at 172° and 176° on either side of the index point. The *theoretical*

electrical travel for the same unit could be 0-340° with *theoretical end points* 170° on either side of the index point.

The three travel ranges described above are defined in terms of end point location. In each case, the lower end point is referred to as the *zero reference* for the particular travel range being considered. Occasionally, it is necessary to refer to some particular wiper position. This is accomplished throughout this book by specifying an *angular travel distance,* θ_W, from the zero reference of the travel range being considered.

Assume the potentiometer whose actual output is plotted in Fig. 2-31B was built to the theoretical function and conformity limits of Fig. 2-30. To evaluate this particular potentiometer's conformity, superimpose the actual output (Fig. 2-31B) on the theoretical function (Fig. 2-30). This composite is shown in Fig. 2-32.

Although the relationship shown in Fig. 2-32 is somewhat exaggerated, it does illustrate the following.

1) The index point by definition is always on the actual curve.
2) Zero output change may occur for a small wiper movement.
3) Output response may be opposite to the expected response.
4) It is also possible that the same exact output could be obtained at two different positions of travel.

At first glance, the upper and lower conformity limits (shown by broken lines in Fig. 2-30) seem to be closer together at the top end. Remember, it is the *vertical* deviation (the change in output voltage) which is being described. Actually, the vertical spacing on the conformity limits is constant.

To summarize this demonstration of conformity:

1) Make a plot of *theoretical* output function of voltage ratio, $\dfrac{E_0}{E_I}$, vs. wiper position, θ_W.
2) Measure the *actual* output of a particular potentiometer and construct a graph of voltage ratio $\dfrac{E_0}{E_I}$ vs. wiper position, θ_W.
3) Evaluate the *conformance* of the actual response to the theoretical response.

If conformity is included in the mathematical relationship previously given, the formula may be written:

$$\frac{E_0}{E_I} = f(\theta) + K$$

where K represents the conformity and is usually specified in terms of a percentage of the total input voltage.

Generally, it is convenient to express the potentiometer's output to input function in terms of a ratio of wiper position, θ_W, to the maximum theoretical electrical travel, θ_T, or:

$$\frac{E_O}{E_I} = f\left(\frac{\theta_W}{\theta_T}\right) + K$$

By its definition, this mathematical relationship is the *transfer function* of the potentiometer.

ABSOLUTE CONFORMITY

Absolute conformity is defined as the greatest *actual* deviation of a potentiometer's output from the specified theoretical transfer function. It is expressed as a percentage of the total applied input voltage and measured over the theoretical electrical travel. An index point of reference is required.

The drawing of Fig. 2-32 illustrates absolute conformity. Note that for some values of travel, the actual output is higher than that predicted from the theoretical curve, while it may be lower for other values of travel. Absolute conformity for this particular potentiometer is the maximum vertical deviation of the actual response from the theoretical curve. This happens to occur at about mid-position for this example, but could happen anywhere along the curve for another unit.

LINEARITY

Linearity is a specific type of conformity where the theoretical function (ideal output curve) is a straight line. A generalized mathematical representation of this function is:

$$\frac{E_O}{E_I} = m\,f(\theta) + b + k$$

Where: E_O is output voltage.
E_I is input voltage.
m is the slope.
b is the slope intercept at zero travel.
θ is the travel.
k is the linearity.

The demonstration method previously described for conformity, is perfectly suited for linearity demonstration.

Linearity is specified in one of 4 ways: *absolute, independent, zero based* or *terminal based*.

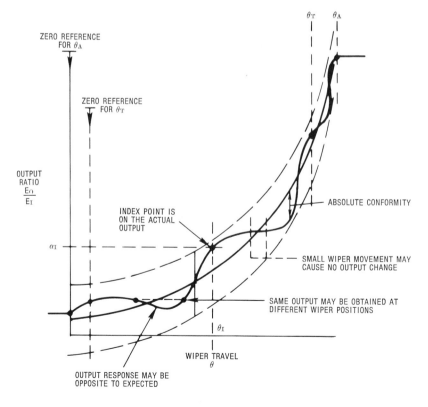

Fig. 2-32 Evaluation of conformity

These specifications differ only in the method of output curve evaluation. Note that the following explanation of each linearity evaluates the actual output curve of one particular potentiometer. This output curve is shown in Fig. 2-33.

Absolute Linearity. Absolute linearity is the maximum permissible deviation of the actual output curve from a fully defined straight reference line. It is expressed as a percentage of the total applied input voltage and measured over the theoretical electrical travel. An index point on the actual output is required.

The straight reference line representing the ideal theoretical output ratio is fully defined by two points. Unless otherwise specified, these points are: (1) Zero travel, $\theta_W = 0$, with an output ratio of 0 and (2) full theoretical electrical travel, $\theta_W = \theta_T$, with an output ratio of 1.

The illustration of Fig. 2-34 shows the conditions necessary to define absolute linearity. In this example, the lower limit of the output ratio (point X, $\theta_W = 0$) is specified as a value slightly greater than zero. The upper limit of the output ratio, (point Y, $\theta_W = \theta_T$) is specified as 1.

The reference line for absolute linearity may be described mathematically as:

$$\frac{E_O}{E_I} = m\left(\frac{\theta_W}{\theta_T}\right) + b$$

In the specific example of Fig. 2-34, the lower limit of the output ratio is specified as 0.05. Therefore, the value of b (intercept) must also be 0.05. In addition, the upper limit of the output ratio is 1 when $\frac{\theta_W}{\theta_T} = 1.0$. To determine the slope, substitute these upper and lower limit values in the general equation and solve for m.

$$\frac{E_O}{E_I} = m\left(\frac{\theta_W}{\theta_T}\right) + b$$

$$1 = m(1) + .05$$

$$m = \frac{1 - .05}{1}$$

$$m = .95$$

For this example, the index point happens to be at an output ratio, $\frac{E_O}{E_I}$, of 0.5 and wiper travel, θ_W, of 170°.

In order to meet the specification, the actual output curve of the potentiometer being evaluated must be within the upper and lower limits defined by absolute linearity. This means the maximum vertical difference (voltage ratio) between the actual output curve and the theoretical reference line must be within the ±k envelope. Typical values of absolute linearity,

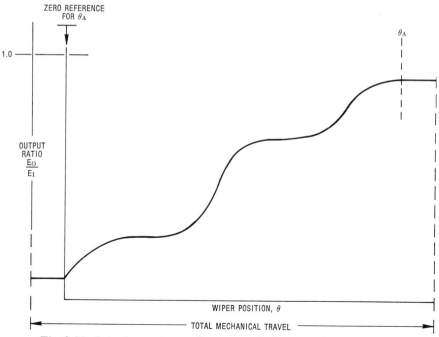

Fig. 2-33 Actual output curve for one particular potentiometer

expressed in percent of total input voltage, range from .2 to 1.0%.

Absolute linearity is the most precise definition of potentiometer output because the greatest number of linearity parameters are controlled. This is the primary advantage of absolute linearity. The methods used to manufacture to these parameters, however, cause *absolute* linearity to be the most expensive of the four linearities.

In Chapter 4, approaches are discussed for achieving absolute linearity performance from more loosely specified (lower cost) linearities by adding adjustment potentiometers. This may be an economical alternative.

Independent Linearity. Independent linearity is the maximum permissible deviation of the actual output curve from a reference line. The slope and position of this reference line are chosen to minimize deviations over all or a portion of the actual electrical travel. In other words, the choice of the values for the slope and intercept are such as to minimize the linearity error. Thus, the reference line is placed for best straight line fit through the actual output curve. Further restrictions may be imposed on the limits of slope and intercept by additionally specifying the range of permissible end output ratios.

Fig. 2-35 illustrates conditions necessary to define independent linearity. The exaggerated wavy line represents the actual output ratio, and is measured over the total *actual* electrical travel. The reference line is positioned on the output curve, without regard to slope and intercept, so the positive and negative deviations or linearity errors are minimized.

The reference line is expressed by the mathematical equation:

$$\frac{E_O}{E_I} = m\left(\frac{\theta_W}{\theta_A}\right) + b$$

where m is an *unspecified* slope, θ_A is the actual electrical travel, b is the *unspecified* intercept value of the output ratio at $\theta_W = 0$.

The independent linearity specification, as shown by the broken lines, are parallel to the reference line and spaced above and below it. These show the allowable output ratio deviation from the theoretical reference line. Typical values of independent linearity, expressed in percent of total input voltage, range from .05 to .20%.

It is more common to specify independent rather than absolute linearity because it gives the tightest tolerance specification for a given cost. The major difference between independent linearity and absolute linearity is that the reference line for independent linearity is positioned to minimize the linearity error. Therefore, the specification of independent linearity should be carefully evaluated to assure interchangeability of devices in a given application.

The determination of actual electrical travel depends upon a clear definition of end points. Generally, there is no problem with wirewound elements, but accurate determination of end points for nonwirewound elements can be quite difficult. In many instances, the output in the region near the end of the nonwirewound element exhibits an abrupt step function. In other cases, the function may be irregular and quite nonlinear with no clearly definable end point.

Irregularities at the end points present little difficulty in most applications where only the middle 80 to 90 percent of travel is used. It becomes necessary, however, to deal with the problem in order to make the linearity specifications meaningful for nonwirewound potentiometers.

There are two possible approaches to characterizing linearity in nonwirewound potentiometers. The first method utilizes an index point of reference while the second merely defines the location of end points.

Fig. 2-36 illustrates the first approach. An index point must be specified as was done for absolute linearity. The travel is presented in terms of a total theoretical electrical travel with respect to the reference index point. Linearity is then determined by constructing a reference line through the actual output curve to minimize the deviations of the actual output from the reference or theoretical line. This best straight line fit is the same as used for independent linearity of wirewound units.

The second approach to specification of independent linearity for nonwirewound potentiometers, Fig. 2-37, defines the end points in terms of specific output ratios. Otherwise, it uses the same basic method as with wirewound potentiometers. A typical set of end points is specified as that travel position where the output voltage ratio is exactly .01 and .99 for the low and high end points respectively. This allows easy measurement of the actual electrical travel, and the independent linearity may be evaluated in the same manner as for wirewound potentiometers.

Zero Based Linearity. Zero based linearity is a special case of independent linearity where the zero travel end of the theoretical reference line is specified. In this case, the theoretical reference line extends over the actual electrical travel. Zero based linearity is the maximum resulting deviation of the actual output from the straight reference line. This straight line is drawn through the specified minimum output voltage

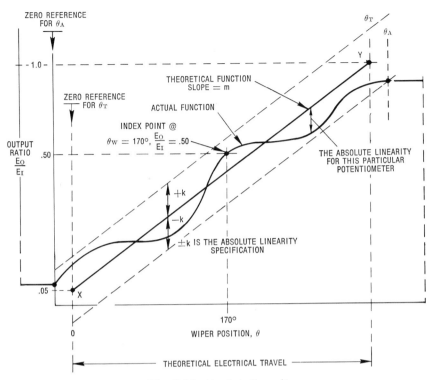

Fig. 2-34 Absolute linearity

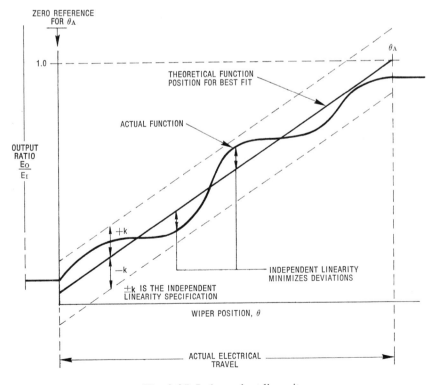

Fig. 2-35 Independent linearity

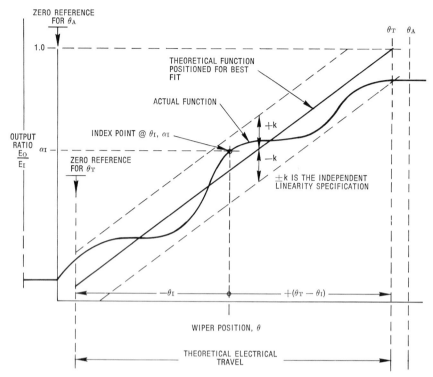

Fig. 2-36 A method to evaluate independent linearity for a non-wirewound potentiometer

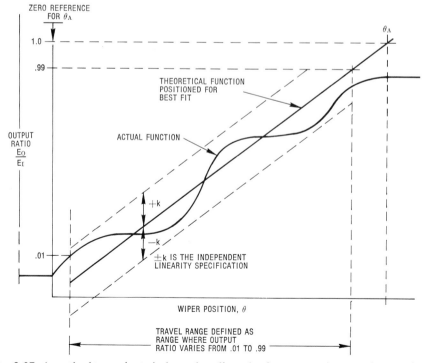

Fig. 2-37 A method to evaluate independent linearity for a non-wirewound potentiometer

ratio with a slope chosen to minimize deviations from the actual output.

Zero based linearity, is expressed as a percentage of total input voltage. Any specified low end output voltage ratio may be used to define the location of the zero travel point of reference. However, unless otherwise stated, the specified value of minimum output voltage ratio is assumed to be zero.

Fig. 2-38 presents the conditions of a zero based linearity specification. For this example, the minimum output voltage ratio is specified as 0%. Note that the transfer functions of both the actual potentiometer output and the theoretical reference line are based upon the *actual* electrical travel. The slope of the reference line is chosen as the best straight line fit in order to reduce the maximum deviations of the actual transfer function from the reference. If an additional specification limits the range of the maximum output voltage ratio, then the range of slope permissible will also be limited.

The mathematical equation describing the actual transfer function is:

$$\frac{E_O}{E_I} = m\left(\frac{\theta_W}{\theta_A}\right) + b + k$$

where m is the unspecified slope whose value is chosen to minimize deviations for a specific potentiometer, b is the specified intercept value determined by the minimum output voltage ratio specification, θ_W is wiper position, θ_A is the actual electrical travel for a specific unit, and k is the linearity.

Zero based linearity is used where: (1) close control of the transfer function is necessary at lower output ratios, (2) greater flexibility of the slope and hence, the transfer function at higher output ratios is permissible.

In many applications, performance very closely resembling that obtained with a costly tight absolute linearity specification may be achieved with a lower cost zero based linearity specification. This is possible when it is used with an adjustment potentiometer to control the overall system gain. Simply stated, an adjustment potentiometer can be used to shift the output (slope) of a precision potentiometer to fit within maximum output limits.

Terminal Based Linearity. A linearity specification sometimes used with wirewound potentiometers is terminal based linearity. It is the maximum deviation of the actual output from a straight reference line drawn through minimum and maximum end points. These points are separated by the actual electrical travel. Unless otherwise stated, the minimum and maximum output ratios are, respectively, zero and

100% of the total applied input voltage. Terminal based linearity is expressed as a percentage of the total applied input voltage.

Terminal based linearity is very much like the absolute linearity except for the definition of reference line end locations as related to travel. With absolute linearity, travel is related to a theoretical movement from a reference index point. The terminal based linearity specification uses actual electrical travel with the end locations on the reference line corresponding to the actual end points of the potentiometer.

Fig. 2-39 shows the requirements for terminal based linearity. For the example here, it is assumed that the minimum and maximum output voltage ratios are given as a basic part of the linearity specification. The 0% and 100% travel limits are implicit.

The reference line for the theoretical output is established by defining two points, X and Y. Point X is the minimum output voltage ratio in the example of Fig. 2-39. It is a travel distance of zero from the lower end point. The second point, Y, is the maximum output voltage ratio in the example and the travel distance is the actual electrical travel. The reference line is constructed with a straight line through the two points, X and Y.

The difference between absolute and terminal based linearity is in the use of *theoretical* electrical travel in the former case and *actual* electrical travel in the latter. There is no significant difference between absolute and terminal based linearity in those applications where the overall system gain may be adjusted to compensate for a variation in the value of the actual electrical travel from one unit to the next. On the other hand, the same degree of interchangeability cannot be expected from a terminal based linearity specification as there would be with an absolute linearity specification.

The actual output function of a given potentiometer purchased under a terminal based linearity specification has the mathematical form:

$$\frac{E_O}{E_I} = m\left(\frac{\theta_W}{\theta_A}\right) + b + k$$

where m is a specified slope of the theoretical reference line, b is the intercept value established by the specified minimum output voltage ratio, θ_W is wiper position, θ_A is the actual electrical travel for a given potentiometer, and k is the linearity error.

POWER RATING

Power rating is the maximum heat that can be dissipated by a potentiometer under specified conditions with certain performance require-

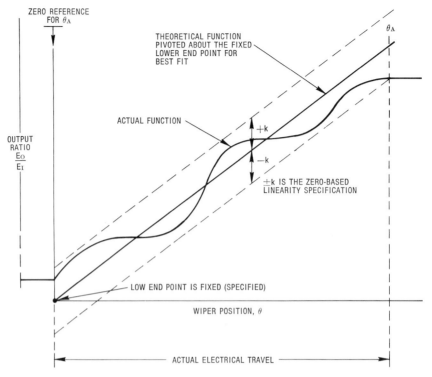

Fig. 2-38 Zero based linearity

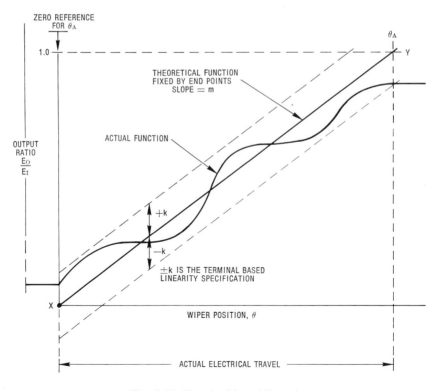

Fig. 2-39 Terminal based linearity

ments. Heat (or power) dissipation is the result of current passing through a resistance. Mathematically:

$$P = I^2R$$

or

$$P = \frac{E^2}{R}$$

where P is the power dissipation in watts and R is the total resistance in ohms. I is the total current in amps flowing through the resistance, R, and E is the total voltage drop expressed in volts, across the resistance, R.

The useful life of a given potentiometer is directly related to the maximum temperature allowed in the interior of the unit. Above a certain internal temperature, insulating materials begin to degrade. A maximum power rating indicates to the circuit designer just how much power may be safely dissipated without harm to the device.

The manner in which a given potentiometer is applied will affect the maximum permissible power dissipation for a given power rating. A detailed explanation of power rating is beyond the scope of this chapter. For a complete analysis relative to applications, refer to Chapter 3.

INSULATION RESISTANCE, IR

Insulation resistance, IR, is the resistance presented to a dc voltage applied between the potentiometer terminals and all other external conducting parts such as shaft, housing, and mounting hardware. It may be thought of as a measure of the electrical leakage between the electrical portion of a potentiometer and other conductive parts of the potentiometer. In the case of ganged (multiple section) units, the insulation resistance specification is also applicable to the resistance between sections.

A commercial megohmmeter with an internal source voltage of the proper value, normally 500v dc, may be used to measure insulation resistance. One lead is connected to all the terminals of the potentiometer and the other lead is connected to the case, shaft, bushing, or other metal parts. Fig. 2-40 illustrates a basic demonstration circuit for insulation resistance.

The power supply must be current limited to prevent damage to it or the electrometer in the case of an unexpected internal short in the potentiometer.

The value of insulation resistance, R_1, is determined by the applied voltage, E, and the resulting current, I:

$$R_I = \frac{E}{I}$$

Typical values of insulation resistance are 1,000 megohms and higher. The insulation resistance parameter, as normally given, refers to bulk leakage resistance under dry operating conditions. Actual equivalent leakage resistance may be much lower (worse) in a given application due to surface leakage paths encouraged by a combination of contaminants and moisture.

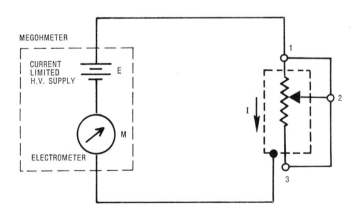

Fig. 2-40 Circuit configuration for demonstration of insulation resistance

POTENTIOMETER ELECTRICAL PARAMETERS SUMMARY

| APPLIES TO: | | | PARAMETER NAME | PAGE | BRIEF DEFINITION | DEMONSTRATION CIRCUIT FIG. | TYPICAL VALUES OR SPECIFICATIONS |
W/W	NON W/W	BOTH					
		●	Total Resistance, TR	18	The resistance between the end terminals	2-2	10Ω nom. to 5 MegΩ nom.
		●	Absolute Minimum Resistance, MR	18	The lowest value of resistance obtainable between the wiper and either end terminal	2-4	0.5Ω max. 1.0% (of TR) max.
		●	End Resistance, ER	20	The value of resistance obtained between the wiper and end terminal with the wiper positioned against the adjacent end stop	2-4	2% or 1Ω to 20Ω
		●	Minimum Voltage Ratio	20	The voltage ratio obtained with the wiper positioned to the Absolute Minimum Resistance	2-6	0.1% to 3.0%
		●	End Voltage Ratio	20	The voltage ratio obtained with the wiper positioned to the End Resistance	2-6	0 to 3.0%
		●	Contact Resistance, CR	22	The resistance of the wiper circuit from the wiper terminal to its ultimate contact with the element	2-8	.1Ω to 2.0% (of TR)
	●		Contact Resistance Variation, CRV	26	The instantaneous change in Contact Resistance that may occur as a result of wiper movement	2-13	±1% (of TR) ±3Ω
●			Equivalent Noise Resistance, ENR	28	Theoretical, lumped parameter resistance reflecting the magnitude of signal loss due to noise	2-19	100Ω max.
	●		Output Smoothness, OS	30	The instantaneous variation in output voltage from the ideal output voltage with wiper in motion	2-21	±0.1% (of TR)
		●	Adjustability of Resistance, A	31	The accuracy to which the wiper can be positioned to an arbitrarily selected resistance value	2-23	±0.1% (of TR)
		●	Adjustability of Output Voltage Ratio	32	The accuracy to which the wiper can be positioned to an arbitrarily selected voltage ratio	2-24	±0.05% (of Full Scale voltage ratio)
		●	Temperature Coefficient, TC	33	The change in total resistance that may occur due to a change in ambient operating temperature	2-25	w/w: ±50ppm/°C Non w/w: ±100 ppm/°C
●			Resolution	34	A measure of the incremental changes in output voltage (or resistance) with wiper travel	2-28	0.01% (of TR) to 1.0% (of TR)
		●	Conformity	37	The allowable deviation of the actual output from a specified theoretical output function	2-30 2-31, 2-32	0.1% to 2.0%
		●	Linearity	40	The allowable deviation of the actual output from a specified, straight line, theoretical output function	2-33 thru 2-38	±.05% to ±1.0% (of Input Voltage)
		●	Insulation Resistance, IR	47	The resistance presented to a d.c. voltage applied between the terminals and all other external, conducting parts	2-40	1000 Meg Ω and higher

Fig. 2-41 A summary of electrical parameters

Notes

Notes

APPLICATION FUNDAMENTALS

Chapter **3**

"Let knowledge grow from more to more."
Alfred, Lord Tennyson

INTRODUCTION

Potentiometers are used in many different applications including calibration adjustments, manual control functions, data input, level and sensitivity adjustments, and servo position feedback transducers. The list is almost endless. However, any specific application can be categorized into one of two operational modes — the *variable voltage divider* or the *variable current rheostat*.

This chapter will look at potentiometer applications from the elemental position. Mathematical derivations of response characteristics are based on a theoretical, ideal model. Ways to modify the two basic operational modes to achieve desired or improved performance are included. The significance of some of the potentiometer parameters are explained. For complete definition of parameters, refer to Chapter 2.

VARIABLE VOLTAGE DIVIDER MODE

A resistive voltage divider provides an output voltage in reduced proportion to the voltage applied to its input. In its simplest form, it consists of two resistances in series. The input voltage is applied across the total circuit and the output voltage is developed across one of the individual resistances.

To construct a *variable* voltage divider, using the potentiometer, the resistive element is substituted for the two resistances of the fixed voltage divider. The wiper provides an adjustable output voltage. The basic circuit is shown in Fig. 3-1A. An input voltage E_I is applied

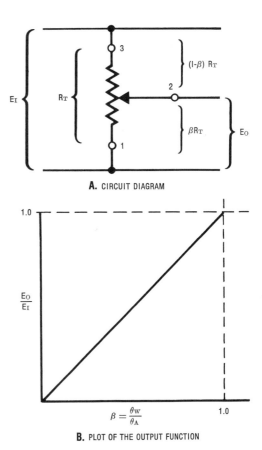

A. CIRCUIT DIAGRAM

$$\beta = \frac{\theta_\mathrm{w}}{\theta_\mathrm{A}}$$

B. PLOT OF THE OUTPUT FUNCTION

Fig. 3-1 The basic variable voltage divider

51

across the entire resistive element. The output voltage, E_O, is developed across the lower portion of the element, between the wiper terminal 2 and end terminal 1. The value of the resistance between terminals 2 and 1 will be βR_T, where β is the ratio of the particular wiper position to the total actual electrical travel or $\beta = \dfrac{\theta_W}{\theta_A}$ and R_T is the mathematical symbol for total resistance. The resistance value for the remainder of the element will then be $(1 - \beta)R_T$.

The unloaded voltage output may be calculated:

$$E_O = \frac{\beta E_I R_T}{\beta R_T + (1 - \beta)R_T} = \beta E_I$$

This relationship is represented by the linear output function shown in Fig. 3-1B.

Adjustability. The adjustability of output voltage ratio A_V is a direct indication of the accuracy with which any, arbitrarily selected, output voltage can be achieved. The A_V specification pertains explicitly to the voltage divider mode. For example, assume a potentiometer having an $A_V = \pm 0.05\%$ is used in an application where $E_I = 100v$. The circuit designer can expect to set any desired output voltage in the range $0 - 100v$ within .05v. Of course, intelligent design procedure requires the selection of a device based on the output voltage accuracy required.

In addition to indicating output accuracy, the A_V parameter is specified together with a maximum time-to-set duration of 20 seconds. This time limitation assures the potentiometer user that the specified accuracy can be obtained without an extended period of trial and error.

Effects of TC. The magnitude of a potentiometer's total resistance may change with variations in ambient temperature. The amount of the change will be proportional to the actual temperature coefficient TC of the particular resistive element. This presents no real difficulty in the simple variable voltage divider whose output voltage is a function of the *ratio* of two resistances. The entire length of the element may be assumed to have an essentially uniform temperature coefficient. If the temperature of the element varies uniformly, then the ratio of the divided portions will remain constant. This means that the output voltage will be unaffected by the TC of the potentiometer.

A slight variation in output voltage can be caused by temperature-change effects on the materials used to fabricate the device. This may result in minor mechanical movements. If the wiper moves relative to the element, then β will change. The same result occurs with a slight temperature difference in two sections of the element. In most actual applications, the voltage input to the potentiometer will not be from a zero output impedance source. Also, the potentiometer output may be driving a load resistance. These conditions are additional sources for output voltage variation with temperature. Although the resulting effects will usually be negligible, they must be considered for critical applications involving significant ambient temperature variations.

Effects of Linearity. Since linearity is a direct indication of the degree to which the output function may deviate from the ideal straight line, the effects of linearity error may have to be considered when the potentiometer is used as a voltage divider.

If the wiper is very accurately positioned, *mechanically*, to some desired output ratio and an electronic measurement is made of the actual output voltage, the output voltage very probably will *not* be as predicted by the relationship:

$$E_O = \beta E_I$$

The error can be caused by any combination of the factors discussed in this chapter. The linearity specification is an indication of the amount of output error that can occur due to linearity factors alone.

When considering a linearity specification, note which of the four linearities is applicable. Refer to Chapter 2 for an explanation of linearity evaluation methods.

Effects of Voltage Resolution. Voltage resolution is the incremental change (steps) in output voltage that occurs as the wiper traverses a wirewound element. As illustrated in Chapter 2, the expanded output voltage function will exhibit a staircase form. The resulting discrete changes in output voltage level make it impossible to adjust the wiper to some values. This limitation is most pronounced in low resistance wirewound potentiometers due to the large diameter resistance element wire used in their construction.

Power Rating. The maximum power rating of most potentiometers assumes that the unit will be operated in the voltage divider mode. Thus a voltage will be applied to the input terminals with an insignificant load current through the wiper circuit.

A typical maximum power rating might be listed as 1.0w at an ambient temperature of 40°C and zero watts at 125°C. This two-point specification is illustrated by Fig. 3-2. The first part of the specification defines the location of point A. The second part of the specification, point B, indicates the operating temperature at which the maximum allowable power dissipation falls to zero.

The permissible power dissipations for temperatures between 40°C and 125°C assume a linear derating curve defined by the straight line connecting the two points.

It is important to realize that it is the resulting *internal* temperature that is critical. It matters little to the potentiometer as to whether heat is caused by current passing through its element, high external ambient temperatures, or a combination of the two.

The manufacturer often views the two-point specification in the manner shown by Fig. 3-3 where the maximum allowable power dissipation is 1.0w for all temperatures 40°C and below. A circuit designer would do well to follow this approach unless he checks with the manufacturer. Extrapolation of the power rating plot to temperatures lower than that of the lowest specification, 40°C in the example, although logical and seemingly practical, is not wise. Other factors such as excessive element current may be involved.

Calculation of a power rating at some intermediate temperature between the two given in the specification is quite easy. First, determine the derating factor, p:

$$p = \frac{\Delta P}{\Delta T} = \frac{P_A - P_B}{T_A - T_B}$$

where P_A and P_B represent the allowable power dissipations at the two temperatures T_A and T_B respectively. For the specific example of Fig. 3-3:

$$p = \frac{1 - 0}{40 - 125} = -0.0118 \text{ watt/}°C$$

Thus, for each degree C of rise in the ambient temperature, the power rating is decreased by .0118w. The power rating for any temperature within the two temperature extremes, T_A and T_B, may now be calculated:

$$P_D = P_A + p(T_D - T_A)$$

where P_D is the allowable power rating at temperature T_D.

Other factors may affect the realistic allowable power dissipation. A full power rating specification should describe the mounting conditions and whether the ambient is still air or forced convection. Generally it is safe to assume that the published rating applies to the standard mounting means for the given potentiometer in still air.

If the unit is mounted in a manner substantially different than the one for which it was designed, consideration should be given to the comparative thermal conductivities. Say that a bushing mounted unit which is normally

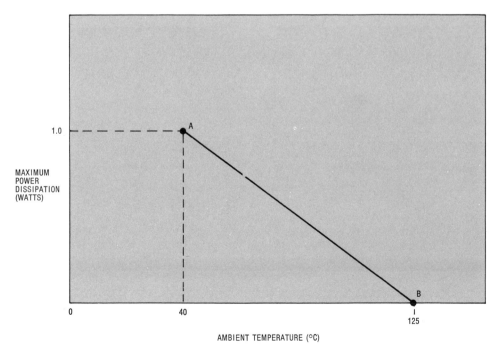

MAXIMUM
POWER
DISSIPATION
(WATTS)

1.0

A

B

0   40   125

AMBIENT TEMPERATURE (°C)

Fig. 3-2 A two-point power rating specification assumes linear operating

mounted to a metal panel will be mounted to a printed circuit board. The thermal conductivity of the epoxy-glass circuit board is much lower than metal, and it may be necessary to reduce the maximum power rating to assure reliable operation.

Be cautious when placing a potentiometer adjacent to a heat producing device such as a power transistor, vacuum tube, power transformer, power resistor, or even another potentiometer. Frequently several adjustment potentiometers may be mounted together with little space between them. With precision potentiometers or controls, more than one unit may be stacked on a common shaft for panel mounting. Derating the allowable power dissipation of the potentiometer may be wise.

Loading Effects. When a load resistance R_L is present in the output circuit of a variable voltage divider, shown in Fig. 3-4A, the output voltage can no longer be represented by the simple relation βE_I. To examine the affect of a load resistance, consider the potentiometer's Thévenin equivalent circuit as shown in Fig. 3-4B. The unloaded open-circuit voltage E', which is equal to E_0 in Fig. 3-1, must be divided between the terminal resistance R' and the load resistance R_L:

$$E'_0 = \frac{E'R_L}{R'+R_L} = \frac{E'R_L}{R_L+(\beta-\beta^2)R_T}$$

The fractional error in the output voltage, as compared with the unloaded ideal value, may be expressed as:

$$\delta = \frac{E'_0-E'}{E'} = \frac{E'_0}{E'} - 1 \qquad (E'=E_0)$$

Inserting the value for the output voltage as given above and simplifying:

$$\delta = \frac{-(\beta-\beta^2)}{\dfrac{R_L}{R_T}+(\beta-\beta^2)}$$

The value of β which yields maximum error may be found by setting the partial derivative of the above expression to zero. That is:

$$\frac{\partial\delta}{\partial\beta} = \frac{-\dfrac{R_L}{R_T}(1-2\beta)}{\left[\dfrac{R_L}{R_T}+(\beta-\beta^2)\right]^2} = 0$$

The only practical solution to this equation is:

$$1-2\beta = 0$$
$$\beta = 0.5$$

Insert this value into the general expression for δ to get the maximum error.

$$\delta_{max} = \frac{-(0.5-0.5^2)}{\dfrac{R_L}{R_T}+(0.5-0.5^2)} = -\frac{1}{1+4\dfrac{R_L}{R_T}}$$

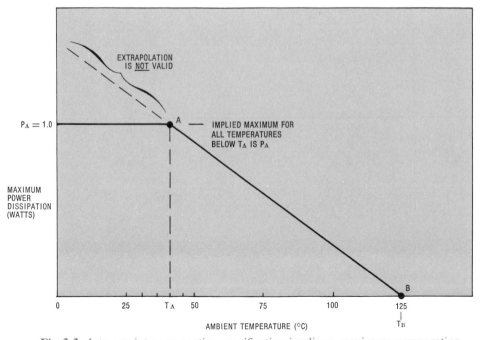

Fig. 3-3 A two point power rating specification implies a maximum power rating

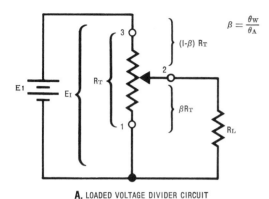

$$\beta = \frac{\theta_W}{\theta_A}$$

A. LOADED VOLTAGE DIVIDER CIRCUIT

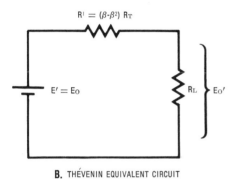

B. THÉVENIN EQUIVALENT CIRCUIT

Fig. 3-4 Variable voltage divider circuit with significant load current

Fig. 3-5 shows the loading error as a function of relative wiper position for two cases. One where the load resistance is only twice the value of the total resistance and another where R_L is 10 R_T. Fig. 3-6 is a tabulation of percentage loading errors over a range of R_L from 0.1 R_T to 100R_T. The solid line of Fig. 3-7 shows the maximum value of loading error (occurring at $\beta = 0.5$). The dashed line indicates the maximum loading error which would occur if the wiper positions were restricted to the end regions of the element.

The required minimum ratio of $\dfrac{R_L}{R_T}$ can be found from Figures 3-5, 3-6, and 3-7 for a given application having a known maximum allowable loading error. In many instances, a certain amount of compromise will be necessary since the large ratio required to assure a low loading error imposes a substantial power loss in the potentiometer element.

Power Rating as Loaded Voltage Divider. If potentiometers are used as variable voltage dividers supplying a significant output current, then power dissipation along the resistive element may be uneven. This is shown in Fig. 3-8.

The portion of resistive element from end terminal 3 to the wiper position conducts a current I_T. It is the sum of the load current I_L and the current through the remainder of the element I_E. The power dissipation per unit length of element is a function of the square of the current passing through it. Therefore, the length of the element supporting I_T will operate at a higher current density, amps per ohm, than that length carrying I_E alone and hence, will be required to support a higher temperature.

The curve of Fig. 3-9 shows how I_T varies as the wiper is moved from zero travel (zero output voltage) to maximum travel (maximum output voltage). For this example, an arbitrary load resistance equal to five times the potentiometer's total resistance was chosen.

Referring to Fig. 3-8, with the wiper at zero travel, I_L drops to zero and the only current through the potentiometer is due to the input voltage and resistive element. This state is analogous to an unloaded voltage divider application. As the wiper travel is increased, the output voltage and load current increase until a maximum value is reached:

$$I_{Lmax} = \frac{E_I}{R_L}$$

Note that the maximum power dissipation per unit length of element actually occurs just as the wiper reaches the end point at terminal 1. At the same time, the total power being dissipated in the portion of the element between terminals 3 and 2 approaches zero. The greatest *spot* temperature rise along the element will occur for a load current slightly below I_L maximum. This condition corresponds with a wiper position near the maximum output voltage setting. The exact values will depend upon the thermal conductivity of the adjacent element core and surrounding structure.

When considering power and load current requirements, the manufacturer's maximum wiper current rating must be respected under all possible conditions of wiper setting, ambient temperature, load resistance and output voltage. Remember, it is power dissipation concentration that can produce a localized elevated temperature that is detrimental to the potentiometer's life. To determine the maximum power rating and maximum wiper current rating, consult the published data sheet or the manufacturer.

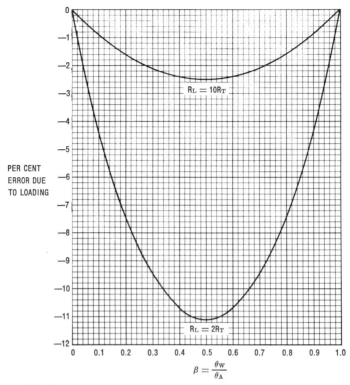

PER CENT
ERROR DUE
TO LOADING

$R_L = 10R_T$

$R_L = 2R_T$

$$\beta = \frac{\theta_W}{\theta_A}$$

Fig. 3-5 Loading error is a function of the relative values of R_L and R_T

$\frac{R_L}{R_T}$ ↓ β→	0.1	0.2	0.3	0.4	0.5	0.6	0.7	0.8	0.9
0.10	-47.37	-61.54	-67.74	-70.59	-71.43	-70.59	-67.74	-61.54	-47.37
0.15	-37.50	-51.61	-58.33	-61.54	-62.50	-61.54	-58.33	-51.61	-37.50
0.22	-29.03	-42.11	-48.84	-52.17	-53.19	-52.17	-48.84	-42.11	-29.03
0.32	-21.95	-33.33	-39.62	-42.86	-43.86	-42.86	-39.62	-33.33	-21.95
0.46	-16.36	-25.81	-31.34	-34.29	-35.21	-34.29	-31.34	-25.81	-16.36
0.68	-11.69	-19.05	-23.60	-26.09	-26.88	-26.09	-23.60	-19.05	-11.69
1.00	-8.26	-13.79	-17.36	-19.35	-20.00	-19.35	-17.36	-13.79	-8.26
1.50	-5.66	-9.64	-12.28	-13.79	-14.29	-13.79	-12.28	-9.64	-5.66
2.20	-3.93	-6.78	-8.71	-9.84	-10.20	-9.84	-8.71	-6.78	-3.93
3.20	-2.74	-4.76	-6.16	-6.98	-7.25	-6.98	-6.16	-4.76	-2.74
4.60	-1.92	-3.36	-4.37	-4.96	-5.15	-4.96	-4.37	-3.36	-1.92
6.80	1-.31	-2.30	-3.00	-3.41	-3.55	-3.41	-3.00	-2.30	-1.31
10.00	-0.89	-1.57	-2.06	-2.34	-2.44	-2.34	-2.06	-1.57	-0.89
15.00	-0.60	-1.06	-1.38	-1.57	-1.64	-1.57	-1.38	-1.06	-0.60
22.00	-0.41	-0.72	-0.95	-1.08	-1.12	-1.08	-0.95	-0.72	-0.41
32.00	-0.28	-0.50	-0.65	-0.74	-0.78	-0.74	-0.65	-0.50	-0.28
46.00	-0.20	-0.35	-0.45	-0.52	-0.54	-0.52	-0.45	-0.35	-0.20
68.00	-0.13	-0.23	-0.31	-0.35	-0.37	-0.35	-0.31	-0.23	-0.13
100.00	-0.09	-0.16	-0.21	-0.24	-0.25	-0.24	-0.21	-0.16	-0.09

PERCENT OUTPUT VOLTAGE ERROR

Fig. 3-6 Loading errors for a variable voltage divider

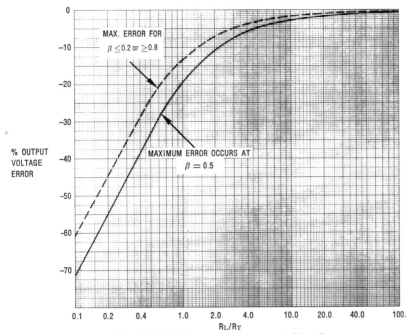

Fig. 3-7 Maximum uncompensated loading error

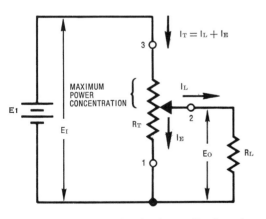

Fig. 3-8 Power dissipation is not distributed uniformly in a loaded voltage divider application

Compensating Loading Errors. The addition of a single compensating resistor, R_1 in Fig. 3-10A, will provide a limited reduction in loading error. Using the equivalent circuit given in Fig. 3-10B, the output voltage and the fractional error can be calculated. The output voltage is:

$$E_O' = E' + \left[\frac{E'' - E'}{R'' + R'}\right]R' =$$

$$E' + \left[\frac{\dfrac{E'}{\beta}\left(\dfrac{1}{1+\eta}\right) - E'}{(\beta - \beta^2)R_T + \left(\dfrac{\eta}{1+\eta}\right)R_L}\right](\beta - \beta^2)R_T$$

The error is:

$$\delta = \frac{E_O - E_O'}{E_O'}$$

$$= \left[\frac{\dfrac{1}{\beta(1+\eta)} - 1}{(\beta - \beta^2) + \left(\dfrac{\eta}{1+\eta}\right)\dfrac{R_T}{R_L}}\right](\beta - \beta^2)$$

$$= \frac{\dfrac{1}{\eta}(1 - \beta^2) - (\beta - \beta^2)}{\left(\dfrac{1}{\eta} + 1\right)(\beta - \beta^2) + \dfrac{R_L}{R_T}}$$

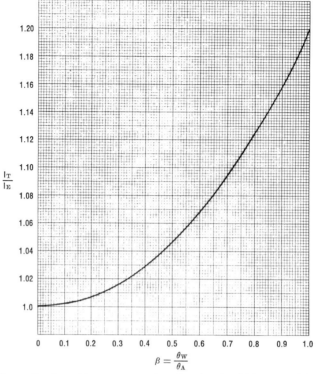

$$\beta = \frac{\theta_W}{\theta_A}$$

Fig. 3-9 Normalized total input current for the circuit of Fig. 3-8 with $R_L = 5R_T$

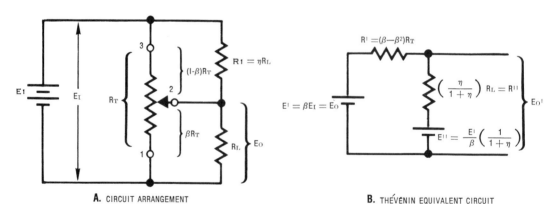

A. CIRCUIT ARRANGEMENT

B. THÉVENIN EQUIVALENT CIRCUIT

Fig. 3-10 A limited degree of compensation for loading effects is possible by use of a single additional resistor

where all terms except η, the ratio of the compensation resistor to the load resistor, have been defined previously. The equivalent circuit of Fig. 3-10B is determined by applying Thévenin's Theorem twice to the circuit of Fig. 3-10A. First, with R_1 and R_L disconnected and then with only the potentiometer disconnected.

Look back at the curves of Fig. 3-5. Notice that the error is greatest when the wiper is at the center position of the potentiometer's resistive element. Intuitively, it might be supposed that if the center region could be forced to zero error, by the proper selection of a compensation resistor, the optimum design would result. Using the formula for compensated output voltage error given above, insert the value $\beta = 0.5$, which corresponds to the center of travel. Then determine the required value for η, to make the error

58

zero:

$$\delta = \frac{\left(\dfrac{1}{\eta}\right)(1-0.5^2)-(0.5-0.5^2)}{\left(\dfrac{1}{\eta}+1\right)(0.5-0.5^2)+\dfrac{R_L}{R_T}}$$

Set the numerator equal to zero:

$$\frac{1}{\eta}(.25)-.25 = 0$$

$$\eta = 1$$

This R_1 to R_L ratio is simple to achieve, but look at the error values for other possible positions of the wiper. Fig. 3-11 presents a plot of the errors resulting from varying degrees of compensation for a specific example where $R_L = 10R_T$. The bottom curve describes the error resulting from no compensation. It indicates a maximum error magnitude of about 2.5%.

The top curve covers the case where the compensating resistor is made equal to the load resistor. Notice that zero loading error is achieved when $\beta = 0.5$ and, for all values of β from 0.33 to 1.0, the error magnitude is lower than for the uncompensated condition. However, for the lower one third of the wiper travel, the error rises sharply and exceeds the worst-case uncompensated error for all values of β. This indicates that the $R_1 = R_L$ degree of compensation is wise only where the active wiper travel will be restricted to the upper portion of the element. Fig. 3-11 also shows other curves for lesser degrees of compensation. Fig. 3-12 is a tabulation of compensated loading errors for several possible degrees of compensation for the condition of $R_L = 10R_T$.

Fig. 3-13 lists the loading errors for moderate ($\eta = 3$) compensation at varying loading ratios. Compare this table with Fig. 3-6. Notice that a substantial reduction in loading error has occurred for a major portion of the wiper travel. Always consider the increased error in the lower regions of wiper travel before employing this method of compensation. In some systems, the errors due to loading or overcompensation may actually be useful as compensation for other possible errors.

Varying the Adjustment Range. The output voltage for the basic variable voltage divider, shown in Fig. 3-1, may be adjusted for any value within the A_V specification between zero and the full input voltage. Many applications do not require this much variation and, in fact, need to have definite limits applied to the adjustment range.

Fixed resistors, placed in series at either one or both ends of the potentiometer's element, as shown in Fig. 3-14, can be used to restrict the minimum and maximum output voltage. The overall effect is equivalent to a potentiometer with a limited wiper travel.

The element of the equivalent potentiometer consists of the two end resistors R_1 and R_2 together with the potentiometer's resistive element. Consider the equivalent parameters of this composite.

The equivalent total resistance is:

$$R_T' = R_1 + R_2 + R_T$$

The minimum equivalent travel position is:

$$\beta_{min}' = \frac{R_2}{R_T'}$$

The maximum equivalent travel position is:

$$\beta_{max}' = \frac{R_T + R_2}{R_T'}$$

The adjustment range is:

$$\Delta\beta' = \beta_{max}' - \beta_{min}' = \frac{R_T + R_2}{R_T'} - \frac{R_2}{R_T'} = \frac{R_T}{R_T'}$$

The formulas given for loading error and maximum loading error are applicable to the composite potentiometer when R_T' is substituted for R_T and β' is substituted for β. Since the equivalent relative travel is restricted to the minimum and maximum limits given above, it may be that the maximum theoretical loading error, occurring at $\beta = 0.5$, may not occur within the adjustment range.

The following text is a step by step generalized design example. The values for E_I, E_{Omin}, E_{Omax}, and R_L are all known. Either the table of Fig. 3-6 or the curve of Fig. 3-7 can be used to determine the minimum $\dfrac{R_L}{R_T}$ ratio that will restrict the maximum loading error to an acceptable limit. For example, if the loading error must be held to less than 2.5%, $\dfrac{R_L}{R_T}$ must be 10 or more.

A tentative value for R_T' can now be computed:

$$R_T' = \frac{R_L}{\left(\dfrac{R_L}{R_T}\right)_{min}}$$

The total resistance of the potentiometer is:

$$R_T = R_T' \cdot \Delta\beta' = R_T'\left[\frac{E_{Omax} - E_{Omin}}{E_I}\right]$$

The value of total resistance obtained from the above formula is unlikely to be a standard value. Choose an available TR value as close as

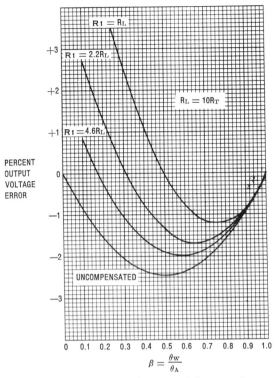

Fig. 3-11 Output error for several degrees of compensation

PERCENT OUTPUT VOLTAGE ERROR									
$\dfrac{R_1}{R_L}$ ↓ β →	0.1	0.2	0.3	0.4	0.5	0.6	0.7	0.8	0.9
NO R_1 COMP.	-0.89	-1.57	-2.06	-2.34	-2.44	-2.34	-2.06	-1.57	-0.89
10.	-0.09	-0.94	-1.57	-1.99	-2.19	-2.18	-1.96	-1.53	-0.88
6.8	0.29	-0.65	-1.35	-1.82	-2.07	-2.11	-1.92	-1.51	-0.88
4.6	0.85	-0.20	-1.01	-1.57	-1.90	-1.99	-1.86	-1.48	-0.87
3.2	1.61	0.39	-0.55	-1.24	-1.66	-1.84	-1.77	-1.44	-0.86
2.2	2.75	1.28	0.12	-0.74	-1.32	-1.62	-1.64	-1.39	-0.84
1.5	4.43	2.60	1.13	0.00	-0.80	-1.28	-1.45	-1.30	-0.82
1.0	7.07	4.65	2.69	1.15	0.00	-0.76	-1.15	-1.16	-0.79
0.68	10.77	7.51	4.85	2.73	1.11	-0.04	-0.74	-0.97	-0.74
0.46	16.24	11.72	8.02	5.04	2.72	1.00	-0.13	-0.70	-0.66
0.32	23.54	17.26	12.16	8.05	4.82	2.37	0.66	-0.33	-0.57
0.22	34.21	25.25	18.07	12.32	7.78	4.30	1.78	0.20	-0.42
0.15	49.67	36.58	26.33	18.24	11.89	6.98	3.36	0.95	-0.22
0.10	72.88	53.06	38.10	26.58	17.65	10.76	5.61	2.04	0.09

Fig. 3-12 Compensated loading error where $R_L = 10R_T$

PERCENT OUTPUT VOLTAGE ERROR									
$\frac{R_L}{R_T}$ $\downarrow$ $\beta \rightarrow$	0.1	0.2	0.3	0.4	0.5	0.6	0.7	0.8	0.9
1.000	16.07	4.40	-3.65	-9.09	-12.50	-14.14	-14.06	-12.09	-7.74
1.500	11.11	3.11	-2.62	-6.59	-9.09	-10.26	-10.11	-8.56	-5.35
2.200	7.76	2.21	-1.88	-4.76	-6.58	-7.41	-7.26	-6.08	-3.74
3.200	5.42	1.56	-1.34	-3.41	-4.72	-5.30	-5.17	-4.30	-2.61
4.600	3.81	1.11	-0.96	-2.44	-3.38	-3.79	-3.69	-3.05	-1.84
6.800	2.60	0.76	-0.66	-1.69	-2.34	-2.62	-2.54	-2.09	-1.25
10.000	1.78	0.52	-0.45	-1.16	-1.61	-1.81	-1.75	-1.44	-0.86

Fig. 3-13 Compensated loading error where $R_1 = 3R_L$

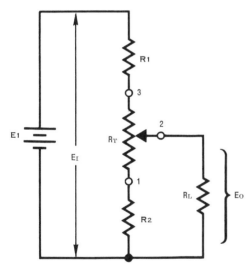

Fig. 3-14 Adjustment range is fixed by resistors placed in series with the potentiometer element

possible to, but not greater than, the calculated value. Now, recompute the value of R'_T:

$$R'_T = \frac{R_T}{\Delta \beta'} = R_T \left[\frac{E_I}{E_{Omax} - E_{Omin}} \right]$$

The end resistors may then be computed from:

$$R_1 = \left[1 - \frac{E_{Omax}}{E_I} \right] R'_T$$

$$R_2 = \beta'_{min} R'_T = \left[\frac{E_{Omin}}{E_I} \right] R'_T$$

As a specific example, assume that the following requirements are given:

$$E_I = 10v \qquad E_{Omin} = 2v \qquad E_{Omax} = 9v$$

$$R_L = 10k \text{ max. loading error} = 2.5\%$$

Fig. 3-6 indicates that $\frac{R_L}{R_T}$ must be 10 or more for a maximum loading error of 2.5%. This gives an initial value for R'_T of $10K/10 = 1K$. Now, compute the value for R_T:

$$R_T = (1k) \left[\frac{9-2}{10} \right] = 700 \text{ ohms}$$

Unfortunately, 700 ohms is not available as a standard value. The nearest ones are, say, 500 and 1000 ohms for the potentiometer type being considered. In order to keep the loading error within the given requirements, choose the lower TR, 500 ohms.

Now, recompute R'_T, using the actual value of TR:

$$R'_T = 500 \left[\frac{10}{9-2} \right] = 714.3 \text{ ohms}$$

Then,

$$R_1 = \left(1 - \frac{9}{10} \right)(714.3) = 71.4 \text{ ohms}$$

$$R_2 = \left(\frac{2}{10} \right)(714.3) = 142.9 \text{ ohms}$$

The resistors used for R1 and R2 should have a TC which matches that of the potentiometer element, if optimum temperature stability is to be achieved. Remember also that the resistance tolerance of the potentiometer will affect the actual range limits. In some applications, it may be wise to use trimming potentiometers for the end resistors to allow precise control of the limits.

The effective resolution of the equivalent potentiometer is improved by a factor equal to $\frac{R'_T}{R_T}$, since the actual resolution of the potentiometer used applies to the adjustment range only.

Load compensation, as described previously, can be applied to the limited adjustment range configuration of Fig. 3-14 using the equivalent composite parameters.

Optimizing Resolution and Adjustability. In many applications, the conformity of the output voltage function is of secondary importance as compared to the accuracy of adjustment over a limited range. A particular application might require an adjustable voltage through a range of only 10% of the total input voltage, most of the time. Occasionally, it may be necessary for the same potentiometer to provide a significantly greater adjustment range. If the basic variable voltage divider approach is used, the normal adjustment range will be a small portion of the potentiometer's actual travel. The full adjustment range will be used only on those occasions where extremes are required.

The output may be intentionally loaded as shown in Fig. 3-15 to alter the output in a way that yields a more desirable adjustment capability. Fig. 3-16 shows the results for three different degrees of this shaping. Note the curve for the condition where R1 and R2 are both

made equal to $0.1R_T$. A major portion of the potentiometer's total travel is required to vary the output voltage from $0.4E_I$ to $0.6E_I$. Specifically, almost 70% ($\beta = .16$ to $.82$) of the wiper total travel is required to produce 20% ($\frac{E_O}{E_I} = .4$ to $.6$) change in output voltage. For most of this range, the output is relatively linear. If required, an output voltage covering the entire range of E_I is possible.

The result is resolution and adjustability are improved by a factor of nearly 6 in the center region. This is compared with the unloaded voltage divider whose output function is indicated by the straight line in Fig. 3-16. Note that the slope for the loaded case is less than the unloaded case in the range $0.1E_I$ to $0.9E_I$. The magnitude of the slope increases rapidly outside this region.

Fig. 3-16 also includes the output function resulting from making R1 and R2 both equal to R_T. With these values, only a very slight improvement is obtained.

The values of R1 and R2 need not be equal except when the output voltage of major interest is around $0.5E_I$. Varying the ratio between the loading resistors will shift the region of improved resolution as shown by the two examples given in Fig. 3-17. Again, the straight line output function of the unloaded case is included for reference.

VARIABLE CURRENT RHEOSTAT MODE

Many applications use the variable resistance between the wiper and one end terminal as a method of current adjustment. This two-terminal method of connection is frequently referred to as simply the *rheostat mode*. The term *variable resistance mode* is also in common usage. However, it is felt that this latter terminology is descriptive of the device's primary characteristic, variable resistance, rather than a particular application mode.

Fig. 3-18A illustrates the basic circuit arrangement. Fig. 3-18B shows three possible load current vs. wiper position curves. The total circuit resistance and applied voltage are equal in all three cases. Only the ratio of potentiometer TR to load resistance R_L has been varied. As the chosen TR becomes larger, compared to R_L, a greater range of load current variability is realized.

The choice of input and output terminals is arbitrary since the potentiometer, when applied as a current control, is a two-terminal bidirectional device. Of course, one of the selected terminals must be the movable contact terminal 2. If end terminal 1 is chosen as the second

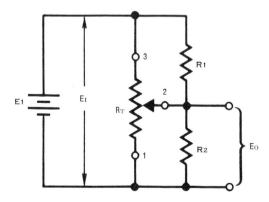

Fig. 3-15 Effective resolution over a limited range can be improved by intentional loading

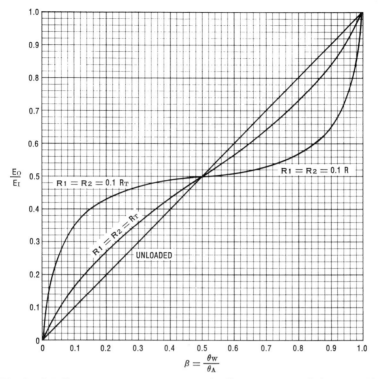

Fig. 3-16 Effective resolution in center of adjustment range is improved by loading

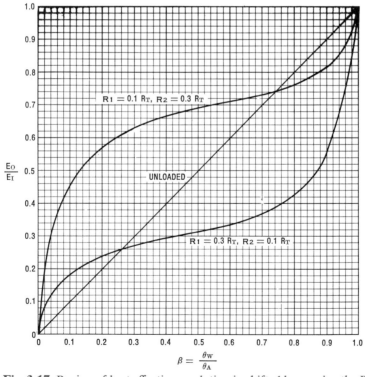

Fig. 3-17 Region of best effective resolution is shifted by varying the R_1/R_2 ratio

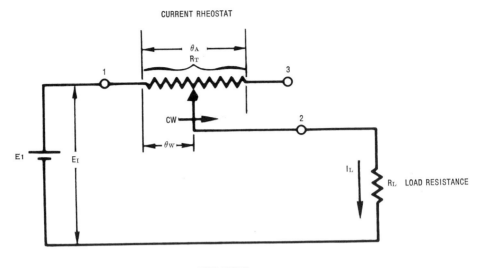

A. ILLUSTRATIVE CIRCUIT

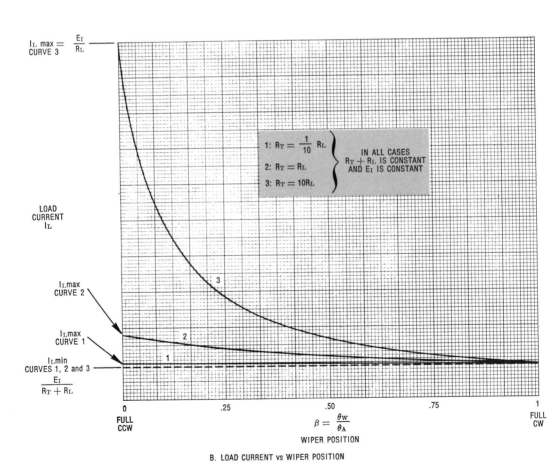

B. LOAD CURRENT vs WIPER POSITION

Fig. 3-18 Potentiometer's variable resistance used to control a load current

terminal, the current control curves will be as shown in Fig. 3-18B. If terminal 3 is chosen as the second terminal, the current control curves will be a mirror image of those shown in Fig. 3-18B. The only criteria for choosing terminal 1 or 3 is the user's preference for current change relative to direction of wiper adjustment rotation.

Importance of Resistance Parameters. The tolerance of the potentiometer's total resistance is frequently not critical in voltage divider applications. This is because proper function depends upon division ratio rather than total resistance. However, in the variable current application, the total resistance becomes significant because it determines the range of resistance adjustment possible. The effects of minimum resistance and end resistance, which include contact resistance, also become more significant in this operational mode. Note, in Fig. 3-19 that the contact resistance R_C is always in series with that portion of the element supporting the current being controlled. Also, the range of resistance available for current controlling purposes lies between the devices minimum resistance R_M and total resistance R_T, subject to the tolerance bands of these parameters.

Fig. 3-20 shows the circuit and output function for a potentiometer whose minimum resistance is not equivalent to its end resistance R_E.

Figures 3-19 and 3-20 are given solely to show those resistance factors which may affect the variable resistance range of a particular device used for current control. In application, sound design procedure requires the assumption that regardless of the type of potentiometer the range of resistance variability will be at least from the absolute minimum resistance value to the minimum R_T value, i.e., $R_T - \%$ tolerance.

Adjustability. The adjustability of in-circuit resistance A_R is a direct indication of the accuracy with which any arbitrarily selected resistance value can be achieved. The A_R specification pertains explicitly to the rheostat mode. For example, if a potentiometer having an $A_R = 0.1\%$ is employed in an application where $R_T = 1000\,\Omega$, the circuit designer can expect to set any desired value of resistance in the range $0{-}1000\,\Omega$ within $1\,\Omega$.

In addition to indicating resistance setting accuracy, the A_R parameter is specified together with a maximum time-to-set duration of 20 seconds. This time factor limitation assures the potentiometer user that the specified accuracy can be obtained without an extended period of trial and error.

Effect of TC. The temperature coefficient of resistance can be quite important when the potentiometer is to be used in the rheostat mode.

As an example, consider a cermet potentiometer having a TC of ± 100 ppm/°C used in an environment which might experience a total temperature variation of 80°C. The *total resistance* could exhibit a variation due to temperature of 8,000 ppm or 0.8%. The negative shift can be compensated by choosing a TR sufficiently high or by choosing a potentiometer whose element construction has a lower TC.

Effects of Resolution. Potentiometer resolution limitations affect rheostat applications directly in a manner similar to its effect in the voltage divider mode. Remember, resolution is a given percentage of a wirewound potentiometer's total resistance, and its effect becomes increasingly important as the total in-circuit resistance is reduced.

Consider an example where a 10,000 ohm potentiometer, having a specified resolution of 0.4%, 40Ω, is used to control a current through a 1000Ω load resistance. A circuit like that shown in Fig. 3-18A could be used. Since $R_T = 10R_L$, curve number 3 in Fig. 3-18B is applicable. In this circuit arrangement, maximum in-circuit resistance is obtained with the wiper in the full clockwise position. With the potentiometer at the maximum resistance, the load current is minimum and if $E_I = 100$v:

$$I_{Lmin} = \frac{E_I}{R_T + R_L} = \frac{10^2}{10^4 + 10^3} = 9.1 \text{ milliamps}$$

The total circuit resistance is 11,000 ohms $(10^4 + 10^3)$ and the chosen potentiometer resolution is 40 ohms. Since 40Ω is .36% of 11,000Ω, the resolution of load current is:

$$.0036 \times 9.1 = .03 \text{ milliamps}$$

When the potentiometer is at its minimum resistance, the load current is maximum or:

$$I_{Lmax} = \frac{E_I}{R_L} = \frac{10^2}{10^3} = 100 \text{ milliamps}$$

The maximum load current must never exceed the manufacturer's maximum wiper current rating.

The total circuit resistance is now 1000 ohms, R_L alone. Since 40Ω is 4% of 1000Ω, the resolution of load current is:

$$.04 \times 100 = 4 \text{ milliamps}$$

The preceding example and curve number 3 in Fig. 3-18B demonstrate that the large ratio of R_T to R_L required for wide range current adjustment is obtained with a corresponding sacrifice in load current resolution at the higher current (lower resistance) settings. Curves 1 and 2 in

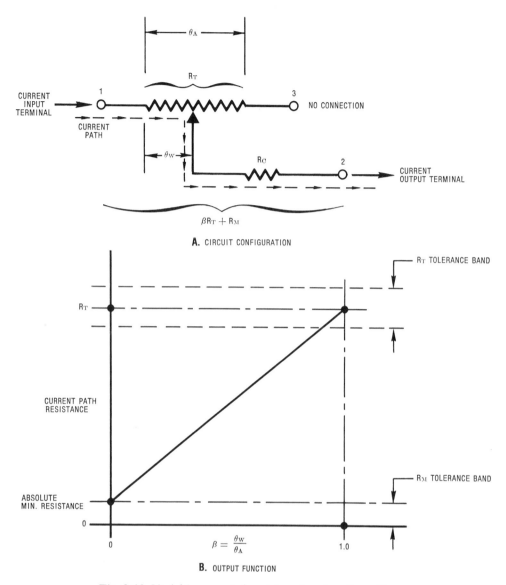

A. CIRCUIT CONFIGURATION

B. OUTPUT FUNCTION

Fig. 3-19 Variable current rheostat mode when $R_M = R_E$

Fig. 3-18B show that a smaller current adjustment range is provided by lower ratios of R_T to R_L but load current resolution is relatively constant over the entire adjustment range.

Power Rating as a Rheostat. The power rating specification given on manufacturer's data sheets applies to the potentiometer in the voltage divider mode as discussed previously. In that application, the power dissipation may be viewed as distributed uniformly along the entire element. When the unit is to be used in the rheostat or two-terminal mode, only a fraction of the total element may be dissipating power for a given setting of the movable contact. That is, as

the wiper is moved from one end of the element to the other, the length of the active portion of the element also changes.

An acceptable method of relating the published power rating to the specific rheostat application is to compute a maximum allowable current. This may be done using the following equation:

$$I_M = \sqrt{\frac{P}{R_T}} \quad \text{(100 ma, absolute maximum)}$$

where P is the allowable maximum power dissipation taken from the manufacturer's data

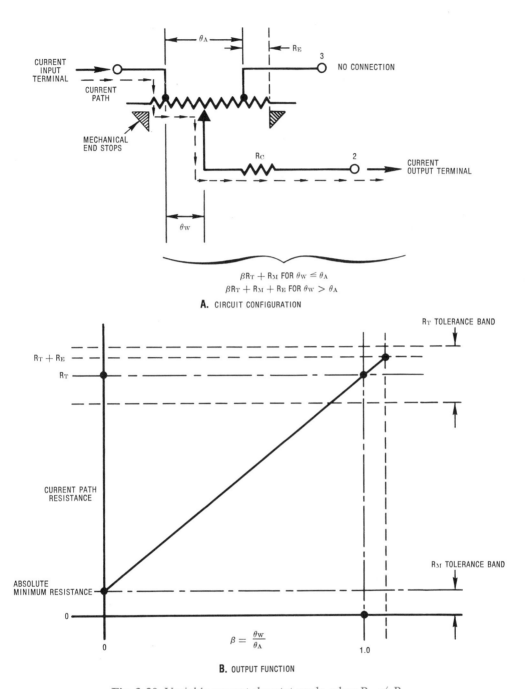

$$\beta R_T + R_M \text{ FOR } \theta_W \leq \theta_A$$
$$\beta R_T + R_M + R_E \text{ FOR } \theta_W > \theta_A$$

A. CIRCUIT CONFIGURATION

$$\beta = \frac{\theta_W}{\theta_A}$$

B. OUTPUT FUNCTION

Fig. 3-20 Variable current rheostat mode when $R_M \neq R_E$

sheet and R_T is the total resistance. If the power is expressed in watts and the resistance in ohms, the current will be given in amps.

A further restriction on maximum current is necessary due to the two-terminal mode of operation. Unlike the voltage divider mode, the rheostat requires the total current flowing

through the resistive element to pass through the wiper circuit. The pressure contact junction of the wiper and element is not always capable of currents as high as the element alone. As already mentioned, the power rating for the voltage divider mode assumes an insignificant wiper current. Therefore, the maximum current in the

67

rheostat mode must be limited to the maximum allowable wiper current for the particular potentiometer being used.

100ma is a common maximum wiper current rating for most wirewound and cermet type units. The manufacturer's data sheet, for the particular unit being considered, should be consulted to ascertain the limiting value of wiper current for rheostat applications.

Once again, refer to the circuit and response curve of Fig. 3-18. The function of the potentiometer is to vary the current through load resistor R_L. When the potentiometer is adjusted fully counterclockwise, the only resistance remaining in the circuit will be that of the load resistor. This is the lowest total circuit resistance condition, hence the high current condition of the circuit. In this state, the total current in the circuit must be limited to the maximum value explained in the previous paragraph. Relating this limitation to circuit voltage and load resistance:

$$\frac{E_I}{R_L} = \sqrt{\frac{P}{R_T}} \quad \text{(100 ma, absolute maximum)}$$

As the wiper is caused to move clockwise, more resistance will be added into the circuit and, therefore, the total current will decrease remaining below the maximum allowable magnitude. The current flowing through the wiper and, hence through the load resistor, is graphically represented in Fig. 3-18B. Applying this maximum current limitation to a rheostat design will insure that the maximum power rating of the potentiometer will never be exceeded.

Using the maximum current limit is only slightly conservative for potentiometers which have rather poor thermal characteristics. For those units which have a good thermal path in the element structure, the maximum power which can safely be dissipated is somewhat larger than that limited by the maximum current calculation. Potentiometers designed specifically for power control or other high power operations have elements wound on an insulated metal core which aids in the distribution of heat. Such potentiometers can have a maximum power limit in the 20 to 30 percent travel range that is twice the 20 to 30 percent of the value indicated by the maximum current calculation.

Some cermet potentiometer designs also have good thermal characteristics, and hence a higher permissible power for limited element applications. Do not assume that the potentiometer will never be adjusted to a particular setting. Always assume that any position is possible and design for that possibility.

Controlling the Adjustment Range. The potentiometer, when used in the rheostat mode, provides a range of resistance from the absolute minimum resistance to the TR. Fixed resistors may be added to alter the adjustment range. Fig. 3-21 shows five basic arrangements and gives formulas for the resulting resistance ranges. Note that the effects of absolute minimum resistance need only be considered in conjunction with minimum settings.

A single series resistor R1 as shown in Fig. 3-21B, provides an effective offset (equal to its value) to the resistance parameters. The resulting output function is still a linear function of relative wiper travel. The effect of R1 is most pronounced at the minimum resistance setting and is often necessary to prevent excess current flow. In all instances, analogous to Figs. 3-21A and B, the total circuit current passes through the potentiometer's wiper circuit.

Placing a fixed resistor in parallel with the potentiometer's element as in Fig. 3-21C, has its most significant effect when the wiper is positioned fully clockwise. The resulting output function is a nonlinear function of travel as illustrated by Fig. 3-22. At the minimum resistance setting, the absolute minimum resistance of the potentiometer is shunted by R2 resulting in a resistance effectively lower than the minimum resistance. This condition is indicated as approximately R_M ($\sim R_M$) on the chart of Fig. 3-21.

Adding a second resistor in the manner shown in Fig. 3-21D, provides the same type of output function shown in Fig. 3-22, but all resistance values are increased by an amount equal to R1. Note that in Fig. 3-21, circuit D is simply the combination of circuits B and C.

In the final arrangement shown in Fig. 3-21E, the shunt resistor R2 is placed in parallel with the series string of the potentiometer TR and R1. The minimum resistance becomes the parallel equivalent of R1 and R2. The maximum terminal resistance is the parallel equivalent of R2 and the sum of R1 and R_T. This configuration permits the control of currents higher than the device's maximum current rating. When the ratio of R1 to R2 is large, most of the total circuit current flows through R2, and only a small portion flows through the potentiometer.

The circuit arrangement of Fig. 3-21E is frequently used where the potentiometer is to provide some small fractional adjustment in the equivalent resistance of a fixed resistor. For example, assume that R1 is equal to 10R2. Fig. 3-23 shows the resulting output functions obtained for two values of R_T. When $R_T = 10R_2$, the total effective circuit resistance varies from about 0.91R2 to a little over 0.95R2.

CIRCUIT ALL CASES CW→	RESISTANCE PARAMETERS (2)		
	GENERAL $f\left(\dfrac{\theta_W}{\theta_A}\right)$	MINIMUM (1)	MAXIMUM
A. 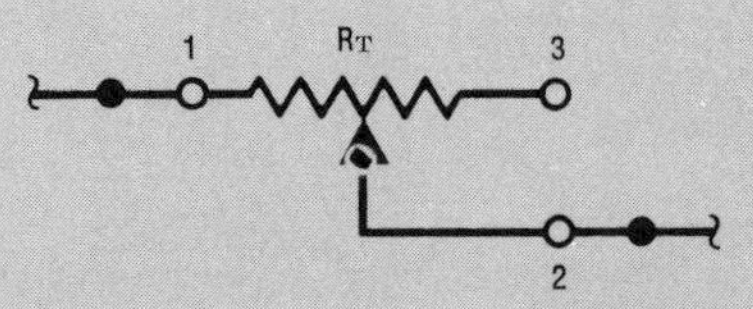	$\beta\,R_T$	R_M	R_T
B. 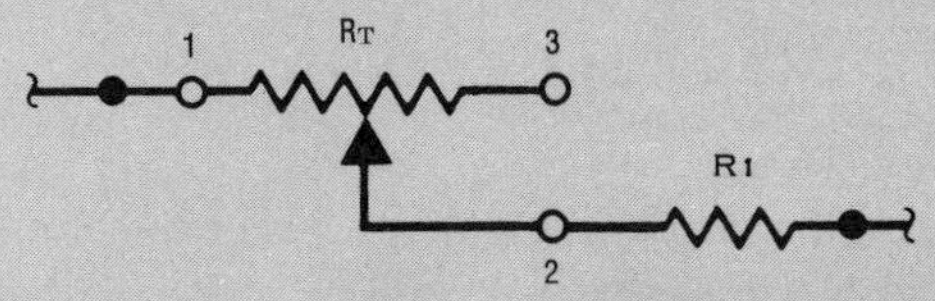	$\beta\,R_T + R_1$	$R_M + R_1$	$R_T + R_1$
C. 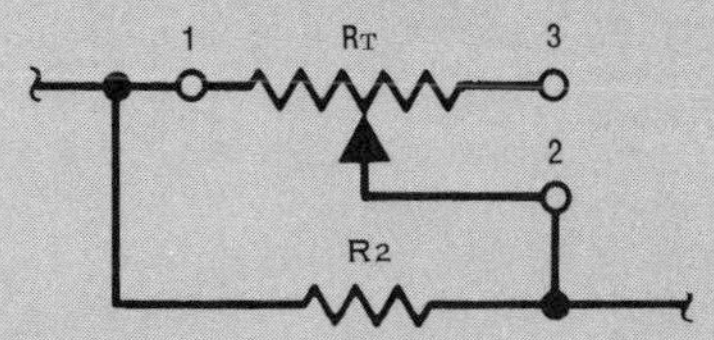	$\dfrac{\beta\,R_T\,R_2}{R_2 + \beta\,R_T}$	$\sim R_M$ LOAD CURRENT CRITICAL	$\dfrac{R_2\,R_T}{R_2 + R_T}$
D. 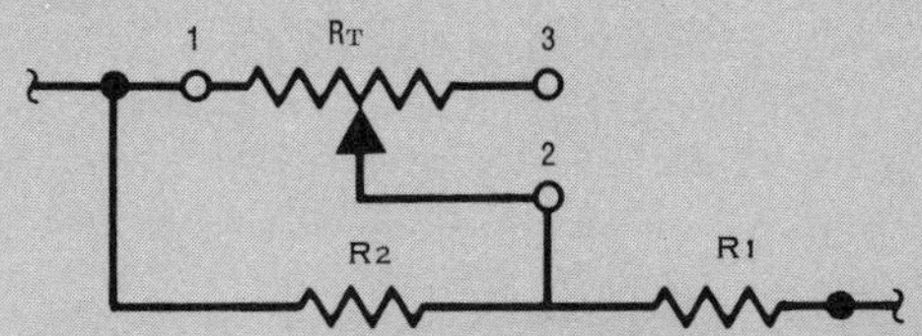	$\dfrac{\beta\,R_T\,R_2}{R_2 + \beta\,R_T} + R_1$	$\sim R_M + R_1$	$\dfrac{R_2\,R_T}{R_2 + R_T} + R_1$
E. 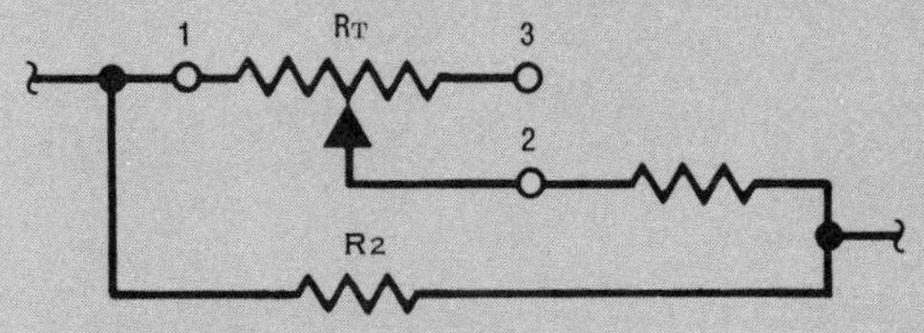	$\dfrac{R_2\,(\beta R_T + R_1)}{\beta\,R_T + R_2 + R_1}$	$\dfrac{R_2\,(R_M + R_1)}{R_M + R_2 + R_1}$	$\dfrac{R_2\,(R_T + R_1)}{R_T + R_2 + R_1}$

1. IN THE MAJORITY OF APPLICATIONS, MINIMUM RESISTANCE CAN BE NEGLECTED
2. IN ALL CASES, THE RESISTANCE PARAMETERS ARE MEASURED BETWEEN THE CIRCUIT-POINTS INDICATED BY THE SOLID DOTS, ● •

LEGEND:
R_M = MINIMUM RESISTANCE
R_T = TOTAL RESISTANCE
$$\beta = \frac{\theta_W}{\theta_A}$$

Fig. 3-21 Fixed resistors vary the adjustment range

If the value of R_2 is made 7.5% higher than the center of the desired adjustment range, then the composite circuit allows an adjustment of about $\pm 2\%$ around the center value. The output function is slightly nonlinear in the end regions, but this does not represent a problem for most trimming applications.

When the relative value of the potentiometer's total resistance is increased to $100R_2$, a greater range of adjustment is obtained. However, the resulting output function becomes even more nonlinear and most of the adjustment will occur in the lower 50% of potentiometer travel, i.e., where $\dfrac{\theta_W}{\theta_A} \leqq 0.5$.

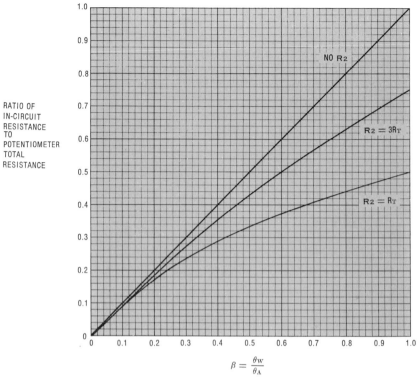

RATIO OF
IN-CIRCUIT
RESISTANCE
TO
POTENTIOMETER
TOTAL
RESISTANCE

NO R_2

$R_2 = 3R_T$

$R_2 = R_T$

$$\beta = \frac{\theta_W}{\theta_A}$$

Fig. 3-22 A fixed resistance in parallel with a variable resistance controls the range
See Fig. 3-21C

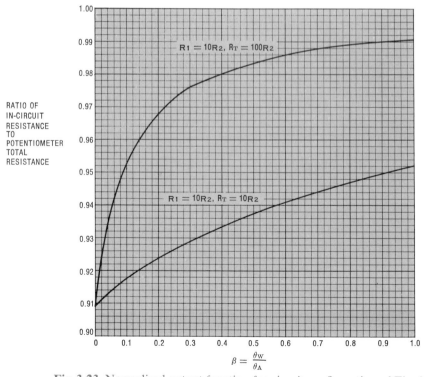

RATIO OF
IN-CIRCUIT
RESISTANCE
TO
POTENTIOMETER
TOTAL
RESISTANCE

$R_1 = 10R_2, R_T = 100R_2$

$R_1 = 10R_2, R_T = 10R_2$

$$\beta = \frac{\theta_W}{\theta_A}$$

Fig. 3-23 Normalized output function for circuit configuration of Fig. 3-21E

70

DATA INPUT

Another basic application of potentiometers is that of data input. Although the actual circuitry may be that of either the variable voltage divider or the variable current mode, there are certain special considerations of data input which deserve discussion.

In this class of application, the potentiometer serves as a means whereby an operator may inject some known value of a given control function into a system by use of some form of dial or scale attached to the potentiometer. Data input applications can vary widely from the simple volume or tone control on an audio amplifier to the high precision stable input for an analog computer.

Dials. Simple dials for single turn potentiometers may be silk-screened or engraved on the mounting panel to provide an easy means for data input with moderate accuracy. The scale might be linear or designed for varying degrees of nonlinearity anywhere from a minor compensation for loading error to logarithmic. Fig. 3-24 illustrate some typical examples.

Various types of dials for use with multiturn potentiometers are also available. One basic style, illustrated in Fig. 3-25, not only displays the number of dial turns but provides a vernier

Fig. 3-25 An example of a turns counting dial for a multiturn potentiometer

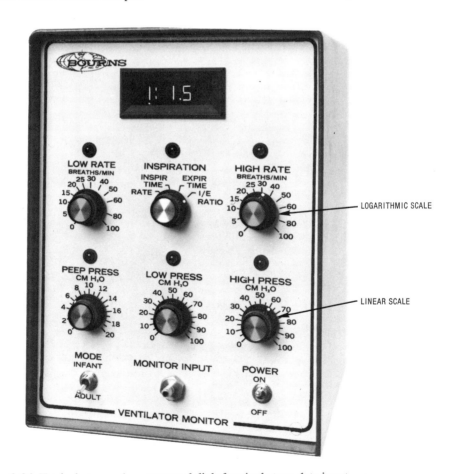

Fig. 3-24 Typical screened or engraved dials for single turn data input

dial for very accurate fractional division of each particular turn. Note the braking feature which allows a particular setting to be locked in and held until the brake is released.

Another style of multiturn dial, shown in Fig. 3-26, resembles a clock face. The short or *hour* hand indicates the number of turns while the long or *minute* hand shows fractional turns.

Fig. 3-26 A multiturn dial with a clocklike scale

For some applications, a *digital* readout dial is more practical. Fig. 3-27 shows several examples.

Some multiturn potentiometers are available with integral dials of either the clock-face or digital type. Use of this style can result in space savings and lower installation costs.

Mechanical Factors. Designs using dials with potentiometers to aid in data input must consider not only the electrical characteristics of the potentiometer but also certain mechanical factors as well.

Readability is an important consideration. For a simple single turn dial, readability is typically 1% to 2% of the total mechanical travel. For multiturn dials, the readability is typically 1% of a single rotation. This results in a readability of 0.1% of full scale for a ten-turn device.

Mechanical backlash can contribute some error if the dial is not directly attached to the potentiometer's shaft. The same dial reading obtained by approaching from different directions can result in slight differences in potentiometer output.

The effective resolution of a potentiometer-dial data input team is a combination of the elec-

Fig. 3-27 Examples of digital type, multiturn dials

trical characteristics of the potentiometer along with the possible errors due to mechanical factors.

Some dials rely on the potentiometer construction to provide end stops as a limit to mechanical travel, while other applications might require the added protection of an additional limit to the maximum travel.

Offsets. Some applications profit from a mechanical offsetting arrangement where the minimum position of the potentiometer may not be zero but some fractional portion of full scale. Consider the voltage divider shown in Fig. 3-28.

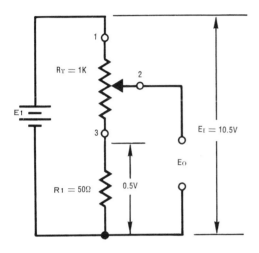

Fig. 3-28 Example of offsetting to modify the covered range of a data input potentiometer

As the potentiometer travel is increased from minimum (1) to maximum (3), the output voltage changes from 0.5V to 10.5V. Some dials will permit mechanical offsetting to display the actual voltage output. Usually these dials are designed to display 15 or more turns and thus can provide up to 50% offset on a ten-turn potentiometer.

Optimum accuracy of offset and adjustment range can be achieved with two potentiometers on each side of the data input potentiometer as demonstrated in Fig. 3-29.

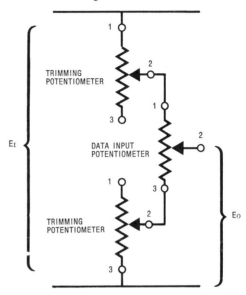

Fig. 3-29 A circuit configuration to optimize offset and adjustment range of data input potentiometer

Logging Charts and Tables. Appropriate dial scales may be designed and applied to the potentiometer mounting panel for single turn devices used where a nonlinear input is required. However, multiturn dials are designed to provide only a linear readout. When a data input application requiring a nonlinear function from a multiturn potentiometer-dial combination, a logging chart or table to relate linear dial readings to the nonlinear variable must be used. The equipment operator is then provided with the proper conversion table and uses it to determine any specific setting. This approach is also useful in making overall system conversions in an expedient manner.

CIRCUIT CONFIGURATION	PAGE	DESIGN CONSIDERATIONS
VARIABLE VOLTAGE DIVIDER MODE E_I INPUT / E_O OUTPUT	51 52 53	FUNCTION: To provide an output voltage in reduced proportion to its input voltage. OUTPUT VOLTAGE: $E_O = \beta E_I$; $\beta = \frac{\theta_W}{\theta_A}$ POWER RATING: Manufacturers data sheet. Linear derating factor between two operating points: $p = \frac{P_A - P_B}{T_A - T_B}$ Where P_A is rated power at temperature T_A and P_B is rated power at temperature T_B Allowable power dissipation at any temperature T_D within the operating temperature range: $P_D = P_A + p(T_D - T_A)$ P_A is specified power at T_A (the low limit of the operating temperature range.)
LOADED VOLTAGE DIVIDER WITH COMPENSATION R_1 COMPENSATION RESISTANCE / R_L LOAD / E_O / E_I	54 54 57 62	MAXIMUM ERROR: Due to output load (no compensation) $\delta = -\frac{R_T}{R_T + 4R_L}$ R_T IS TOTAL RESISTANCE MAXIMUM LOAD CURRENT: $I_{Lmax} = \frac{E_I}{R_L}$, never exceed manufacturer's absolute maximum rating COMPENSATION: Refer to figures 3-11, 3-12, 3-13 OPTIMIZE RESOLUTION: Intentional loading, figures 3-15, 3-16 and 3-17
VARYING THE ADJUSTMENT RANGE R_T' / R_L / $R_T - R_2$ / E_O / E_I / R_2 / R_3	59	EQUIVALENT PARAMETERS: Total Resistance $R_T' = R_1 + R_2 + R_3$ Minimum travel position β' S.S. MIN. $= \frac{R_3}{R_T}$ Maximum travel position β' S.S. MAX. $= \frac{R_T + R_3}{R_T'}$ Adjustment range $\Delta\beta' = \beta'$ MAX. $- \beta'$ MIN. $= \frac{R_T}{R_T'}$
VARIABLE CURRENT RHEOSTAT MODE NO CONNECTION / R_T / R_L / I_L / R / E_I	62 66 69	FUNCTION: To control current Total circuit resistance is changed to vary current, resolution of current changes as total circuit resistance is changed. $I_{Lmin} = \frac{E_I}{R_T + R_L}$; $I_{Lmax} = \frac{E_I}{R_L}$, never exceed manufacturer's absolute maximum rating POWER RATING: Active portion of element is varied as wiper moves $P = \frac{E_I^2 R}{(R + R_L)^2}$ $I_{Lmax} = \sqrt{\frac{P}{R_T}}$ never exceed manufacturer's absolute maximum rating Various circuit configurations for particular applications are summarized in Figure 3-21.

Fig. 3-30 Summary of application fundamentals

Notes

Notes

APPLICATION AS A CIRCUIT
ADJUSTMENT DEVICE

Chapter 4

"Design is revealed in terms of a number of fundamental principles and relationships . . . One must realize that design focuses more attention on the individual to structure his thinking along guided and hopefully productive paths. . ."

Percy H. Hill
The Science of Engineering Design

INTRODUCTION

Potentiometers were considered from their elemental circuit functions in the previous chapter. A broader look at applications shows their use as circuit adjustment, control, and precision devices. The latter two include the man-machine interface function; that is, communications between man and machine in the form of electronic input and output data. This chapter covers the first of these three important functions — adjustment devices.

Adjustment potentiometers can provide the means for compensation of various error sources that are not predictable quantities during the design phase, i.e., currents, resistances and voltages. The adjustments are made during final checkout and may never be needed again. These applications are commonly described as *set and forget* or *trimming* functions. The adjustment capability also permits correction for

long-term variances, e.g., component replacement or aging.

It is these adjustment capabilities — either set and forget or correction for long term variances — that have been most responsible for the potentiometer being called a *cost-effective component*. For it is often found by cost-benefit analysis that proper application of potentiometers *is* the most cost-effective alternative.

It would be a massive task to describe all of the possible potentiometer applications in this category and no attempt to do so is made here. A wide variety of applications will be presented to indicate typical areas where trimming potentiometers provide a valuable function.

Although the application descriptions will be brief, enough information will be included to illustrate basic adjustment techniques which may be adapted to many other circuits.

POTENTIOMETER OR FIXED RESISTORS?

For those applications having a low probability of ever needing readjustment, the *set and forget* class, it is well known that a selected fixed resistor will yield a stable performance for a longer time duration than a rheostat connected potentiometer. Even a voltage divider potentiometer could be replaced by two fixed resistors. The basic cost of two fixed resistors may be less than that of a good adjustment potentiometer, but there are other factors which should be considered carefully.

First, compare the problems of inventory. Fine adjustment capability requires that many different values of fixed resistors be available during final checkout. If the additional cost of ordering, stocking, and handling many fixed resistors is considered, the economics of the potentiometer begins to look better.

In addition, 1% precision resistors are only readily and economically available in discrete values approximately 2% apart. If the application requires better adjustment resolution than that, then two values will have to be chosen in a two-step selection process.

Proper selection of fixed resistors requires some form of test substitution, which must be temporarily attached to the circuit in order to determine the exact value required, e.g., a decade resistance box. Care must be taken not to induce noise or stray capacitance. Proper values must be chosen by a process of reading the dials on the decade box carefully, calculating the nearest available value, then obtaining a part and installing it in the assembly by hand. This means that the instructions regarding the selection process will have to be more detailed as the skill of the operator must be higher. Also the selection operation consumes more time.

Compare this with possible automatic insertion and wave soldering of a potentiometer during assembly and a simple screwdriver adjustment during check-out. The labor cost of potentiometer installation and adjustment will be lower, and finer adjustment is practical. Negligible noise or capacitive loading is induced, and final adjustment may be made when the full assembly is complete and even inserted in a case! Thus, serious consideration of the cost-effective component — a potentiometer — is well worthwhile.

Some set and forget applications suddenly need *to be remembered and reset* when a field failure occurs. If a critical component must be replaced, the original adjustment must be modified to fit new conditions. The cost for field selection and installation of a precision fixed resistor will be very high. If an adjustment potentiometer had been placed in the original design, the field service technician could instantly set the right value with very little effort. Also, no unsoldering of the old resistor and soldering of the new one would be required . . . a further saving and elimination of a risky rework operation.

POWER SUPPLY APPLICATIONS

Many applications of adjustment potentiometers are found in power supplies. Certain parameters must be adjusted to compensate for tolerance variations of components used in the power supply assembly.

Precise Output Voltage Adjustment. Even in the simplest form of regulated power supply, there are usually several components whose value will influence the final value of the output voltage. Consider the simple voltage regulator circuit of Fig. 4-1. The output voltage is determined primarily by the breakdown voltage of the voltage reference (VR) diode, D_1, and the two resistors R_3 and R_4. But E_0 is also influenced, to a lesser degree, by the base to emitter voltage of transistor Q_1 and the values of R_1 and R_2.

Assume that the purchase tolerance on the VR diode is $\pm 5\%$ at a test current value which is likely to be different from the required bias value. If 1% resistors are used for R_3 and R_4, the worst-case error in the voltage divider is $\pm 2\%$. This gives a total worst-case error of more than $\pm 7\%$. Adding all of the other error possibilities will show that the possible output voltage variation due to component tolerances can be as much as $\pm 10\%$. This is too much for most applications.

It is possible to replace either of the two divider resistors R_3 or R_4 with a potentiometer connected as a rheostat, but the preferable application mode is shown in Fig. 4-2. An adjustment potentiometer R_5 is used as a voltage divider to permit variation of the output voltage. The adjustment range is determined by the TR value of R_5 compared to R_3 and R_4. The potentiometer resistance can be chosen to provide enough adjustment range for a precise setting of the output voltage with optimum resolution, or chosen to permit a much larger variation, thus making the circuit more versatile.

If the proper trimmer is selected, economical 5% resistors can be used for R_3 and R_4, provided they are relatively stable, e.g., metal film. Further economic advantage can be achieved by purchasing VR diodes with a $\pm 10\%$ tolerance.

Always attempt to limit the adjustment range

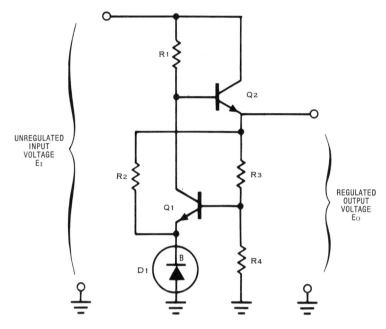

Fig. 4-1 The output voltage of a simple voltage regulator is affected by tolerances of several components

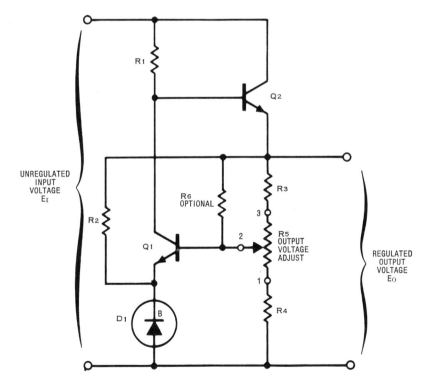

Fig. 4-2 A single potentiometer compensates for component tolerances and permits precise adjustment of output voltage

to those values anticipated. This will optimize resolution as well as restrict output excursions. An excessive range might cause damage to other circuitry if the potentiometer is not pre-adjusted before application of power. Even if pre-adjustment is accomplished, someone is sure to set the potentiometer to any given point within its range at some time or another!

If the adjustment potentiometer is located remotely from the rest of the circuit for accessibility, consider adding an extra resistor such as R_6, shown in Fig. 4-2. Its value is not critical but should be chosen to be an order of magnitude higher than the divider resistors. The function of R_6 is to provide feedback to limit the regulator output voltage in those cases where the circuit connection to the wiper terminal becomes electrically open. This can occur because an assembler failed to install a wire or, for a multitude of possible reasons, the wire breaks.

Other more elaborate power supplies may use similar adjustment schemes to facilitate a limited and controlled range of output voltage adjustment.

Current Limit Adjustment. Fig. 4-3 illustrates another power supply regulator that utilizes an integrated circuit. Potentiometer R_2 permits output voltage adjustment, as explained in the previous paragraphs, with range limited by resistors R_3 and R_4.

A second adjustment capability is included in Fig. 4-3 to permit control of the short-circuit current limit. The IC regulator will limit the output current to a value necessary to establish approximately 0.6V between the *limit* and *sense* terminals. The voltage across resistor R_5 is proportional to the output current. Potentiometer R_1 provides a fractional part of the developed voltage to the current limit input.

Although R_5 could be replaced directly by the potentiometer, it is generally not practical because of the very low ohmic values required. It is more reasonable to choose a value for R_5 that will force it to carry the output current, than select a practical value for R_1 which has adequate resolution.

It may be required to place another fixed resistor, R_6 in Fig. 4-3, in series with R_1 to prevent

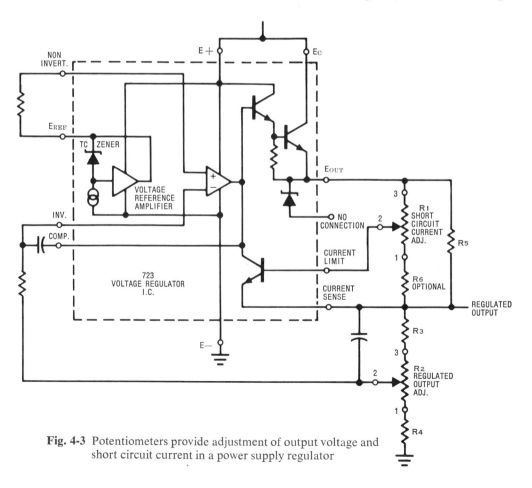

Fig. 4-3 Potentiometers provide adjustment of output voltage and short circuit current in a power supply regulator

current limit from exceeding a given value.

TC VR Diode Reference Supply. A *tempera-ture compensated* VR diode is frequently used as a *voltage reference* supply. A greatly expanded characteristic curve for a typical TC VR diode is shown in Fig. 4-4A. Note that the diode voltage is a function of temperature at any bias current above or below an optimum value.

By providing a trimming potentiometer for the bias current, adjustment of the overall temperature coefficient to its optimum value is possible. In the circuit diagram of Fig. 4-4B, potentiometer R1 provides adjustment of the current generator by varying the total resistance in the emitter circuit of transistor Q1.

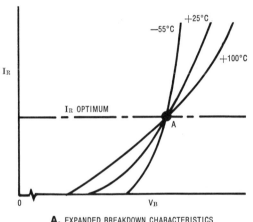

A. EXPANDED BREAKDOWN CHARACTERISTICS

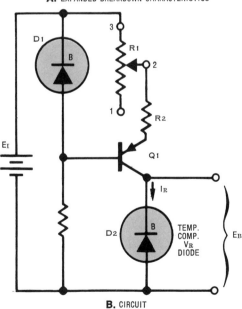

B. CIRCUIT

Fig. 4-4 Adjustment of bias current through a temperature compensated VR diode optimizes its temperature coefficient

R2 is a current limiting resistor necessary for the condition when R1 is adjusted to its minimum value. Without R2, this condition could result in excessive current flow and severe damage to R1, Q1, and the TC VR diode, D2. Choose the values of R1 and R2 to provide the design value of bias current through D2 when the potentiometer is set to its center travel position. R1 will typically be about one fifth the value of R2 to provide an adjustment range of roughly ±9%. Remember, R1's minimum setting is the high current condition and must not exceed the potentiometer's maximum wiper current rating.

Temperature Compensating Voltage Supply. It is often desirable to develop a temperature compensating voltage for correction of a temperature induced error within a system. Fig. 4-5 illustrates one possible circuit design.

Diode D1 is forward biased and has a forward voltage drop E_F which decreases about 2mV/°C. Potentiometer R2 provides a nulling adjustment to cause the output voltage E_0 to be zero at a given temperature, typically 25°C.

As the operating temperature of D1 increases, E_F drops and the value of E_0 rises by an amount controlled by trimmer R5. Thus E_0 may be used as a correction voltage whose increment is adjustable for a given temperature change.

OPERATIONAL AMPLIFIER APPLICATIONS

Integrated circuit operational amplifiers are very common and extremely useful components. Potentiometers are used to adjust for an equivalent zero offset voltage or to set the overall gain in the op-amp's feedback circuit.

Offset Adjustment. Many IC operational amplifiers provide access to the internal circuitry for the purpose of nulling the offset voltage with an external potentiometer. Fig. 4-6 illustrates three methods for common IC types.

In Fig. 4-6A, the potentiometer is tied directly between pins 1 and 5 with the wiper connected directly to the negative dc supply. The other two circuit arrangements, Fig. 4-6B and 4-6C, are more complicated.

The circuit arrangement given on some op-amp data sheets calls for a potentiometer having a TR of 5 megohms. Cermet or wirewound potentiometers are preferable for stability, even though any variations in potentiometer resistance will produce only second order effects in the actual drift performance of the amplifier. However, 5 megohms is beyond the range of wirewound and is available in only a few cermet models.

The simple circuit arrangement of Fig. 4-6B accomplishes the offset nulling requirement, with

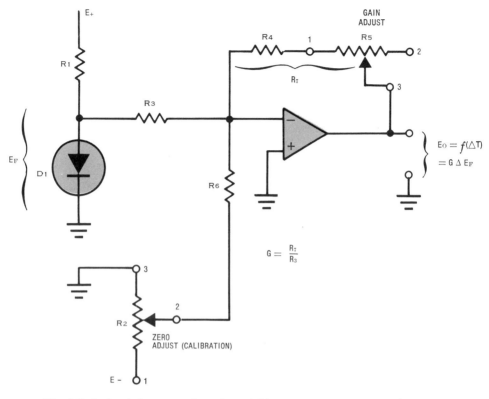

Fig. 4-5 A circuit for generation of a variable temperature compensating voltage

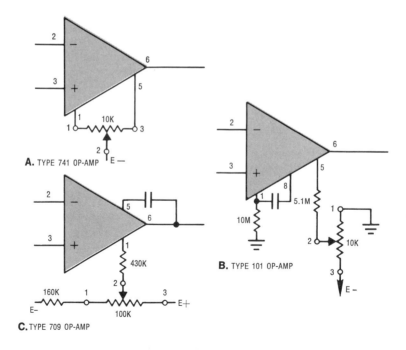

Fig. 4-6 Offset adjustment for operational amplifiers with internal balance access.

a more practical (lower) potentiometer value. Note that the actual value of the adjustment potentiometer is of little significance, since the loading is 5 megohms.

Offset voltage compensation can be accomplished on operational amplifiers that do not provide the internal access feature of those shown in Fig. 4-6. Four methods are presented in Fig. 4-7. Note that a different arrangement is required for each basic amplifier configuration.

For all of these offset adjustment methods the actual value of the trimmer from a performance standpoint is of minor significance. Lower values, however, will cause a greater power supply drain and yield somewhat poorer resolution with wirewound units. If a wirewound potentiometer is required, 10KΩ, which has a typical resolution of about 0.2%, is a practical total resistance value. Higher resistance values generally cost more. For cermet units 20K to 100K is the preferable choice.

In the offset compensation arrangements of Fig. 4-7, the compensation voltage is fed by a low output resistance voltage divider to prevent resistance level variations which might change the operating gain.

Gain Adjustment. Potentiometers are also useful in providing a means of adjusting the voltage gain of an operational amplifier circuit by modifying the feedback ratio.

Several gain adjustment arrangements for non-inverting amplifiers are shown in Fig. 4-8. In the configuration of Fig. 4-8A, the adjustment potentiometer is used to vary the value of R4. Operating voltage gain G_E is given by:

$$G_E = 1 + \frac{R_4}{R_1}$$

which is much less than the open-loop gain of the basic operational amplifier.

The minimum gain is obtained when the potentiometer R3 is adjusted to its minimum resistance. Maximum gain occurs when the TR of R3 is adjusted into the feedback circuit or:

$$G_{E\,min} = 1 + \frac{R_M + R_2}{R_1}$$

$$G_{E\,max} = 1 + \frac{R_T + R_2}{R_1}$$

Although the presence of resistor R2 is not

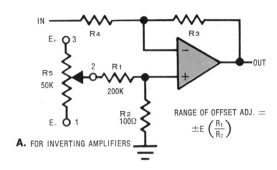

A. FOR INVERTING AMPLIFIERS

RANGE OF OFFSET ADJ. $= \pm E\left(\dfrac{R_1}{R_2}\right)$

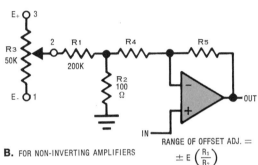

B. FOR NON-INVERTING AMPLIFIERS

RANGE OF OFFSET ADJ. $= \pm E\left(\dfrac{R_1}{R_2}\right)$

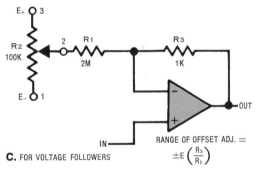

C. FOR VOLTAGE FOLLOWERS

RANGE OF OFFSET ADJ. $= \pm E\left(\dfrac{R_3}{R_1}\right)$

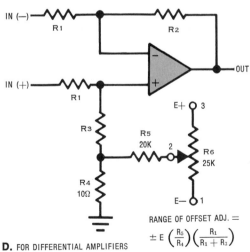

D. FOR DIFFERENTIAL AMPLIFIERS

RANGE OF OFFSET ADJ. $= \pm E\left(\dfrac{R_3}{R_4}\right)\left(\dfrac{R_1}{R_1 + R_3}\right)$

Fig. 4-7 Offset adjustment for various operational amplifier configurations

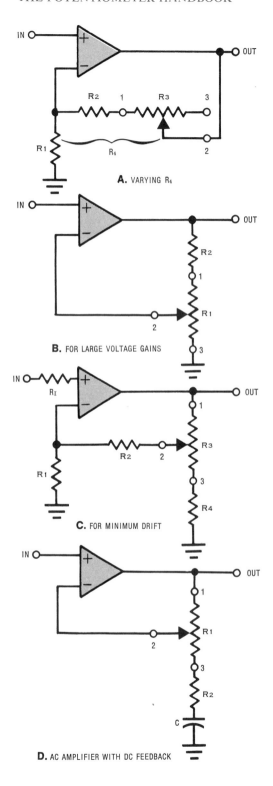

A. VARYING R₄

B. FOR LARGE VOLTAGE GAINS

C. FOR MINIMUM DRIFT

D. AC AMPLIFIER WITH DC FEEDBACK

Fig. 4-8 Gain adjustment for non-inverting amplifiers

absolutely necessary, it is advisable for several reasons. First, an absolute minimum gain is generally desired with a certain amount of gain increase possible. As in all adjustment arrangements, any excess in the adjustment range is wasted and results in reduced adjustability and some loss of stability. R_2 effectively establishes the minimum voltage gain, the potentiometer's minimum resistance being negligible, and the adjustment range is provided by the variable resistance.

Where the necessary adjustment range is a small fraction of the overall gain, R_2 results in some additional benefits. If R_3 must be remote from the operational amplifier circuitry, the possible noise picked up by the potentiometer leads is reduced. Note that this is not true if the relative position of R_2 and R_3 are interchanged.

The circuit arrangement of Fig. 4-8B is particularly useful when voltage gains of very large magnitude are required. With the wiper of the adjustment potentiometer set at the ground end, the gain is equal to the open-loop gain of the basic amplifier. An additional fixed resistor could be placed in series with the ground end of the potentiometer element in order to limit the maximum gain to a lower value.

The configuration shown in Fig. 4-8C uses a voltage divider between the output of the adjustment potentiometer and the feedback input. This serves several purposes. First, it can reduce the required total resistance value of the potentiometer. For critical dc amplifier circuits it is necessary that the equivalent resistances seen by the two differential inputs be matched for minimum drift. The value of the input source resistance R_I might be 10KΩ, thus requiring the output resistance of the total feedback circuit to likewise be 10KΩ. Assuming a desired voltage gain of about 1000 ±200, the values of R_2 and R_3 in Fig. 4-8A would be 8 and 4 megohms, respectively. A more practical value of potentiometer total resistance is certainly needed and is developed below.

Consider the arrangement given in Fig. 4-8C with $R_1=10KΩ$, $R_2=8.06MegΩ$, $R_3=10KΩ$, and $R_4=20KΩ$. When the adjustment potentiometer is set to the terminal 1 end, the full output voltage is fed to R_2 and the voltage gain is approximately 800, the desired minimum value. Then, when the potentiometer is set to the terminal 3 end, only 1.5 times the output voltage is applied to R_2. The resulting voltage gain will be three-halves the previous value or 1200, the desired maximum. The 10KΩ resistance for the adjustment potentiometer is much more practical than 4 megohms and the end result meets all of the desired requirements.

The voltage divider arrangement of Fig. 4-8C

also provides improved noise immunity if the potentiometer must be remotely located. The relative signal level is high at the location of the potentiometer and the resistance level is fairly low. Both of these factors will reduce noise pickup.

Fig. 4-8D illustrates a gain adjustment arrangement for an ac amplifier with 100% dc feedback. The impedance of capacitor C must be very low in comparison with R2.

Voltage gain adjustment configurations for inverting amplifiers are given in Fig. 4-9. The feedback signal is a current, unlike the non-inverting amplifiers of Fig. 4-8 which utilize a voltage feedback signal.

The basic arrangement of Fig. 4-9A has a voltage gain of:

$$G_E = \frac{R_4}{R_1}$$

R4 is again composed of a fixed value, which establishes the minimum gain, and an adjustment potentiometer which provides the desired adjustment range.

Where potentiometer resistance values for the circuit of Fig. 4-9A become unreasonable, the circuit arrangement of Fig. 4-9B provides benefits analogous to those realized in the arrangement of Fig. 4-8C discussed previously.

Fig. 4-9C illustrates a circuit capable of achieving a wide range of gain variation with practical values.

Filters. Operational amplifiers are frequently used in active filter circuits. Trimmers are used to adjust both the Q and the operating frequencies.

Fig. 4-10 illustrates one simple bandpass filter which has a variable Q (and gain) controlled by the adjustment potentiometer R2.

The center frequency of this filter may be changed (without affecting the Q) by varying C1 and C2 or R1 and R3.

Each of the frequency determining resistors may be replaced by a fixed resistor in series with a trimmer. Dual trimming potentiometers, which allow easy adjustment of both resistors simultaneously, are not readily available. This is because demand for them is low. Their cost is relatively high when compared with two separate potentiometers. When two separate trimming potentiometers are used they must be adjusted individually, varying each a little at a time, while trying to change each by an equal amount.

Variable Capacitance. Operational amplifiers may be used to multiply the effective values of either resistive or reactive elements. Fig. 4-11 illustrates one configuration which can be used

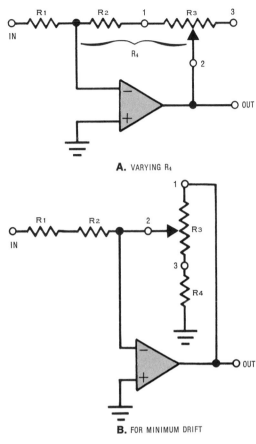

A. VARYING R4

B. FOR MINIMUM DRIFT

$$G_E = \frac{R_4}{R_1}$$

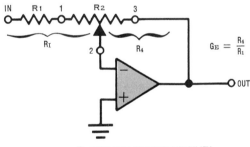

C. WIDE GAIN VARIANCE CAPABILITY

Fig. 4-9 Gain adjustment for inverting operational amplifiers

to develop an equivalent variable capacitor with a range from 0.1 to 1.0 microfarads using a fixed capacitor C1.

This application illustrates how a trimming potentiometer may be used to vary a parameter other than current or voltage ratio alone. R2 varies the relative currents fed from the outputs of the two operational amplifiers. This signal is fed to the inverting input of the second unit and thereby adjusts the amount of multiplication which occurs.

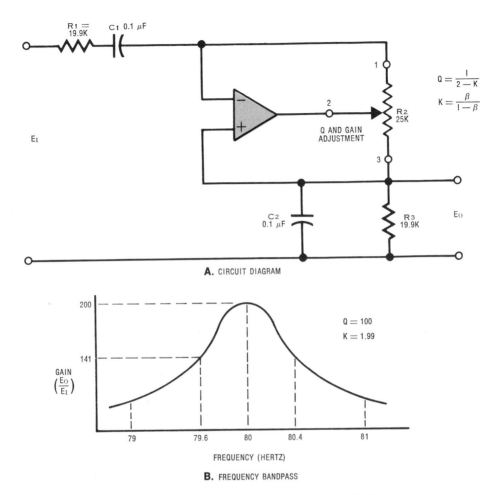

A. CIRCUIT DIAGRAM

$$Q = \frac{1}{2-K}$$

$$K = \frac{\beta}{1-\beta}$$

$Q = 100$
$K = 1.99$

B. FREQUENCY BANDPASS

Fig. 4-10 Active band pass filter with variable Q

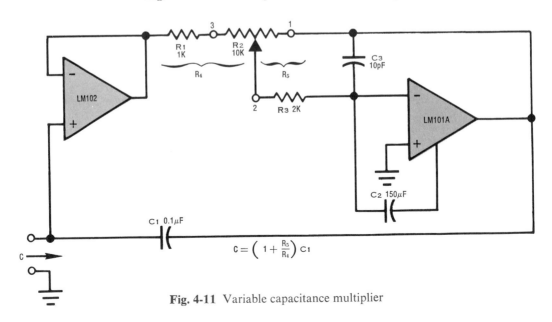

$$C = \left(1 + \frac{R_5}{R_4}\right) C_1$$

Fig. 4-11 Variable capacitance multiplier

Trimming potentiometers and IC operational amplifiers make good teammates. The wise designer will make full use of both of them.

DIGITAL CIRCUITS

Trimmers may be used in digital applications to provide adjustment for common characteristics such as time delay, clock frequency, and threshold levels. They are available in dual in-line packages (DIP). In addition to conventional solder mounting they may be inserted in IC sockets, thus permitting popular digital system wiring techniques to be used.

When using potentiometers in digital circuits, where fast rise or fall times are required, choose a cermet type. A wirewound device may exhibit a significant inductance and can result in undesirable behavior.

A few typical applications are presented to illustrate the possibilities.

Monostable Timing. One of the most common digital applications of adjustment potentiometers is the control of time delay in an integrated circuit monostable. Fig. 4-12 illustrates one of the commonly used monostable types.

The amount of time delay is directly proportional to the product of capacitor C and the sum of R2 and R3 or:

$$\text{Time Delay} = C[R_2 + R_3] = R_4C$$

R2 serves to limit the minimum value of the timing resistance as required for the given IC. R2 can be selected to cause the potentiometer to adjust the time delay around a given nominal value.

Fig. 4-13 illustrates the circuit for another type of monostable using a 555 IC timer. Once again, a trimmer is used to vary the RC time constant to control the time delay interval.

The additional circuitry, consisting of resistor R3 and potentiometer R4, provides a calibration adjustment where the timing resistor may be a precision potentiometer with a dial. The nominal delay time is given by:

$$t_I = 1.1R_5C$$

Some variation exists from one IC to the next causing the factor 1.1 to vary over a small range. In order to make several circuits yield the same time delay, R4 is adjusted to vary the voltage appearing at pin 5. This voltage is nominally two-thirds of the supply voltage E+. Adjustment of R4 will compensate for timing capacitor tolerance variations as well as IC differences.

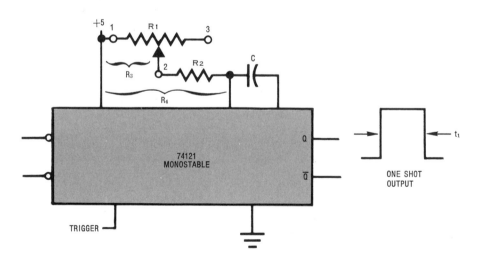

Fig. 4-12 A trimming potentiometer is frequently used in digital circuits to adjust timing of monostable

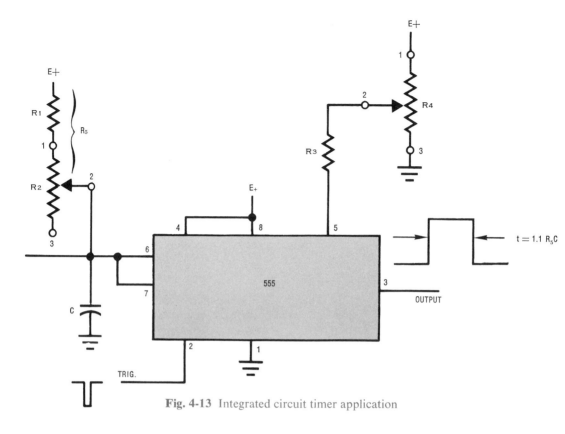

Fig. 4-13 Integrated circuit timer application

Clock Generator. Where an accurately controlled clock is not required, a single monostable IC may be used in a somewhat non-standard mode to yield an astable clock generator as shown in Fig 4-14.

The basic width of the output pulse is relatively constant and depends upon delays within the IC. The interval between output pulses, and hence the clock frequency, is controlled by the RC time constant which is easily adjusted by the trimmer.

Photocell Sensitivity. All photocells, whether resistive or voltaic, exhibit some variation in sensitivity from one unit to the next. Trimming potentiometers may be used to adjust the sensitivity of each cell to a uniform value or compensate for other variations in the optical system. Fig. 4-15 shows the circuit diagram for the photocell amplifier of one channel in a paper tape reader.

With no light falling on the cell (no hole in the tape) the photocell conducts little current. Transistor Q_1 is turned on and Q_2 is off. When light strikes the photocell through a hole in the paper tape the photocell conducts enough bias current away from the base of Q_1 to turn Q_2 off. This turns Q_2 on and the output is pulled to ground. The feedback path through R_6 produces a slight amount of regeneration.

The trimmer R_1 controls the base bias current to Q_1 and hence the amount of light required on the photocell to activate the circuit. Each channel has its own sensitivity adjustment to compensate for differences in individual photocells, minor position errors, and other circuit variations.

Resistive photocells are frequently used in a bridge circuit where the cell is in one branch and a trimmer is located in the opposite branch to adjust the balance at a given light level.

Data Conversion. Analog to digital and digital to analog converters are common system interface functions. Trimming potentiometers are necessary to provide adjustment of offset errors and scaling values.

Fig. 4-16 illustrates the arrangement for a typical analog to digital module. One potentiometer R_2 is necessary to adjust the input offset voltage such that digital zero will result from zero input voltage. Another potentiometer R_1 is used to control the sensitivity such that all digital outputs will be "1" when the desired full-scale voltage is applied to the analog input.

In some high precision analog to digital and digital to analog conversion systems, individual adjustment may be provided for several of the

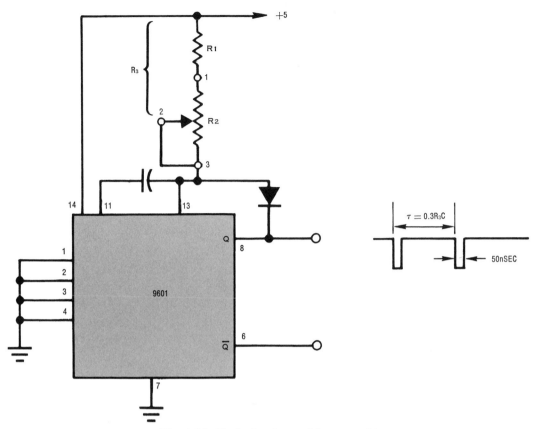

Fig. 4-14 Clock circuit uses IC monostable.

high order bits.

Digital Magnetic Tape Deck. In the typical digital tape deck, which usually has 9 tracks, there are 26 adjustment potentiometers. They are required to provide the degree of uniformity needed, from one deck to the next, for tape interchangeability. This is not only a cost-effective application but also an essential one.

One potentiometer adjusts the logic power supply voltage and another controls the photocell sensitivity for the beginning and end of tape marker detectors.

Individual gain controls are provided for each of the nine read amplifier channels. This compensates for possible variations in read-head sensitivity as well as component tolerance differences in the amplifiers.

Another adjustment potentiometer is used to vary the timing in a read strobe delay monostable, while nine more potentiometers are used to individually adjust the write deskew monostable circuits for each track.

The remaining potentiometers are used for adjustment of the capstan servo system and are critical in assuring uniformity from one tape deck to the next. One adjusts for capstan servo offset while two more allow precise setting of the forward and reserve tape speed. Finally, trimming potentiometers permit adjustment of the forward and reverse stop ramps.

INSTRUMENTS

Trimming potentiometers play an important part in the electronic instrument field. They assist in making economical assembly and checkout possible as well as facilitating easy calibration in day to day use. They also provide adjustment for recalibration when repairs require the replacement of a critical component or where normal aging of components has caused a loss in accuracy.

Applications for electronic instruments are ever widening in scope and include such diverse fields as communications, computer, medical, manufacturing and process control and automotive performance analysis. Examples of trimmer applications from some of these fields are the subject of the following paragraphs.

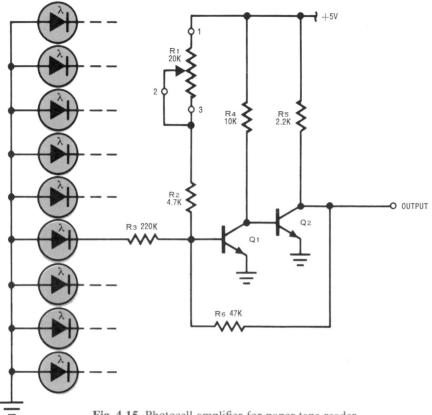

Fig. 4-15 Photocell amplifier for paper tape reader

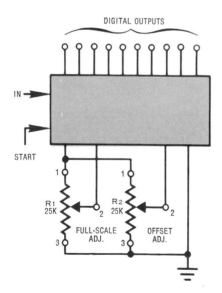

Fig. 4-16 Analog to digital converter modules use potentiometers for offset and full-scale adjustment

Digital Voltmeters. Adjustment potentiometers are used in digital voltmeters to provide compensation for component tolerance variations and permit proper calibration.

Typical applications include power supply control, both for the operating supply and the precision reference voltage, zero adjustment, amplifier gain control, and ramp speed adjustment. Separate calibration adjustments are often provided for each range, especially if the instrument includes a preamplifier to allow low-level measurement.

Generators. Signal generators require adjustment of oscillators, timing circuits, trigger circuits, linearity controls, and duty cycles. Where a precision control dial is used on the front panel, a trimmer is often used to adjust for proper calibration.

Oscilloscopes. Precision instrumentation oscilloscopes rely on adjustment potentiometers for control of their power supplies, amplifiers, timing and sweep circuits, and triggering circuits.

In some cases, access to some of the potentiometer trimmers is provided through small holes

90

in the front panel. While adjustment may not be needed often it may be easily performed when necessary.

Portable Electronic Thermometer. Fig. 4-17 shows one of the major advancements made in the medical instrumentation field in recent years. The electronic thermometer offers significant improvements in body temperature measurement. It is safer, easier to use, faster and provides greater accuracy than the standard mercury thermometer.

Inside the instrument case, an adjustment potentiometer is used to balance a bridge circuit which compensates for component and probe tolerances. In this application, a single turn, low cost, cermet device provides a cost effective alternative to selecting fixed resistors during construction. The potentiometer reduces assembly costs and simplifies calibration and maintenance procedures.

MISCELLANEOUS APPLICATIONS

There are many additional applications where trimming potentiometers are useful. A few are briefly outlined below.

Phase Locked Loops. Phase locked loops consist of a voltage controlled oscillator, a phase detector, and a low pass filter connected in a servo system arrangement. Adjustment potentiometers are often used to control the free-run frequency of the internal oscillator in the manner shown in Fig. 4-18.

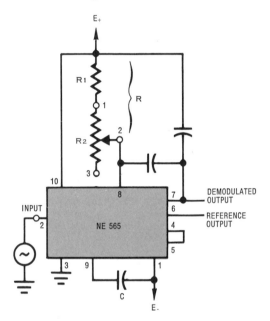

Fig. 4-18 Trimming potentiometers adjusts free-run frequency in a phase locked loop FM demodulator

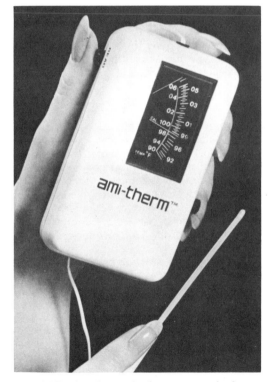

Fig. 4-17 The electronic thermometer is the first major improvement in the fever thermometer in over 100 years. (AMI Medical Electronics, Div. of LMC Data, Inc.)

Potentiometers might also be used to set the levels of various threshold detectors used in conjunction with phase locked loop circuits.

Linearity Optimization. In applications where a precise conformity between a precision potentiometer function and relative wiper travel is required, it is common to specify a potentiometer with an absolute linearity specification. It is possible in many cases to save money and possible delivery delay by using a lower cost precision potentiometer purchased to an independent linearity specification; then using trimmers to optimize the operating linearity in your application.

Fig. 4-19 illustrates this cost-effective circuit arrangement. In Chapter 2, the basic difficiency in an independent linearity specification, as compared with absolute linearity, was shown to be a lack of control for the intercept and slope of the best straight line reference function.

Trimmer R_1 in Fig. 4-19 acts primarily to control the slope of the transfer function. It is necessary that the input voltage E_I be slightly larger than the maximum full-scale output voltage required.

The second trimmer R_3 permits adjustment of the effective intercept point. An index point of some kind is necessary. For this application, an index point near the low end will be best for proper adjustment of R_3.

There is a certain amount of interaction between the two adjustments, so it may be necessary to repeat the calibration procedure one or more times. When the adjustment is completed, the performance obtained will be identical to the performance of a precision potentiometer purchased to an absolute linearity specification. The circuit of Fig. 4-19 provides added flexibility to compensate for minor errors in the dial or linkage controlling the wiper travel position.

Nonlinear Networks. Trimmer potentiometers can be used with VR (voltage reference)

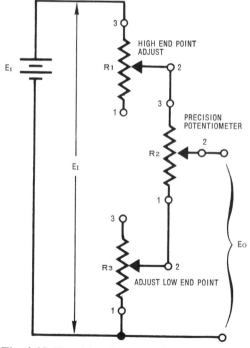

Fig. 4-19 Two trimming potentiometers may be used to optimize linearity error in a precision potentiometer

diodes in the manner shown in Fig. 4-20 to produce a nonlinear resistance network. The voltage breakpoints are set by the breakdown voltages of the VR diodes, while the slope of the incremental resistance is adjusted by the trimmers.

Since all the lower voltage branches will affect the higher ones, the adjustment procedure should begin with R_1 and proceed in order through R_4.

Replacing the VR diodes with clamping diodes and variable voltage sources, results in additional flexibility over the shape of the characteristic curve.

RF Tuning. Usually a trimmer would not be considered for adjustment of an RF tuned circuit. There *is* a tuned circuit, which may prove to be cost-effective, that *does* use an adjustment potentiometer. In this design, which is becoming increasingly popular, particularly in TV receivers, a variable capacitance diode (varactor) tunes the RF circuit. The required tuning capacitance is achieved by adjusting the voltage applied to the diode with a trimmer.

Fig. 4-21 illustrates this simple arrangement. The dc bias control circuit may be located at a remote point. This method is also very useful when the circuit to be tuned is in an inaccessible location, such as within a temperature-controlled oven.

Custom designs. Applications of potentiometers chosen from manufacturers standard product line will usually produce the optimum in electrical and economical design. However, the circuit designer is not limited to these standard devices. Occasionally, electrical, mechanical or environmental application requirements will demand a variable resistance device of unique design. The following paragraphs describe a few of the components created by potentiometer manufacturers to meet the custom application demands of the electronic industry.

The potentiometer shown in Fig. 4-22 is a low resistance ($< 1\Omega$), high power (25 watts) rheostat. It is used in the regulator of an AC/DC converter in a computer system. It functions as a current control through parallel power semiconductors that carry load currents to the central processor unit. In this application, the custom designed rheostat proved to be the most economical alternative due to the low cost of field maintenance as compared with other methods.

When the circuit function requires many variable and fixed resistors to accomplish a task, the most cost effective approach could be the multipotentiometer network. Fig. 4-23 illustrates a network that was custom designed for multi-channel varactor tuning of a television receiver. The advantages of this thick-film module are: 1) less space required for packaging, 2) fewer parts to stock, inventory and install in the system, 3) the lowest cost per variable function in high volume production quantities.

A very ingenious method of adjusting the electrical output of an implanted heart pacer from outside a patient's body, without the need for surgery or through-the-skin leads has recently been developed.

The technique utilizes a tiny magnetically driven mechanism inside the pacemaker module that can be made to rotate (adjust) by spinning a precisely configured and positioned magnetic field — outside the body, directly over the implanted pacemaker.

At the core of the mechanism is a single-turn cermet adjustment potentiometer. See Fig. 4-24. The potentiometer/mechanism is installed in a tiny, magnetically-transparent metal can embedment. The potentiometer adjustment slot is linked to a small clock-like precision gear train. At the input end of the gear train is a miniature wheel with two rod magnets installed parallel and on either side of the wheel centerline. The gear ratios and the fine balance of the mechanism are such that very little torque is required to spin the mechanism. The relationship of gear turns to movement of the potentiometer element wiper enables extremely precise adjustments. This also protects the patient from any detrimental effects due to movement of the mechanism as a result of vibration, inertia or stray magnetic energy.

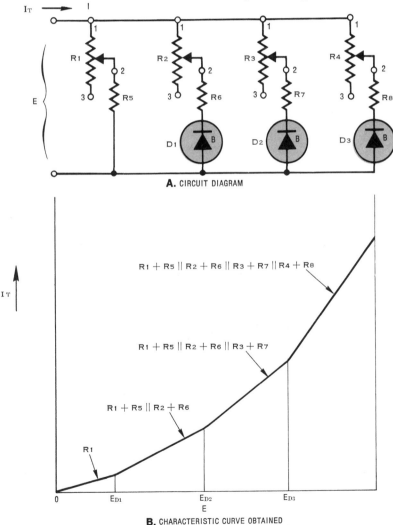

A. CIRCUIT DIAGRAM

B. CHARACTERISTIC CURVE OBTAINED

Fig. 4-20 A nonlinear resistance network synthesized with VR diodes and trimmer potentiometers

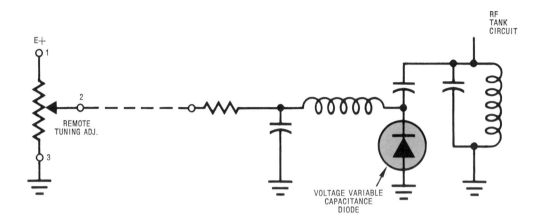

Fig. 4-21 Trimming potentiometers may be used to perform RF tuning

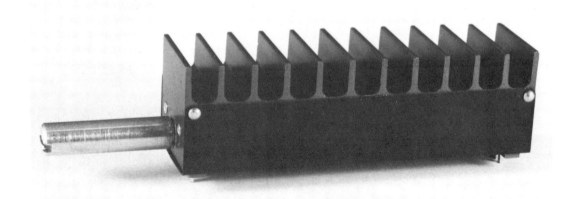

Fig. 4-22 A high power, low resistance custom designed rheostat

Fig. 4-23 A custom designed multi-potentiometer network

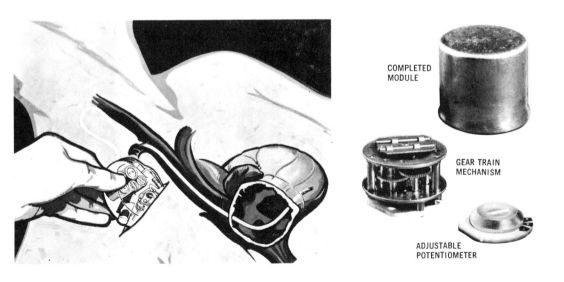

Fig. 4-24 A potentiometer provides adjustment of electrical output of implanted heart pacer

Notes

APPLICATION AS A
CONTROL DEVICE

~~~~~~

Chapter

5

Double Control Unit

Electrad, Inc. announces a Model B Super Tonatrol which is particularly adapted for use by manufacturers on account of its arrangement whereby if desired, two completely isolated circuits may be controlled by one shaft. The contact is a pure silver multiple type which floats over the resistance element with amazing smoothness...

From the New Products section of
Electronics Magazine, April 1930.

INTRODUCTION

Potentiometers can provide a means for frequent adjustment of an electrical circuit or system where operator control is desired. Chapter 4 presented applications requiring only an initial or occasional adjustment. These are usually best served by a trimmer type of potentiometer.

In this chapter, the focus is on applications in which more frequent and convenient adjustment is anticipated. These applications are described generally as control functions. Many of them are the man-machine interface that provides selectivity, versatility and variability to a circuit or system. The type of potentiometer used may vary according to specific needs, but usually the wiper travel is manually controlled by the turning of a knob or turns counting dial. Cost-effective circuit design and application of the potentiometer depends on the designer having a broad knowledge of the economical options available.

Applications for *precision* control devices are sometimes electromechanical and usually re-quire greater mechanical accuracy, better stability, and longer life. They may also be subjected to severe environments. These applications are discussed in the next chapter.

Control functions can generally be classified as one of the following:

Calibration — *How much correction?*
Level — *How much?*
Rate — *How fast?*
Timing — *How soon?*
Position — *Where?*

Many of the techniques described in Chapter 4 are directly applicable to control applications. The sections regarding gain adjustment of operational amplifiers, filters, and frequency control are of particular interest. One difference between Chapter 4 applications and those here is *how often* the adjustment may be needed.

These application examples represent a very small sample of the huge number possible. The brief descriptions give a general idea of the type of control functions which use potentiometers.

97

BASICS OF CONTROL

Before looking at specific applications, some fundamental guidelines of good control function design should be considered. Although some of the ideas may seem very obvious, each one should be carefully considered as a factor in the design of a control scheme for an instrument or system.

Control the actual function of interest. It is usually possible to control a function in a number of ways. Some may be rather direct while others may require an indirect approach. Where practical, the control function scheme should provide the operator with a direct relationship between the position of the control dial and response of the controlled variable. This is in line with good human engineering and should be followed where practical.

An example will make this clearer. Suppose that the *on* time of a simple oscillator must be controlled. One possible circuit is shown in Fig. 5-1. As shown by the operational equations, the setting of the potentiometer wiper is affective in determining the charging (t_1) and discharging (t_2) time constants of the circuit. Varying R_2 will change the amount of time t_1 that the output is high. R_2 also varies the time t_2 during which the output is low. The controlled function is actually the *frequency* of the oscillator.

In Fig. 5-2, two potentiometers are used to allow independent control of t_1 and t_2. A few additional components allow one potentiometer R_2 to be active only during the charging cycle, thus controlling t_1. The other potentiometer R_3 is active only during the discharging cycle, thus controlling t_2. If only t_1 control is needed, R_3 may be replaced by a fixed resistor of appropriate value.

Note in Fig. 5-2 that a fixed resistor is included in series with R_2. This prevents possible component damage in case both potentiometers are set to their minimum resistance values. A fixed resistor could be placed in series with R_3 to achieve greater values of t_2.

In some applications, it is required to control t_1 while the oscillator frequency remains constant. This can be done with the circuit of Fig. 5-2 by decreasing t_2 each time t_1 is increased. Although the circuit arrangement achieves the requirements, the overall adjustment procedure is more complicated and requires greater operator care and skill than necessary.

Direct control of t_1 is easily accomplished by the circuit shown in Fig. 5-3. The oscillator frequency is a function of the potentiometer's total resistance only, and is not affected by wiper position. When the control is actuated, the division ratio of R_1 changes. Since the total resistance of R_1 is constant, the output frequency will remain constant while the variable division ratio provides a variable duty cycle. The duty cycle is directly proportional to the relative wiper travel.

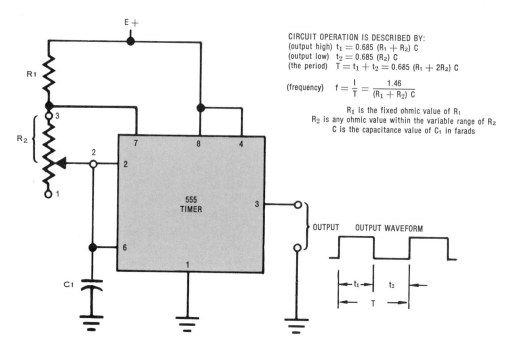

CIRCUIT OPERATION IS DESCRIBED BY:
(output high) $t_1 = 0.685\ (R_1 + R_2)\ C$
(output low) $t_2 = 0.685\ (R_2)\ C$
(the period) $T = t_1 + t_2 = 0.685\ (R_1 + 2R_2)\ C$

(frequency) $f = \dfrac{1}{T} = \dfrac{1.46}{(R_1 + R_2)\ C}$

R_1 is the fixed ohmic value of R_1
R_2 is any ohmic value within the variable range of R_2
C is the capacitance value of C_1 in farads

Fig. 5-1 A potentiometer provides control of frequency in an oscillator circuit

It can be indicated on a read-out dial.

Control requirements should be carefully analyzed to make certain that the circuit chosen satisfies those requirements in the most *direct* manner. This will result in the most cost-effective approach with a logical man-machine interface.

Provide adequate range and resolution. The control arrangement must provide adjustment of the variable over the required range for the life of the system. Adequate adjustment margin must be provided to compensate for electronic component tolerances and aging effects. It may be necessary to restrict the control range somewhat or shape the control function in some

manner. Adjustment resolution must be adequate for the specific application. Methods for achieving various responses are discussed in Chapter 3.

Choose a logical direction of control sense. The control sense refers to the direction of change in the controlled function compared to the direction of change in mechanical input rotation or wiper movement. The criteria for choosing a control sense are those factors dictated by good *human* engineering. For example, a clockwise rotational input should cause the controlled function to increase while counterclockwise causes a decrease. In the case of linear

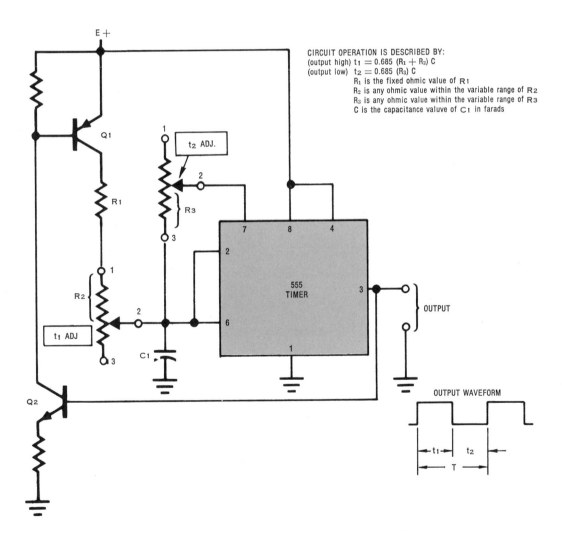

CIRCUIT OPERATION IS DESCRIBED BY:
(output high) $t_1 = 0.685 (R_1 + R_2) C$
(output low) $t_2 = 0.685 (R_3) C$
R_1 is the fixed ohmic value of R_1
R_2 is any ohmic value within the variable range of R_2
R_3 is any ohmic value within the variable range of R_3
C is the capacitance valuve of C_1 in farads

Fig. 5-2 Potentiometers provide independent control of ON and OFF times in an oscillator circuit

motion potentiometer controls, movement upward or to the right should increase; down to the left, decrease. Position controls should provide an upward or left-to-right movement for a clockwise rotation of the control knob.

As an illustration, suppose that a potentiometer is used to control a current through a low resistance load. A simple rheostat connection will accomplish the desired control function. The choice of end terminal should be such that a clockwise rotation of the adjustment shaft will produce an increase in current.

In control sense selection, primary consideration should be given to how the operator will view the control function. Say a control is provided for changing the period of an oscillator but the operator will be interested in the resultant frequency change. Clockwise rotation should cause a *decrease* in period so that the operator will experience *increase* in frequency for clockwise rotation.

Changing the control sense after final circuit assembly is a simple task. This may be necessary where, after check out, it is discovered that the man-machine interface *seems* backwards or unnatural. Reversing the wires connected to the end terminals will invert the control sense for a voltage divider. For a rheostat, simply remove all connections from the end terminal being used and connect them to the other end terminal.

Assume worst case conditions. When a potentiometer is designed into a system as a control device, assume that the wiper will be set to all possible positions. Don't be satisfied and feel safe with a warning contained in an instruction manual which might say, *Do not turn the gain control more than 75 percent of the full clockwise position.* If there is a possibility of circuit failure beyond a *safe* limit, *design in* a control range restriction. Remember Murphy's Law: The instruction manual will not be read until all else fails, a control knob will be inadvertantly bumped and the skill level of the operator will be much lower than required.

Make controls independent. Whenever possible, make all controls independent so that adjustment of any one will have no affect on the setting of another. If this is not practical, attempt to cause the dependence to be restricted to one direction. If this is done, the operator first adjusts the independent control, then the affected dependent control, without having to go back

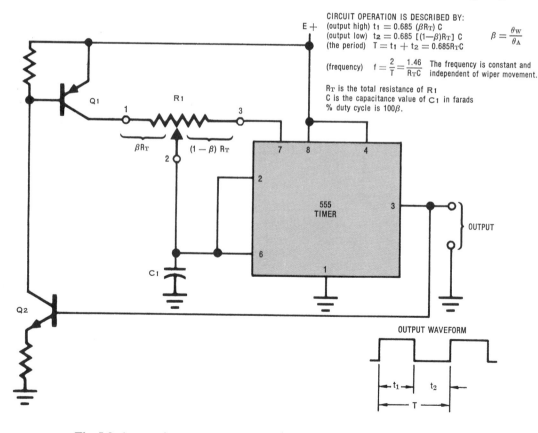

CIRCUIT OPERATION IS DESCRIBED BY:

(output high) $t_1 = 0.685 \ (\beta R_T) \ C$

(output low) $t_2 = 0.685 \ [(1-\beta)R_T] \ C$

(the period) $T = t_1 + t_2 = 0.685 R_T C$

$\beta = \dfrac{\theta_W}{\theta_A}$

(frequency) $f = \dfrac{2}{T} = \dfrac{1.46}{R_T C}$ The frequency is constant and independent of wiper movement.

R_T is the total resistance of R_1

C is the capacitance value of C_1 in farads

% duty cycle is 100β.

Fig. 5-3 A potentiometer controls the percent duty cycle in an oscillator circuit

and forth. When dependent controls cannot be avoided, adjustment instructions should clearly indicate the proper sequence of adjustment for minimum interaction.

Consider the shape of the controlled function (output curve). Many control requirements are satisfied by the characteristics of a linear function potentiometer. Some applications, however, require a potentiometer with a non-linear function characteristic. This is easily accomplished for applications that permit the use of carbon element potentiometers which are available in a wide variety of functions. If stability requirements will not permit the use of a carbon element potentiometer, then consider a potentiometer with a cermet or wirewound element.

In some applications, the control function may be shaped using the methods described in Chapter 3. It is possible to change the effective shape of the control function by proper arrangement of the control circuit.

Suppose the current through a resistive device is to change linearly with respect to the adjustment of a control potentiometer. Adjusting the current by changing the resistance in the circuit loop (rheostat) will produce a hyperbolic function whereas adjusting the voltage across the resistance in a linear fashion will satisfy the current linearity requirement.

Consider environmental and stability requirements. The potentiometer, when properly designed and applied, will not respond to temperature, vibration or shock, beyond its established tolerance limits. Choose an element type and a mechanical construction style that will yield sufficient stability for the application.

If high vibration may be present during circuit operation, choose a potentiometer model that provides a means for mechanically locking the wiper in position. A simple friction brake may be added to the shaft in many instances. Some potentiometer designs have inherent friction which results in a high torque to actuate the wiper. This high torque provides greater stability under vibration.

Additional precautions against a harsh environment include a water-tight seal to the control panel or protection of the control devices with a cover which must be lifted when adjustments are necessary. There are many combinations of environmental factors possible. The most expedient and cost-effective approach is to discuss a particular application with a potentiometer manufacturer.

Choose a proper location. Controls which must be adjusted often should be easily accessible. This seems obvious but is sometimes overlooked and difficult to correct after a system is built. Other influences on control location, e.g.,

noise susceptibility and stray capacitance, may require that the control potentiometer be located deep within the equipment. A rigid or flexible shaft extension connected to a front-panel knob can be used.

INSTRUMENT CONTROLS

Potentiometer controls serve many functions on various instruments including those for test and measurement. A few examples will give an idea of typical control possibilities.

Oscilloscopes. A modern test oscilloscope has many potentiometric controls as indicated in the photograph of Fig. 5-4. Controls are provided for focus, beam intensity, beam and graticule illumination, and beam positioning. Other controls allow adjustment of triggering level and polarity. Even the normally fixed calibration switches controlling the input voltage sensitivity and sweep speed employ potentiometers to provide some degree of variable control between ranges.

Function Generators. Control potentiometers are used for many functions in both digital pulse and analog function generators. Fig. 5-5 shows a simplified schematic diagram of a function generator with the potentiometers emphasized. Note that most of the front panel control funcions from triggering level to output level use control potentiometers. Trimmer potentiometers, also shown in Fig. 5-5, are used for many of the calibration functions.

The block diagram of a typical pulse generator is shown in Fig. 5-6. In the case of clock frequency, delay time, and pulse width, capacitors are switched to provide the typical decade range changing. Potentiometers provide the necessary fine adjustment within a given range.

Additional control potentiometers are indicated for adjusting the trigger sensitivity and output level.

Power Supplies. Adjustments on fixed voltage power supplies are usually made with a trimmer as discussed in Chapter 4. Laboratory power supplies, on the other hand, require frequent adjustment of output voltage and output current limit. These use control potentiometers with knobs easily accessible to the operator. Here the results are monitored for control with a meter rather than calibrating the input using a pointer or indicator line.

Fig. 5-7 gives the schematic of a versatile laboratory power supply. Control potentiometers permit both coarse and fine adjustment of either the output voltage or output current. These control potentiometers need no calibration dial since meters on the panel indicate the resulting current or voltage same as explained above.

Some power supplies use a multiturn poten-

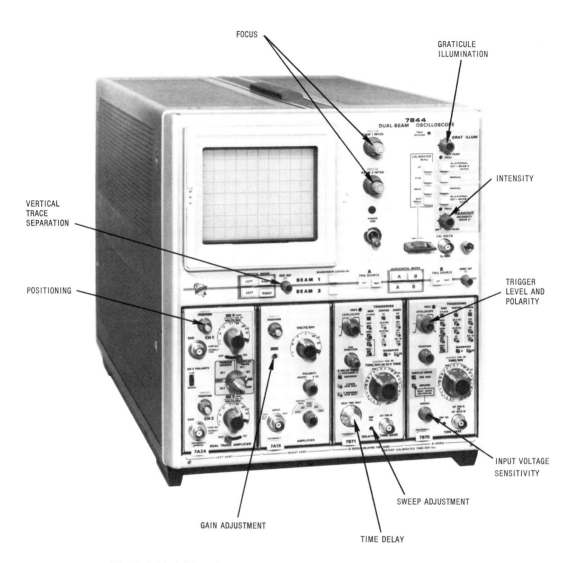

Fig. 5-4 Variable resistance controls contribute to the versatility of the modern test oscilloscope (Tektronix, Inc.)

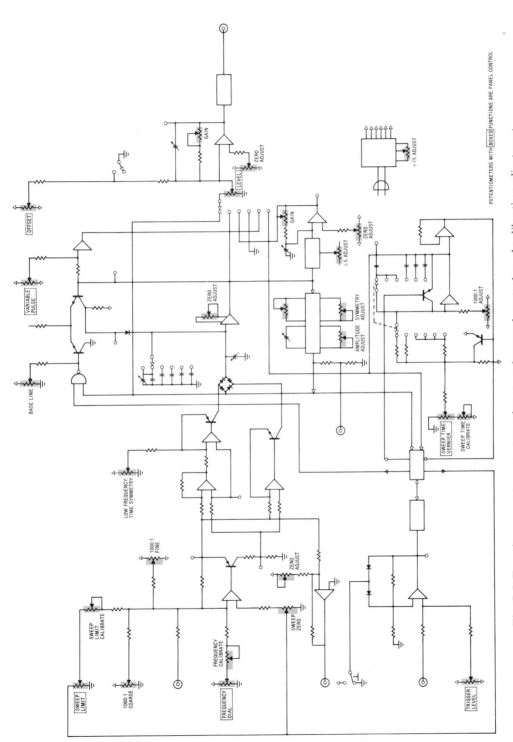

Fig. 5-5 Function generator uses potentiometers for panel controls and calibration adjustment (Interstate Electronics Corp.)

103

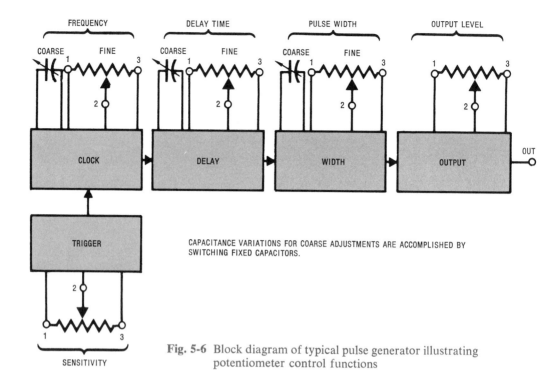

Fig. 5-6 Block diagram of typical pulse generator illustrating potentiometer control functions

tiometer, rather than a single turn, to provide adjustability of the control function. This results from the better resolution provided by multi-turn devices.

Photometers. An example of a correction function performed by a control potentiometer is shown in Fig. 5-8. The dark current in a pho-tomultiplier tube varies from unit to unit and over the life of the tube. In addition, it is temperature sensitive. Proper operation of the photometer requires frequent adjustment to compensate for dark current variations.

Recorders. Strip-chart and X-Y recorders use control potentiometers for pen reference positioning. A voltage signal is injected into the servo system to produce an adjustable error to compensate for other possible errors. This variable voltage moves the pen zero reference to any desired position within its normal operating range. Usually, the position signal is fed in at a high-level point in the system after the preamplifiers and range attenuators.

In some units, a very wide input offset adjustment is provided to allow an expanded scale display at some level above ground. Good resolution and stability in this application are absolute necessities.

Another control potentiometer is frequently included in order to provide a variable sensitivity. Earlier recorders also included a servo gain control on the front panel, but better designs have permitted this control to be delegated to an infrequent adjust trimmer.

Meters. Control potentiometers are used for meter zeroing on dc analog meters as illustrated by the control labeled *zero* in Fig. 5-9. Although contemporary solid-state designs are much more stable than the older vacuum tube models, a certain amount of operator adjustable zero control is necessary at very low voltage levels. The total zero adjustment range provided for the instrument illustrated is only ±15 microvolts.

Another control, labeled *null* in Fig. 5-9, is provided to adjust an internal voltage supply in order to produce an input zero offset. Good resolution and stability are required. Note that an adjustment potentiometer is available to set the output level for an optional external recorder.

AUDIO

Perhaps the most frequently adjusted type of potentiometer control is the volume on radios, audio amplifiers, and television sets. Carbon element potentiometers are generally used, and the function is usually logarithmic, to more closely match the nonlinear response of the human ear. This logrithmic resistance variation is commonly referred to as resistance *taper*.

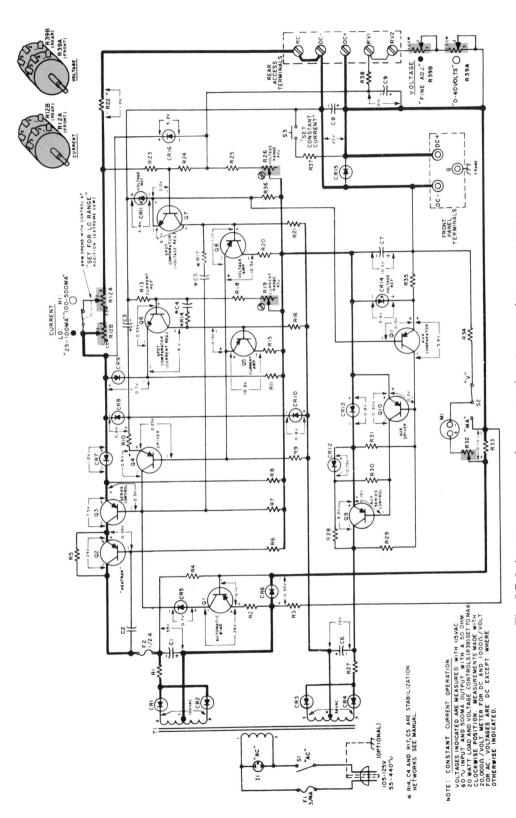

Fig. 5-7 Laboratory power supply uses potentiometer controls to vary output voltage and current limits over a wide range (Power Designs, Inc.)

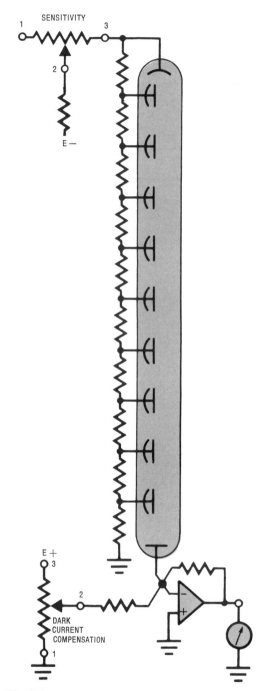

Fig. 5-8 Photometer circuit uses potentiometer control to compensate for photomultiplier dark current

Resistance Taper. Resistance taper is the output curve of resistance measured between one end of the element and the wiper. It is expressed in percent of total resistance *versus* percent of effective rotation. Three resistance tapers have

been established as standard by military specifications and by industry usage. These three standard tapers — linear, clockwise audio, and counter-clockwise audio — are shown in Chapter 7, Fig. 7-13. Most manufacturers list other standard resistance tapers and produce special tapers on request.

In recent years linear slide potentiometers have become popular in audio applications and may some day surpass rotary control usage. Some rotary type potentiometers are actually actuated by a linear motion via a mechanical linkage. Other audio controls include ones for tone and balance. On stereo systems, a set of controls is usually provided for each channel.

The master mixer board Fig. 5-10 found in recording studios uses sliding type potentiometer controls to adjust the level of each input channel. Since these units are frequently adjusted during the recording session, smooth operation and a low noise level are required. Rotary potentiometers are used for control of special effects, i.e., output to an echo chamber and returning input to the console.

MISCELLANEOUS CONTROLS

Potentiometer controls are used in many forms both in the home and in industry. This section contains a few typical applications.

Model aircraft remote control system. The ingenious *single stick* positioner in Fig. 5-11A uses two space saving, conductive plastic potentiometers to provide the output that controls two independent model aircraft functions.

The complete RC transmitter shown in Fig. 5-11B uses two of the dual potentiometer assemblies to control a model's throttle, ailerons, rudder and elevator. The unit shown is actually a six-channel transmitter. The extra two channels are used for special controls such as landing gear retraction.

The *pilot* controls the model much like he would if he were at the controls of the real thing.

Phase Shift Control. In Fig. 5-12 dual ganged potentiometers are used to provide an adjustable phase shift from about 10 to 165 degrees for an input signal frequency of 400 Hz. By proper circuit configuration, a phase shift control is achieved without changing the output amplitude.

Control potentiometers are available in multiple ganged units and thus may be used to simultaneously change voltage, current, or resistance levels at different parts of the circuit. Even if tracking of these variables is imperfect, the availability of ganged controls adds a great flexibility to the designer's resources.

The equation included in Fig. 5-12 shows that phase shift is a nonlinear function of the resist-

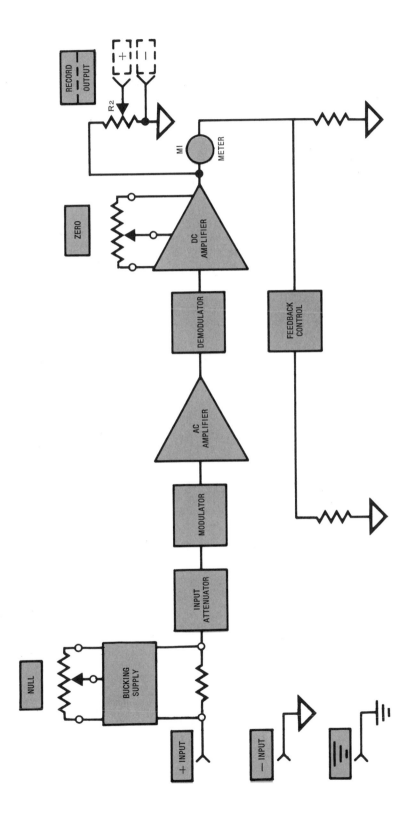

Fig. 5-9 Potentiometer controls are utilized for zero and offset null on this voltmeter (Hewlett-Packard)

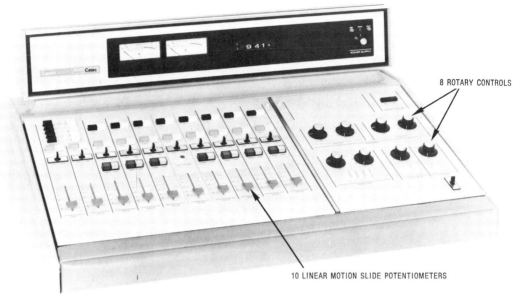

8 ROTARY CONTROLS

10 LINEAR MOTION SLIDE POTENTIOMETERS

Fig. 5-10 Master mixer board for recording studio
(Cetec, Inc.)

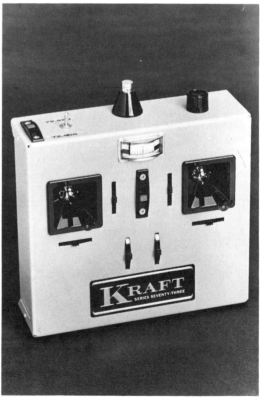

A. SINGLE STICK POSITIONER

B. SIX-CHANNEL TRANSMITTER

Fig. 5-11 Remote control system for model aircraft
(Kraft Systems, Inc.)

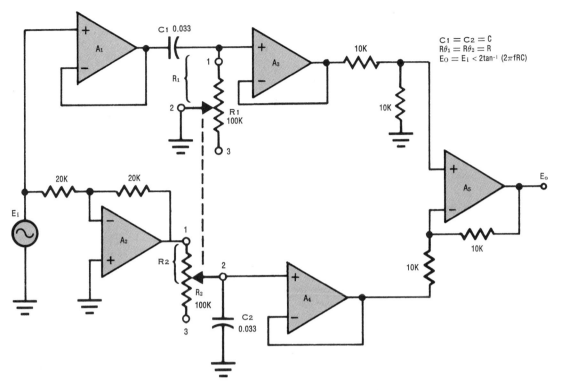

Fig. 5-12 Ganged potentiometers yield phase shift control

ance value. If the potentiometer elements have a logarithmic transfer function, a smoother phase shift control is obtained.

Attenuators. A very common control used in communications equipment is the constant impedance variable attenuator. Fig. 5-13 shows five typical circuit configurations for these attenuators. The unique characteristic of all these configurations is to maintain the input impedance and output impedance at an equal and constant level as the amount of attenuation (from input to output) is varied.

All five of the circuits shown in Fig. 5-13 perform an identical function. The difference in the configurations is the accuracy with which it performs that function. The circuits are arranged in relative order of accuracy. Fig. 5-13A is the least accurate and Fig. 5-13E is the most accurate.

For the bridged T configuration of Fig. 5-13B, to keep the impedances constant requires maintaining the relationships:

$$R_1 = Z \ (K\text{-}1) \quad \text{and} \quad R_2 = \frac{Z}{K\text{-}1}$$

where $K = \text{antilog} \left[\dfrac{A}{20} \right]$ and A is the

attenuation in decibels (db).

Solving the above relationships for K and setting them equal to each other:

$$R_1 R_2 = Z^2 = \text{constant}$$

This condition can be achieved by constructing R_1 and R_2 to produce a logrithmic output function. R_1 must be counterclockwise logrithmic and R_2 clockwise logrithmic. R_1 and R_2 must be mounted in a common shaft.

Motor Speed Control. Modern drill motors have great versatility because of adjustable speed control. A simple circuit is included within the case. This circuit uses a potentiometer and a triac in the manner shown in Fig. 5-14. The operator squeezes a trigger that is mechanically linked to the potentiometer. The potentiometer setting determines the point in the input voltage cycle where the triac is turned on and, hence, controls the average voltage applied to the motor. This allows a very large range of usable motor speeds.

Relatively small potentiometers teamed with modern solid-state circuitry are used to control the speed of very large motors. The control point can be right at the motor, as in the case of the drill motor or a blender, or it can be at some remote point more convenient to the opera-

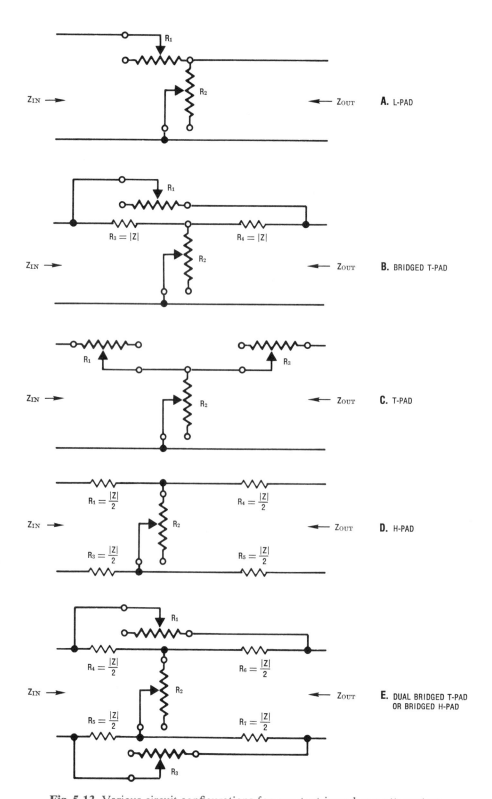

Fig. 5-13 Various circuit configurations for constant impedance attenuators

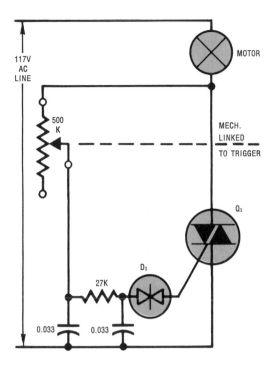

Fig. 5-14 Simple motor speed control

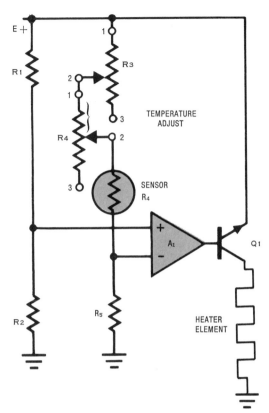

Fig. 5-15 Temperature control circuit uses a balanced bridge

tor. Even with great separation, the man-machine interface capability of potentiometers is effective.

Temperature Control. Control potentiometers may be used to adjust the amount of power supplied to a heater, or they may be included in a temperature servo control to adjust the set point.

The same basic circuit shown in Fig. 5-14 may be adapted to control the amount of ac power supplied to a heater element. By adjusting heater power with a potentiometer the operator controls the operating temperature indirectly.

For *precise* temperature control, a servo feedback system could be used to adjust the amount of power and thus the temperature of the heater. In operation, the desired temperature is preset by a control potentiometer. When the operating temperature, as monitored by a temperature sensor, is below the setpoint more power is applied to the heater. Usually, the sensing transducer and the control potentiometer are part of some form of bridge circuit such as that shown in Fig. 5-15.

In this circuit, the bridge is balanced when the sum of R_3 and R_4 bear the same ratio to R_5

as R_1 does to R_2 or:

$$\frac{R_3 + R_4}{R_5} = \frac{R_1}{R_2}$$

If the sensor resistance is high, causing an imbalance in the bridge, it indicates the temperature is low. Then the error produces a positive output voltage from amplifier A_1 and heater power is increased an amount determined by the temperature error and the voltage gain of A_1.

When the resistance of the sensor drops below the balance point in the bridge, indicating that the temperature is too high, then the output voltage from A_1 is negative. This will turn off transistor Q_1 and no heater power is supplied to the heater element.

Making the gain of A_1 very high will result in a system in which heater power goes from off to full on with a very small temperature change. On the other hand, a moderate voltage gain will yield a more or less proportional control in which the amount of power supplied will be proportionate to the temperature error.

The relative positions of the sensor and control potentiometer in the circuit may be changed if the temperature coefficient of the sensor is positive rather than negative or the same bridge configuration may be used with the inputs to the amplifier reversed. The sensor and potentiometer could be relocated to opposite branches of the bridge, but the configuration shown always brings the bridge back to the exact same operating conditions with the same power requirement.

Lighting Level Control. Potentiometer controls may be used to vary light levels by adjusting the power applied to the lamps in a manner similar to that for heaters described in the preceding paragraphs. The control may be a direct one as is common for *mood* lighting control in homes or in stage lighting. It may be used in an overall servo system to control the exact set point of the light level using a photoelectric sensor.

Once again, a small, unimposing, low power control potentiometer may be used to control huge banks of high power lamps when it is applied with modern solid-state circuitry.

Multifunction Control. In fields such as tests and measurement, there is a great need for a variety of front panel controlled switches and potentiometers. Often multiple functions are used on a single adjustment shaft or multiple functions are controlled by concentric shafts from the same front panel control. Also modest quantities of specials that vary from circuit-to-circuit are sometimes needed. An economical and versatile assembly with many options is manufactured in the configuration shown in Fig. 5-16. These are modular components in a standard, expandable package with a variety of functions available in each section at relatively low cost.

SUMMARY

Control devices are used in applications in which frequent manual adjustment is anticipated and convenient adjustment is desired. Many of these applications involve man-machine interface. Cost-effective application of the control potentiometer depends on the designer's knowledge of economical options available.

Factors in the design of control applications and examples of typical control possibilities are listed in Fig. 5-17.

Fig. 5-16 Multifunction control with modular construction provides a variety of functions including potentiometers and switches

Page	DESIGN FACTORS FOR CONTROL APPLICATIONS	Page	SOME TYPICAL CONTROL APPLICATIONS
			INSTRUMENT CONTROLS
98	Control the actual function of interest	101	Oscilloscopes
99	Provide adequate range and resolution	101	Function Generators
99	Choose a logical direction of control sense	101	Power Supplies
100	Assume worst case conditions	104	Photometers
100	Make controls independent	104	Recorders
101	Consider the shape of the controlled	104	Meters
	function (output curve)	104	AUDIO
101	Consider environmental and stability		
	requirements		MISCELLANEOUS CONTROLS
101	Choose a proper location	106	Model Aircraft Remote Control
		106	Phase Shift Control
		109	Motor Speed Control
		111	Temperature Control
		109	Attenuators
		112	Lighting Level Control
		112	Multifunction Control

Fig. 5-17 Design factors and some typical control applications

Notes

APPLICATION AS A PRECISION DEVICE

You will find me wherever men strive to attain still higher levels of accuracy, and place no petty premium upon perfection. In the laboratory I am ever present amidst the chemist's test tubes and pipettes. In the observatory I am always at the astronomer's elbow. Each day I guide the fingers of a million pairs of hands, and direct the destinies of countless busy machines . . . I am Precision.

From an early advertisement by
HERBERT H. FROST, INC.
(Now CTS Corp.)

INTRODUCTION

Precision potentiometers find application where there is interest in the relationship between the incremental voltage level and the incremental displacement of a mechanical device. Often precision potentiometers are used in the simple control functions discussed in Chapter 5.

In this chapter the emphasis is on applications where a higher degree of accuracy is required than has been previously discussed. The importance of power rating, the effects of frequency, linear and nonlinear functions, and other electrical parameters are examined through the use of application examples. The examples include servo systems, coarse-fine dual level controls, and position indication/transmission systems. Since precision devices are electro-mechanical in nature, a discussion of them is not really complete with only electrical application data. Therefore Chapter 6 includes important mechanical parameters such as mounting, torque, stop strength, mechanical runouts and phasing.

OPERATIONAL CHARACTERISTICS

The resulting performance of precision potentiometers is dependent not only on the wisdom of their design and the accuracy of their construction but also on the conditions under which they must perform. Some of the operational factors which effect the quality of performance or the duration of life include the excitation (input) and wiper currents, the excitation frequency, the heat conductivity of the potentiometer mounting, and the temperature, pressure, and humidity of the surrounding environment.

In evaluating performance, the best method is to study the effects of each of the above factors separately. In any given application, however, the potentiometer is influenced by a combination of these factors. The resulting performance cannot be predicted by considering a mere linear summation of these effects. Recognizing this places more and more importance on the methods of environmental testing which simulate the actual conditions of application. With these methods the true reaction of any physical component to the particular environment can be reliably evaluated.

POWER RATING

Power rating is an indication of a maximum power that can be safely dissipated by the device when a voltage (excitation) is applied to the end terminals. It is most often determined by a temperature-rise method. This prevents any part of the potentiometer from exceeding the maximum operating temperature at full rated power. For extremely accurate low-noise units the power rating should take into consideration noise, life, electrical and mechanical angles, the number of sections, and other functional characteristics.

The maximum power that can be dissipated is dependent upon the capability of the mounting structure to get rid of (sink) its heat by conduction or convection. This capability must be sufficient to keep the operating temperature of critical parts below levels which can cause permanent, physical damage. In certain less critical applications, such as those discussed in Chapter 5, potentiometers can function in the presence of minor physical deterioration. If the resistance element remains unbroken and the wiper continues to make satisfactory contact with the element, operation is not affected. In precision instrumentation, however, a minor increase in noise level or a slight dimensional shift in the resistance element may be prohibitive. Therefore, a meaningful power rating of a precision potentiometer should include a definition of the level of physical and electrical deterioration that can be tolerated in the particular application.

Potentiometers are not often fortunate enough to operate under laboratory mounting and ideal controlled environmental conditions. The tendency is to tightly package them with other components that may create greater temperatures. In Fig. 6-1 typical power derating curves are used to illustrate the relationship between power rating and basic potentiometer size. The curves shown are based on metallic cases and wirewound resistance elements. These curves are for single section devices. When sections are added the result is a power rating (for multicup units) that is 75% of the single section rating. The curves clearly indicate that operation at temperatures below 70°C is conducive to longer life. Voltage

excitation limits are usually determined by the insulation resistance of the resistance wire, in other words, the allowable voltage drop per turn of the resistance winding.

Fig. 6-2 shows how the power dissipation capability of wirewound potentiometers changes with diameter. The trend varies roughly as a square law because of its relationship to the device's surface area. Single turns with metal cases and ten turns with plastic cases are illustrated. The curves do not go to zero for the hypothetical potentiometer of zero size because any plate on which a potentiometer is mounted will have some heat sink capability.

Special considerations are necessary when the potentiometer is connected as a rheostat. The power capability depends upon how much of the resistance element is employed and thus upon the position of the wiper. Fig. 6-3 is a power derating curve for rheostats with wirewound resistance elements and metallic cases. This curve pertains to all sizes and is presented in terms of the percent of maximum power rating. The shape of the derating curve is interesting and shows the surprising fact that 50% of the total rated power can be dissipated satisfactorily by only 20% of the resistance element. This is due to the fact that the case and the remainder of the winding serve to conduct heat away from the small portion of the winding actually being used.

Fig. 6-4 shows a power derating curve for potentiometers with plastic cases. Because of

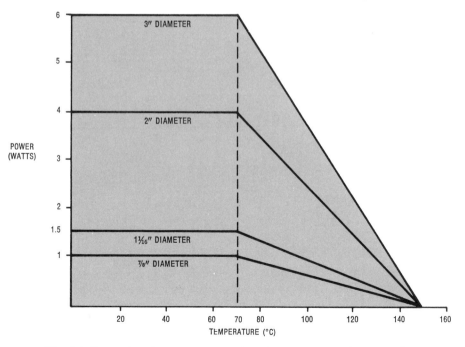

Fig. 6-1 Power derating curves for typical single turn precision potentiometers

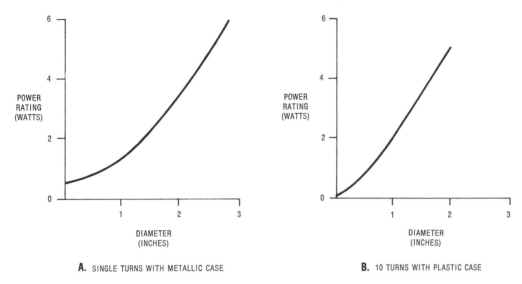

A. SINGLE TURNS WITH METALLIC CASE   **B.** 10 TURNS WITH PLASTIC CASE

Fig. 6-2 Trend in power rating with change in diameter

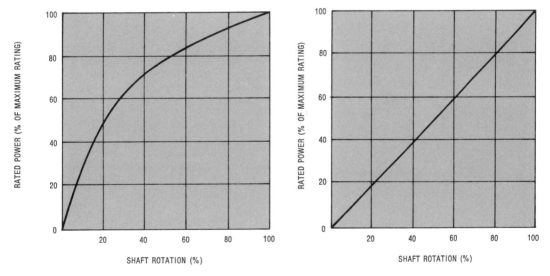

Fig. 6-3 Power derating curve for a rheostat connected, metal case potentiometer

Fig. 6-4 Power derating curve for a rheostat connected, plastic case potentiometer

the poor thermal conductivity of the plastic material, the advantages of cooperative heat dissipation are not enjoyed, and a more linear derating characteristic results.

Continuing advancements in the state of the art (materials and processes) are having a marked effect on the high temperature capabilities of potentiometers. In the past fifteen years, maximum operating temperatures of precision potentiometers have increased from 80°C to 150°C for most devices. Special design considerations have increased some devices to 200°C and higher.

FREQUENCY CHARACTERISTICS

The resistance element of a wirewound precision potentiometer acts as a pure resistive load for direct current and the common line frequency of 60 Hz. When used in the kilohertz frequency range, the reactances of the distributed inductance and capacitance become significant. When combined with the resistance of the element, these reactances form a complex impedance characteristic load.

The following paragraphs describe four important potentiometer alternating current (ac) parameters: *input impedance, output impedance, quadrature voltage and phase shift*. These parameters are industry standard definitions used to characterize the ac response of potentiometers. Typical values of these parameters are not presented due to the wide range of voltage and frequency possible. In addition, these characteristics vary considerably from design to design due to the well known effects of physical construction and geometry on ac response. They may be applied to any potentiometer, wirewound or nonwirewound, but are most pronounced in units constructed with wirewound elements.

Input Impedance. The total impedance (ac reactive and dc resistive) measured between the potentiometer's end terminals is the input impedance. It is always measured with the wiper circuit open (no load). The voltage and frequency at which the impedance is measured must be specified and the wiper must be positioned to a point that results in the largest impedance value.

Output Impedance. The total impedance (ac reactive and dc resistive) measured between the potentiometer's wiper terminal and either end terminal is the output impedance. This characteristic is always measured with the end terminals connected together (electrically shorted). As with input impedance, output impedance must be specified together with a voltage and frequency.

Quadrature Voltage. When an ac voltage of sufficiently high frequency is applied as the input voltage to a potentiometer, the resistive element exhibits a characteristic impedance. This impedance is composed of a capacitive reactance X'_C, inductive reactance X'_L, and a resistance R. The reactive components are 180° out of phase and therefore, the larger will cancel the effects of the smaller. The resultant impedance seen by the input voltage (or looking back from the output) will consist of a single reactive component (X_C or X_L) and a resistive component R. These resultant components can be represented by two vectors in quadrature, i.e., separated in phase by 90°. One is the voltage across the real (resistive) impedance component, the other is across the imaginary (reactive) impedance component. The *quadrature voltage* for a potentiometer refers to that voltage across the reactive component of the output voltage. Fig. 6-5 summarizes the quadrature voltage parameter for a potentiometer whose X_L is much larger than X_C.

Fig. 6-6 is the industry standard test circuit for quadrature voltage. In this configuration, a standard potentiometer having a negligible reactive component is used to null the real (resistive) component of output voltage, leaving the reactive voltage displayed on the meter M_1. The nulling procedure is performed several times until a maximum reading on M_1 is obtained. The quadrature voltage specification for the particular potentiometer being tested is then calculated using the formula given in Fig. 6-6.

Phase Shift. The reactive component of the potentiometer's characteristic impedance will cause a *phase shift* between the input and output voltages. The phase shift of a potentiometer refers specifically to sinusoidal inputs. The input frequency, voltage and wiper position must be specified. Mathematically, phase shift may be written:

$$\Phi = \sin^{-1}\frac{E_X}{E_O}$$

(E_X and E_O must be in like terms, i.e., RMS, Peak, or Average)

Where:

Φ is the phase shift in degrees.

E_X is the quadrature voltage as measured in Fig. 6-6.

E_O is the output voltage.

A very complicated circuit condition exists if the wiper is connected to a complex impedance load and is allowed to move along the resistance element. The analysis of such a circuit configuration depends critically upon the nature of the external load. The following text assumes that

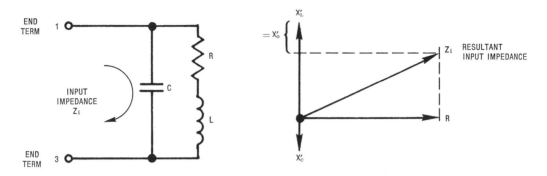

A. RESULTANT INPUT IMPEDANCE FOR THIS UNIT IS EFFECTIVELY INDUCTIVE AND RESISTIVE

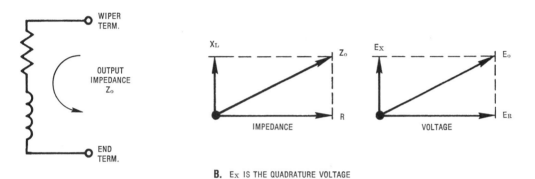

B. E_X IS THE QUADRATURE VOLTAGE

Fig. 6-5 Quadrature voltage for a particular potentiometer at a specified input voltage and frequency

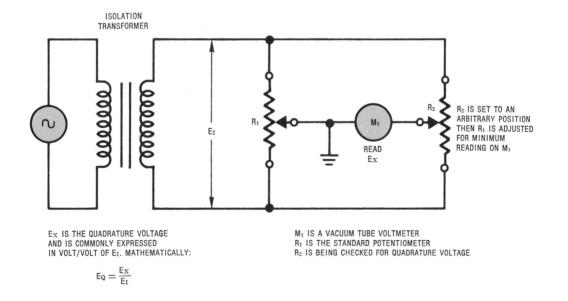

E_X IS THE QUADRATURE VOLTAGE
AND IS COMMONLY EXPRESSED
IN VOLT/VOLT OF E_I. MATHEMATICALLY:

$$E_Q = \frac{E_X}{E_I}$$

M_1 IS A VACUUM TUBE VOLTMETER
R_1 IS THE STANDARD POTENTIOMETER
R_2 IS BEING CHECKED FOR QUADRATURE VOLTAGE

Fig. 6-6 Industry standard for quadrature voltage measurement

the wiper is unloaded and set to the low voltage end of the resistive element.

A wirewound resistance element can be analyzed as a uniform transmission line when excited at high frequency. The resistive wire provides the resistance R of the line and the coiled turns of wire result in inductance L. Closeness of one winding to the next and winding to case produce capacitance C. For elements wound on insulated copper mandrel or other conductors there is also capacitance between the mandrel and winding. These three parameters, R, C and L, are spread along the entire length of resistance wire and are approximated by the equivalent circuit of Fig. 6-7.

For a wirewound potentiometer with conductive mandrel the resistance and inductance of each turn are shown as Rw and Lw. Cw represents capacitance between adjacent turns of wire. Capacitance between winding and housing is identified as Cc. The capacitive coupling Cw between individual turns increases when a conductive mandrel is involved. To better understand the total electrical state of a wirewound potentiometer refer again to the lumped parameter circuit of Fig. 6-7. Measurements of an actual potentiometer will yield the most reliable performance data with respect to R, C, and L.

The use of nonlinear windings, shunt loading, and variable pitch wire spacing can be expected to alter the frequency performance. The use of an enameled copper mandrel to support the resistance winding has been found to lower the potentiometer's frequency range capability.

LINEAR FUNCTIONS

In this chapter, linear refers to an electrical response rather than a mechanical style. Many of the concepts discussed apply, regardless of mechanical design; but the chapter is dealing with rotary style potentiometers.

A knowledge of the general design characteristics of linear precision potentiometers can help the user in selecting a device that will best satisfy the demands of a particular application. The curves in Fig. 6-8 and Fig. 6-9 are intended to indicate general design trends rather than specific design values.

Fig. 6-8 shows how the range of achievable voltage resolution varies with the potentiometer's diameter. The upper limit of the resolution zone pertains to low values of total resistance (approximately 1,000 ohms). The lower boundary of the resolution zone is for relatively high total resistance (approximately 50K ohms).

Fig. 6-9 illustrates the general trend in linearity with a change in potentiometer diameter. The solid curves represent common linearity figures. The broken line curves represent linearities achievable with special design and construction techniques on potentiometers whose total resistance is 5K ohms or greater.

The concepts of linearity and resolution are related. Certainly the 1/N resolution figure places a limit on the achievable linearity. The linearity deviations are usually from two to five times greater than the 1/N resolution value. Linearity and resolution depend on the value of

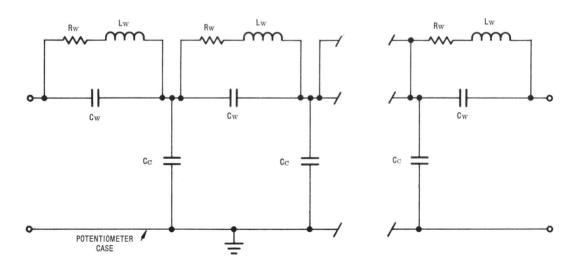

Fig. 6-7 Lumped-parameter approximation for wire wound potentiometers

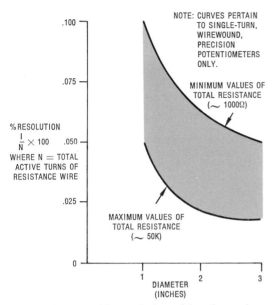

% RESOLUTION

$$\frac{1}{N} \times 100$$

WHERE N = TOTAL ACTIVE TURNS OF RESISTANCE WIRE

NOTE: CURVES PERTAIN TO SINGLE-TURN, WIREWOUND, PRECISION POTENTIOMETERS ONLY.

MINIMUM VALUES OF TOTAL RESISTANCE (~ 1000Ω)

MAXIMUM VALUES OF TOTAL RESISTANCE (~ 50K)

DIAMETER (INCHES)

Fig. 6-8 Trend in resolution with a change in potentiometer diameter

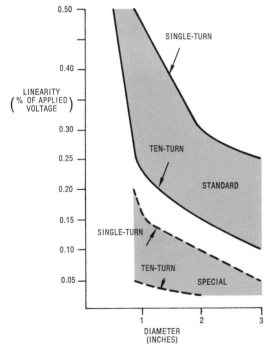

LINEARITY $\left(\begin{array}{c} \% \text{ OF APPLIED} \\ \text{VOLTAGE} \end{array} \right)$

SINGLE-TURN

TEN-TURN

STANDARD

SINGLE-TURN

TEN-TURN

SPECIAL

DIAMETER (INCHES)

Fig. 6-9 Trend in potentiometer linearity with a change in diameter

total winding resistance. As winding resistance increases, linearity and resolution improve (decrease).

Any discussion of potentiometer electrical functions is not complete without stressing the interdependence of the accuracy and the

methods of resistive element construction. The construction of resistive elements is discussed in detail in Chapter 7. The following list outlines those construction factors which influence the accuracy of the assembled wirewound potentiometer:

1. The uniformity of the supporting mandrel.
2. The tension of the resistance wire.
3. The spacing and number of turns of resistance wire.
4. The degree of cleanliness of the resistance element.
5. The concentricity of the mounted winding relative to the rotational axis of the wiper.

The angular length of a linear resistance winding for a single turn potentiometer is usually about 350°. For special applications, this angle can be reduced. By careful construction, the winding can be made to accommodate more than 350° of wiper rotation. Obviously, it would be undesirable to short circuit the two ends of the resistance windings; but it is possible to place the two ends of the resistance element in proximity and still control the design such that the wiper will not bridge the gap between the ends. When such precaution is taken, the potentiometer is described as *non-shorting*.

For some special applications, such as the sine/cosine functions presented in the next section, it may be desirable to rotate the slider continuously to produce a repetitive voltage waveform. In such instances, the two ends of the resistance winding are joined together. Other forms of continuous winding are discussed in later sections of this chapter.

NONLINEAR FUNCTIONS

The accuracy with which a nonlinear function can be produced is difficult to generalize because of the many controlling factors and the variety in design approaches available to the potentiometer designer. The following paragraphs will explain some of the nonlinear techniques available. Expecting certain variations to occur in manufacture, the potentiometer designer usually designs within a band that is equal to one-third the required conformity band. This conservative approach gives assurance that the final performance will be well within the conformity specification.

To create workable specifications the user should be aware of the producibility of a given nonlinear function, the minimum size into which it can be built, the ease of manufacturing a large quantity with a minimum of production difficulty, and the cost. The table in Fig. 6-10 shows the common nonlinear functions available from most manufacturers. It is important to note the

FUNCTION DESCRIPTION		SPECIFICATIONS	CONDUCTIVE PLASTIC SINGLE TURN		
			7/8" DIAMETER	1¹⁄₁₆" DIAMETER	2" DIAMETER
SIN-COS 360° 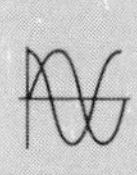	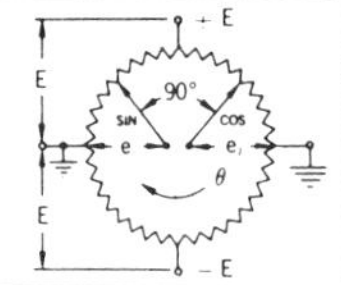 $\dfrac{e_1}{E}=\sin(\alpha)$ $\dfrac{e_2}{E}=\cos(\alpha)$ $0°\le\alpha\le360°$ $0°\le\theta\le\theta_T$	Resistance Range Ω Conformity (1) Std. ±% Best θ_T	1K-50K 60K 2.0 2.5 1.0 1.25 360°	1K-50K 60K-75K 1.25 1.5 0.5 0.75 360°	1K-50K 60K-100K 1.0 1.25 0.25 0.3 360°
SINE 360° 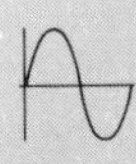	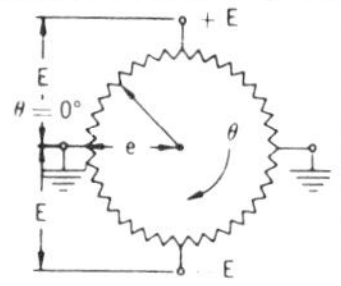 $\dfrac{e}{E}=\sin(\alpha)$ $0°\le\alpha\le360°$ $0°\le\theta\le\theta_T$	Resistance Range Ω Conformity (1) Std. ±% Best θ_T	1K-50K 60K 2.0 2.5 1.0 1.25 360°	1K-50K 60K-75K 1.25 1.5 0.5 0.75 360°	1K-50K 60K-100K 1.0 1.25 0.25 0.3 360°
COSINE 360° 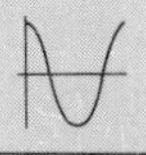	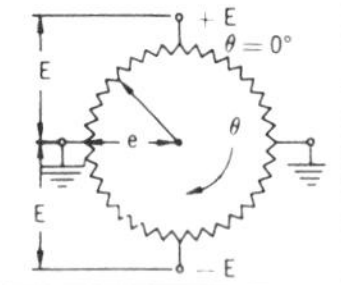 $\dfrac{e}{E}=\cos(\alpha)$ $0°\le\alpha\le360°$ $0°\le\theta\le\theta_T$	Resistance Range Ω Conformity (1) Std. ±% Best θ_T	1K-50K 60K 2.0 2.5 1.0 1.25 360°	1K-50K 60K-75K 1.25 1.5 0.5 0.75 360°	1K-50K 60K-100K 1.0 1.25 0.25 0.3 360°
SINE 180° 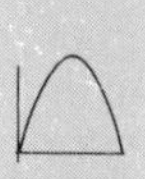	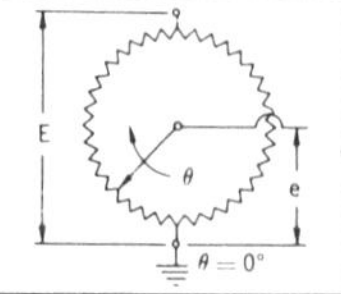 $\dfrac{e}{E}=\sin(\alpha)$ $0°\le\alpha\le180°$ $0°\le\theta\le\theta_T$	Resistance Range Ω Conformity (1) Std. ±% Best θ_T	1K-50K 2.0 1.0 360°	1K-50K 1.25 0.75 360°	1K-50K 1.0 0.5 360°
COSINE 180° 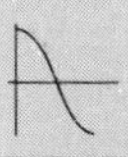	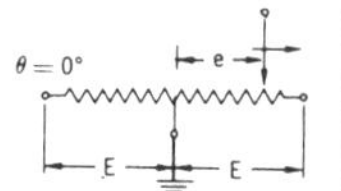 $\dfrac{e}{E}=\cos(\alpha)$ $0°\le\alpha\le180°$ $0°\le\theta\le\theta_T$	Resistance Range Ω Conformity (1) Std. ±% Best θ_T	1K-100K 2.0 1.0 320°	1K-100K 1.25 0.5 340°	1K-100K 1.0 0.25 350°
SINE 90° 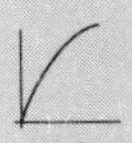	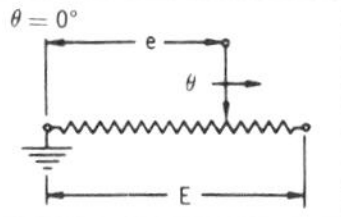 $\dfrac{e}{E}=\sin(\alpha)$ $0°\le\alpha\le90°$ $0°\le\theta\le\theta_T$	Resistance Range Ω Conformity (1) Std. ±% Best θ_T	1K-100K 1.7 1.0 320°	1K-100K 1.25 0.5 340°	1K-100K 0.75 0.25 350°
COSINE 90° 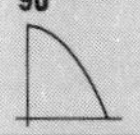	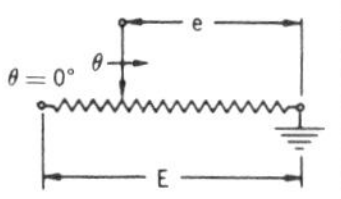 $\dfrac{e}{E}=\cos(\alpha)$ $0°\le\alpha\le90°$ $0°\le\theta\le\theta_T$	Resistance Range Ω Conformity (1) Std. ±% Best θ_T	1K-100K 1.7 1.0 320°	1K-100K 1.25 0.5 340°	1K-100K 0.75 0.25 350°
SINGLE SIDED SQUARE 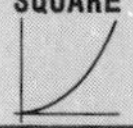	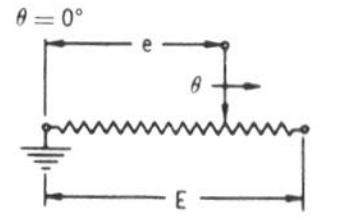 $\dfrac{e}{E}=X^2$ $0\le X\le1$ $0°\le\theta\le\theta_T$	Resistance Range Ω Conformity (1) Std. ±% Best θ_T	1K-50K 1.25 1.0 320°	1K-75K 1.0 0.5 340°	1K-100K 0.6 0.25 350°
DOUBLE SIDED SQUARE 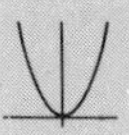	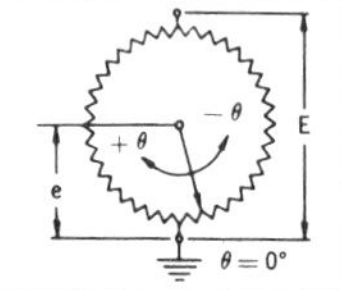 $\dfrac{e}{E}=X^2$ $-1\le X\le+1$ $-\theta_T\le\theta\le+\theta_T$	Resistance Range Ω Conformity (1) Std. ±% Best θ_T	1K-50K 2.0 1.0 ±180°	1K-50K 1.25 0.75 ±180°	1K-50K 1.0 0.5 ±180°
20 db LOG 	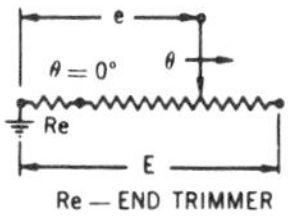 $\dfrac{e}{E}=10^{(x-1)}$ $0\le X\le1$ $0°\le\theta\le\theta_T$ Re — END TRIMMER	Resistance Range Ω Conformity (1) Std. ±% Best θ_T	1K-75K 1.25 0.75 320°	1K-100K 1.0 0.5 340°	1K-150K 0.6 0.5 350°
40 db LOG 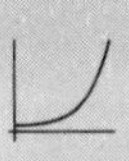	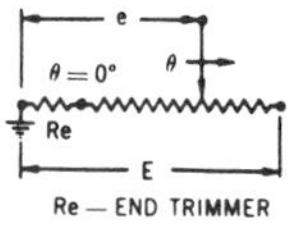 $\dfrac{e}{E}=10^{2(x-1)}$ $0\le X\le1$ $0°\le\theta\le\theta_T$ Re — END TRIMMER	Resistance Range Ω Conformity (1) Std. ±% Best θ_T	1K-50K 3.0 2.0 320°	1K-100K 2.0 1.0 340°	1K-100K 1.0 0.75 350°

Fig. 6-10 Standard nonlinear functions

WIREWOUND SINGLE TURN				⅞″ DIAMETER WIREWOUND MULTITURN		
⅞″ DIAMETER	1 1/16″ DIAMETER	2″ DIAMETER	3″ DIAMETER	3 TURN	5 TURN	10 TURN
N/A N/A N/A N/A N/A N/A	N/A N/A N/A N/A N/A N/A	250 16K 1.0 1.0 0.6 0.4 360°	250 30K 1.0 0.5 0.5 0.25 360°	N/A N/A N/A	N/A N/A N/A	N/A N/A N/A
250 4K 2.0 1.5 1.1 0.9 360°	250 5K 1.5 1.0 1.0 0.75 360°	250 16K 1.0 1.0 0.6 0.4 360°	250 30K 1.0 0.5 0.5 0.25 360°	500 7K 1.0 1.0 0.7 0.5 1080°	1K 12K 0.8 0.8 0.5 0.35 1800°	1K 25K 0.7 0.7 0.25 0.2 3600°
250 4K 2.0 1.5 1.1 0.9 360°	250 5K 1.5 1.0 1.0 0.75 360°	250 16K 1.0 1.0 0.6 0.4 360°	250 30K 1.0 0.5 0.5 0.25 360°	500 7K 1.0 1.0 0.7 0.5 1080°	1K 12K 0.8 0.8 0.5 0.35 1800°	1K 25K 0.7 0.7 0.25 0.2 3600°
250 4K 1.8 1.5 1.1 0.9 360°	250 5K 1.5 1.25 1.1 0.9 360°	250 16K 1.5 1.1 0.6 0.4 360°	250 30K 1.25 1.0 0.5 0.25 360°	500 8K 1.5 1.1 0.6 0.5 1080°	500 14K 1.2 1.0 0.5 0.35 1800°	1K 25K 1.1 0.9 0.25 0.2 3600°
500 15K 1.8 1.5 1.1 0.9 350°	500 18K 1.5 1.0 1.0 0.8 350°	500 75K 1.0 1.0 0.6 0.4 350°	500 125K 0.75 0.5 0.4 0.25 350°	1K 25K 1.0 1.0 0.7 0.6 1080°	1K 45K 0.8 0.8 0.5 0.4 1800°	1K 100K 0.7 0.7 0.3 0.2 3600°
1K 13K 1.6 1.3 0.6 0.5 350°	1K 18K 1.5 1.2 0.5 0.4 350°	1K 75K 1.0 0.9 0.4 0.2 350°	1K 125K 1.0 0.9 0.35 0.2 350°	1K 25K 1.2 1.0 0.4 0.3 1080°	1K 45K 1.0 1.0 0.25 0.2 1800°	1K 100K 1.0 1.0 0.2 0.15 3600°
1K 13K 1.6 1.3 0.6 0.5 350°	1K 18K 1.5 1.2 0.5 0.4 350°	1K 75K 1.0 0.9 0.4 0.2 350°	1K 125K 1.0 0.9 0.35 0.2 350°	1K 25K 1.2 1.0 0.4 0.3 1080°	1K 45K 1.0 1.0 0.25 0.2 1800°	1K 100K 1.0 1.0 0.2 0.15 3600°
1K 13K 1.5 1.0 0.7 0.6 350°	1K 18K 1.2 1.0 0.65 0.5 350°	1K 50K 1.0 0.5 0.4 0.25 350°	1K 100K 0.6 0.4 0.3 0.15 350°	1K 20K 0.9 0.5 0.5 0.35 1080°	1K 35K 0.8 0.5 0.3 0.2 1800°	1K 70K 0.7 0.5 0.2 0.15 3600°
250 3.5K 1.8 1.5 1.4 1.0 ±180°	250 4K 1.7 1.3 1.2 1.0 ±180°	500 13K 1.2 1.0 0.65 0.45 ±180°	500 25K 1.0 0.75 0.5 0.25 ±180°	500 7K 1.2 1.2 0.75 0.65 ±540°	500 8K 1.1 1.1 0.5 0.4 ±900°	1K 20K 1.0 1.0 0.3 0.25 ±1800°
1K 12K 1.5 1.0 0.8 0.7 350°	1K 14K 1.2 1.0 0.7 0.6 350°	1K 50K 0.75 0.75 0.5 0.3 350°	1K 75K 0.75 0.5 0.5 0.15 350°	1K 20K 0.75 0.75 0.45 0.35 1080°	1K 33K 0.65 0.65 0.35 0.25 1800°	1K 70K 0.5 0.5 0.2 0.15 3600°
500 5K 2.5 2.0 1.5 1.2 350°	500 7K 2.0 1.75 1.4 1.1 350°	1K 25K 1.5 1.2 0.7 0.45 350°	1K 40K 0.8 0.75 0.5 0.25 350°	1K 10K 1.2 1.2 0.75 0.65 1080°	1K 16K 1.1 1.1 0.5 0.4 1800°	1K 35K 1.0 1.0 0.35 0.3 3600°

NOTES: (1) All conformities listed are peak-to-peak conformities, i.e., the absolute difference between the maximum and minimum voltage applied.
(2) e = Output Voltage   E = Total Applied Voltage θ = Shaft Position θ_T = Theoretical Travel $X = \theta \div \theta_T$

relationship between size, total resistance range, and conformity tolerances for each of these common functions. Many other nonlinear functions can be achieved. To order special functions, supply the potentiometer manufacturer a mathematical equation representing the function or a series of data points and a graphical representation of the desired function.

It is possible to make an estimation of the resolution achievable for a given nonlinear function with a wirewound potentiometer. Approximate portions of the function by straight line segments and estimate the resolution as though these linear segments were part of a continuous linear potentiometer. In the case of linear potentiometers, it is possible to use wire size, resistivity, and spacing to minimize the 1/N resolution. However, it is usually not possible to achieve as low a 1/N resolution figure for a similar linear portion of a nonlinear potentiometer due to other design factors. The following method will give an approximate resolution figure:

1. Plot a curve of percent voltage (or percent resistance) *vs.* percent rotation.
2. Approximate the nonlinear curve with joined straight-line segments.
3. Measure the approximate slope of a linear segment occupying a given region of the function.
4. Specify the desired total resistance of the nonlinear potentiometer.
5. Construct a resolution curve for a linear potentiometer of the same size, type of construction, and total resistance using data from a manufacturer's catalog sheet. Read the approximate 1/N resolution for the region of the function under consideration.

As might be expected, this method yields small 1/N resolution values in the regions of relatively high slope. This is because higher slopes require a greater number of turns of fine, high resistance wire.

Loading. There are two methods of generating nonlinear functions by loading. They are:

Shunt loading the resistive element.

Shunt loading the wiper.

The principal feature of shunt loading techniques is its flexibility.

Loading the Resistive Element. If the wirewound element is constructed with several taps, a variety of nonlinear curves can be apppproximated by connecting the appropriate shunt resistors. The element taps are usually accessible via terminals on the potentiometer case. The resistance element is excited by a voltage applied across the end terminals. The resistivity, and therefore the output function, may be different in various portions of the element depending on the values of fixed shunt resistors across each

section as selected by the system designer. This technique has considerable benefit to the potentiometer user and manufacturer because of its versatility.

When the nonlinear function requirement is precisely defined *before* potentiometer construction, the manufacturer can supply a unit complete with shunt resistors for optimization of the electrical function. Potentiometers can be designed so shunt resistors can be attached internally or externally.

Some users maintain a stock of tapped potentiometers that they can quickly shunt to provide a variety of custom nonlinear functions. In this case, the exact nonlinear function is unknown when it is ordered but a linear potentiometer is specified with taps at selected locations. The user can then attach external shunt resistors that produce the desired function.

Loading the Wiper. In Chapter 3, output error due to wiper circuit loading was presented in detail. In certain precision potentiometer applications, this error may be used to an advantage. By proper selection of the load, the output can be shaped to vary (the error) in some desired nonlinear manner. Fig. 6-11 shows a schematic example. Proper selection of the shunt-to-element total resistance ratio can produce smooth nonlinear functions such as tangent, secant, square root, square, and reciprocal. Fig. 6-12 illustrates a possible nonlinear function with a loaded wiper circuit.

Anytime current is caused to flow through

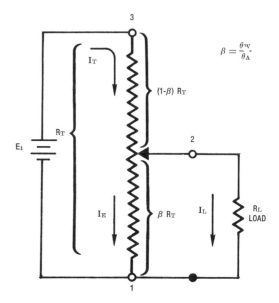

Fig. 6-11 A potentiometer with a loaded wiper circuit

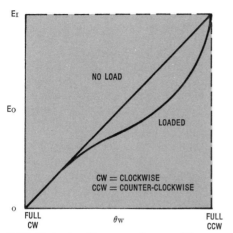

Fig. 6-12 Output voltage vs wiper position with and without wiper load

the wiper, special care must be taken. Every potentiometer has a maximum wiper current that it is capable of handling without degrading its operational life. Consult the manufacturer's data sheet or the manufacturer directly to determine maximum wiper current capability.

If a particular application requires the wiper to be loaded to a degree which causes excessive nonlinearity, then some type of compensation will be necessary. Chapter 3 discusses some com-

pensation methods. Another method utilizes an isolation amplifier inserted in the circuit between the wiper and the load. The high input impedance of the isolation amplifier draws negligible wiper current.

Compensation can also be affected by the use of a loaded linear potentiometer. The resistance element is compensated — made nonlinear — during manufacture to correct the loading effect. Given the resistance value of the load, the potentiometer manufacturer can calculate the deviation from the theoretical function. This information is then used to manufacture a compensated element.

Voltage Clamping. Another technique for obtaining nonlinear functions from linear element potentiometers involves clamping (electrically holding) various taps at the *voltage levels* of the desired function. This technique is shown in Fig. 6-13. End terminals or tap points are connected to a voltage divider. If the voltage divider is of sufficiently low resistance relative to the potentiometer element, each of the tap points can be set independently.

The result is an approximation of the desired function by straight-line segments. When approximating a nonlinear function by voltage clamping, the clamping voltages are generally established with the wiper circuit load connected. This provides compensation of loading errors and

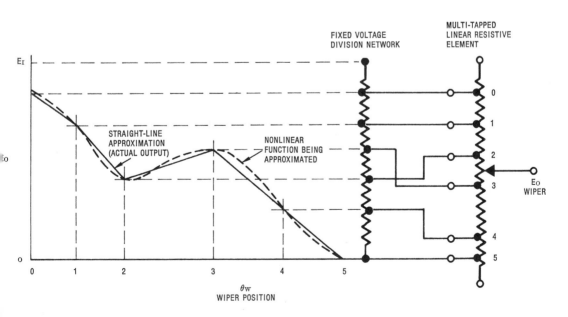

Fig. 6-13 Use of voltage clamping to obtain a straight-line approximation of a nonlinear function

assures function coincidence at the tap points. If the wiper is loaded by a low impedance circuit, the straight-line segments may sag between the tap points. This effect can be compensated by setting the clamping voltage slightly above the theoretical tap point values.

Nonlinear functions with steep slopes require many taps if a high degree of accuracy is desired. These steep slopes must have adequate power rating to dissipate the heat resulting from the voltage drop at established resistance levels. The steepest slope, the number of taps in this slope and the power rating of the entire resistance element must be considered when determining the output voltage scale.

Split windings can be used if small irregularities (flats) in the output are permissible. For this construction, pairs of taps are used to terminate isolated lengths of resistance element. Very steep slopes are practical when these sections are connected to proper voltage sources.

Voltage clamping provides versatility and also offers the advantage of not requiring a precision linear resistance element. The small flats in the output curve at each tap may be a limitation for some applications. For this reason, the distance between taps are of concern. The cause of flats in the output curve is the effect of the wiper contacting more than one turn of wire on each side of the tap at the same time it touches the tap.

Cascaded, Ganged Linear. Nonlinear functions with extremely steep slopes can be made using ganged, linear potentiometers. This is done by connecting the potentiometers in series (cascading) with the wiper output of one providing the excitation voltage for the next. The precision of such an arrangement depends on just how close a mathematical power series expansion can approximate the desired nonlinear function. Use of poly-nominal curve fitting methods will usually result in satisfactory power series.

When applications require various functions relative to shaft rotation, the ganged network may be used. Some of these relationships can be achieved with nonlinear potentiometers or shunting techniques. Then the power series expansion with appropriate resistors can be supplemented with nonlinear terms.

Potentiometer loading characteristics, discussed earlier, can be used to develop nonlinear functions. Complex loading characteristics result from series ganged potentiometers. By careful selection of resistances, nonlinear functions can be approximated. The required design procedure, especially in situations with polarity changes between terms of the initial power series, is often complex and prohibitive. Functions with low order power series can occasionally be designed by overlooking the effects of loading and then using summation network design to compensate. Calculations and empirical methods are used to determine the values of the summing resistors to obtain the necessary correction factors.

Designing the ganged arrangement using conventional network synthesis techniques results in a more versatile method of determining component values in the network. The ganged circuit can be handled as a two element network simply by setting some limitations on its characteristic polynominal. A longer trial and error method can be avoided if the above conditions can be met for a specific application. Then, using RL transfer function techniques, the ganged circuit and weighing network can be designed.

VOLTAGE TRACKING ERROR

Voltage tracking error is the difference between the actual output voltages of commonly actuated (i.e., common shaft) potentiometers at any point within their total electrical travel. It is expressed as a percentage of the total voltage applied. Tracking is a conformity specification that compares the outputs of ganged units. Tracking error is checked continuously by rotating the shaft at a slow, constant speed and recording the difference in voltage between each section and a reference section. No angle of rotation scale is used.

In a typical tracking application, the shaft is positioned by the output voltage, unlike the case where the application requires that the potentiometer shaft be positioned by gearing from some outside control. The most precise conformity in the outputs can be obtained by means of a tracking arrangement. Instead of trying to achieve simultaneous terminal conformity between the geared potentiometers or by checking each section against a theoretical angle scale, tracking compares the output voltage of one section with the output of each other section and uses the difference to describe simultaneous conformity between the outputs of the ganged sections. The voltage difference signal drives the potentiometer by means of a motor and gear linkage. The other sections then provide outputs that track the reference section with extremely high conformity due to the matching of error curves. Tracking error will always be less than the sum of the terminal conformities of each section. Where manufacturers' techniques permit combination of nonwirewound and wirewound sections on the same shaft, it is possible to gain the particular advantages of each type in the various sections.

The accuracy of a tracking potentiometer is inherent in its design and construction. A ganged combination, built to the particular tracking tolerance, operates with an accuracy that cannot be matched by single units geared together or by available conformity tolerance methods.

Specifying tracking eliminates errors due to an intermediate angle scale, angle differences between sections, gearing defects, voltage divider or test equipment errors, and end resistance. Repeatable error patterns are nullified by the self-correcting effect of the voltage difference signal. This voltage comparison technique eliminates dependence on terminal conformity. In nonlinear functions, this provides much greater checking accuracy throughout the range of the function. In single turn potentiometers, it is feasible to select sections (cups) by matching their curve, reducing differences, and thereby achieving much closer tolerances. Precision tracking units are subjected to functional testing which permits the system designer to specify tolerances closer to the system requirements. Each of these methods does have an effect on the unit price, and this should be taken into consideration when determining the necessity of a tracking potentiometer.

CLOSED LOOP FUNCTIONS

A closed loop (electrical) function is a function that is active over 360° of rotation. Some of the most significant closed loop functions have already been reviewed in the previous discussions on nonlinear functions. The sine and cosine functions are two of the most common mathematically repetitive functions. There are many other applications, however, that utilize the concept of a closed loop function. One of the most popular type is a synchro-resolver. This is a 360° electrical function with three equally spaced taps, that is, at each 120° of electrical rotation. A function such as this with multiple taps is much more difficult to evaluate electrically than a simple series resistance. Since a closed loop function is much more accurately described by a series-parallel network, the potentiometer user must realize the importance of completely specifying the total resistance required. Any closed loop function must have the total resistance specified as the value of total resistance over a specific electrical rotation. In addition, it must be known whether the resistance is measured with the electrical loop closed or open.

MECHANICAL PARAMETERS

In this modern age of electronics, it is easy for the potentiometer user to delay mechanical considerations until the final system design phase. In recent years, the importance placed on end product package size continues to remind engineers and designers of the relatively great importance of mechanical parameters. The potentiometer is an electro-mechanical device and obviously has mechanical limitations in addition to the electrical ones previously discussed. Chapter 8 includes further details on mounting and packaging not included here.

A presentation of mechanical parameters requires a discussion of *mounting* methods including the effects of starting and running *torque, overtravels, backlash,* shaft, lateral, and pilot diameter *runouts,* end and radial *play, stop strength,* and mechanical *phasing.*

Mounting. There are two basic mounting styles of precision potentiometers—bushing and servo mount. Each is characterized by its particular application.

Bushing — Manual Adjust. Applications, such as those discussed in Chapter 5, all use a manually set bushing mounting style convenient for hand adjustment. The shaft is generally available with plain, slotted, or flatted end. Some bushings incorporate a self-locking feature. Many bushing mount styles have an anti-rotation pin extending from the mounting surface. Suitable drilling or punching of the mounting panel allows the engagement of the anti-rotation pin such that the housing is firmly restricted from rotating.

Servo-Motor Driven. Many precision potentiometers utilize the servo mount or screw mount style for motor driven applications. Either of these mounting styles can be recognized by the flanged, flat mounting face. The operating shaft extends through the mounting face. Fig. 6-14 shows typical examples of servo mount and screw mount potentiometers. The machining tolerances on the pilot diameter are held extremely close. These tolerances, generally less than ±.001 inch, are required to insure proper fit and concentricity with adjacent components such as servo drive motors. Additional measures to insure concentricity include close machining of the shaft diameter, the servo mounting flange diameter, and the mounting flange thickness.

Another significant feature in the design of motor or gear driven potentiometers is the use of ball bearings in the front and rear of the device. The ball bearings insure a longer life, better concentricity and closer mechanical interface match with adjoining components. Since the majority of precision potentiometer applications use servo mount units, the mechanical parameters discussed in the following sections are related directly to the servo mount style.

Torque. In many applications, the torque of a precision potentiometer is a critical design consideration. There are two types of torque to consider.

1. *Starting torque* is the maximum moment (of inertia) in the clockwise or counter-

Fig. 6-14 Servo mount and screw mount potentiometers

clockwise direction required to initiate shaft rotation regardless of wiper position on the element.

2. *Running torque* is the maximum moment (angular force) in the clockwise or counterclockwise direction required to sustain uniform shaft rotation at a specified speed throughout the total mechanical travel.

Generally, starting torque for a precision unit is less than 2 ounce-inches. The running torque is usually 75% to 80% of the starting torque. The actual torque values are dependent on the diameter of the potentiometer and the total number of sections on a common shaft. Fig. 6-15 is a table of starting and running torques. The values shown are for single section units only.

Overtravels. In Chapter 2, the various travel ranges for potentiometers are presented in detail. These ranges are total mechanical travel, actual electrical travel, and the theoretical electrical travel. In precision potentiometer applications, the intimate relationship between electrical and mechanical parameters necessitates the use of overtravel terminology to describe or control the relationship of the various travel ranges. There are two overtravels used throughout the industry —mechanical overtravel and electrical overtravel.

Mechanical overtravel refers to the range of wiper travel between the end point (or theoretical end point) and its adjacent end stop or limit of total mechanical travel. It is common to ex-

DIAMETER (Inches)	STYLE (Mounting and Rotation)	STARTING (oz.-in.)	RUNNING (oz.-in.)
7/8	BUSHING/SINGLE-TURN	.15	.15
1 1/16	BUSHING/SINGLE-TURN	.20	.20
2	BUSHING/SINGLE-TURN	1.50	1.00
7/8	SERVO/SINGLE-TURN	.10	.10
1 1/16	SERVO/SINGLE-TURN	.10	.10
2	SERVO/SINGLE-TURN	1.00	.60
1/2	BUSHING/MULTI-TURN	.60	.60
7/8	BUSHING/MULTI-TURN	.50	.50
1 13/16	BUSHING/MULTI-TURN	2.00	2.00
1/2	SERVO/MULTI-TURN	.50	.30
7/8	SERVO/MULTI-TURN	.40	.30
1 13/16	SERVO/MULTI-TURN	1.20	.80

Fig. 6-15 Maximum torque values for single section units only

press mechanical overtravel in degrees of shaft rotation.

Electrical overtravel refers to the range of wiper travel between the end of the actual electrical travel (or theoretical electrical travel) and the adjacent end stop or the point at which electrical continuity between wiper and element ceases.

Backlash refers to the maximum allowable difference in actuating shaft position that occurs when the wiper is positioned twice to produce the same output ratio but from opposite directions. When a backlash test is made, the actual wiper position on the element is obviously identical for each measurement. Any difference in mechanical position of the shaft (backlash) is due to mechanical tolerances of the total actuating system.

Mechanical Runouts. To insure proper fit and function with adjacent mechanical components, the precision potentiometer is designed to conform to specific mechanical runouts with respect to the actuating shaft. Five common mechanical runout parameters are described in the following paragraphs. Refer to Fig. 6-14.

Shaft runout refers to the eccentricity of the shaft diameter with respect to the rotational axis of the shaft. It is measured at a specified distance from the end of the shaft. The body is held fixed and the shaft is rotated with a specified load applied radially to the shaft. The eccentricity is expressed in inches of total indicator reading (TIR). Control of shaft runout insures that the potentiometer will run true and not cause uneven wear in the mating component or the potentiometer itself.

Lateral runout refers to the perpendicularity of the mounting surface with respect to the rotational axis of the shaft. It is measured on the mounting surface at a specified distance from the outside edge of the mounting surface. The shaft is held fixed and the body is rotated with a specific load applied radially and axially to the body. The lateral runout is expressed in inches of total indicator reading.

Pilot diameter runout refers to the eccentricity of the pilot diameter with respect to the rotational axis of the shaft. It is measured on the pilot diameter. The shaft is held fixed and the body is rotated with a specified load applied radially to the body. The eccentricity is expressed in inches of total indicator reading. For many servo applications the pilot diameter is extremely critical. Its relationship to adjacent surfaces is also critical. Therefore, the allowable pilot diameter runout is controlled to insure minimum build up of tolerances.

Shaft radial play refers to the total radial excursion of the shaft. It is measured at a specified distance from the front surface of the unit. A specified radial load is applied alternately in opposite directions at the specified point. Shaft radial play is specified in inches.

Shaft end play refers to the total axial excursion of the shaft. It is measured at the end of the shaft with a specified axial load applied alternately in opposite directions. Shaft end play is expressed in inches.

Shaft radial play, end play, and the runouts are controlled by the manufacturer to provide optimum mechanical life and the highest accuracy possible for interfacing with adjacent mechanical components. The potentiometer user should recognize that any mechanical misalignment of adjacent components with respect to the operating shaft can result in a load on the shaft that will degrade the potentiometer's maximum rotational life.

Stop Strength. The stop strength specification means *static* stop strength. It is the maximum static load that can be applied to the shaft at each mechanical stop. The force is applied for a specified period of time and no permanent change of stop position, greater than specified, is allowed. Single turn precision potentiometers usually do not have stops. They are continuous rotation devices with a nonconductive bridge between ends of the resistance element. The contact sweeps across the bridge and returns to zero without a change in the direction of rotation.

Some single turn and all multiturn precision potentiometers have mechanical stops at each end of rotation. The stop strength of the unit is dependent on its physical size. In general, the larger the diameter, the higher the stop strength. Most motor and gear driven potentiometers have higher rotational forces applied to the operating shaft than the stop strength rating. Unless the manufacturer is made aware of special requirements, the stop strength of the potentiometer is usually *not* sufficient to function as a stop for the whole system.

PHASING

Phasing is a parameter used to describe the relationship of one potentiometer output function to another. Although phasing can be used to describe the relationship between two separate devices, it is generally used to describe the relationship of one section to another in a multiple section precision device. In most applications, it is extremely difficult to discuss an electrical phasing requirement without discussing mechanical effects. The phase relationship in most nonlinear output functions, built in multisection assemblies, is established by the physical location of the wiper in one section with respect to the wiper position in another section. The common location is referred to as the *phasing point*. Phasing is often done internally so the potentiometer user may not be aware of a physical difference from outside appearance.

Another method used to establish the phasing of multisection devices is rotating the body of one section with respect to another. This is possible through the use of clamp ring sections as shown in Fig. 6-16. The phase relationship of multiple sections is normally specified by the user when the potentiometer is designed. The ability

NOTE THAT TERMINALS ARE OFFSET
BETWEEN SECTIONS. THIS INDICATES
THAT THE ASSEMBLY **MAY** BE
A PHASED ASSEMBLY.

CLAMP RING

Fig. 6-16 Clamp ring style, phased potentiometer

to change phase relationships of multiple sections and to vary electrical and mechanical angles has significant benefit to the potentiometer user. However, once the device has been built, the clamp ring set screws are usually secured in some fashion (ie., with an adhesive). This sealing may make changes impossible without violation of manufacturer's warranty.

NONWIREWOUND PRECISION POTENTIOMETERS

Nonwirewound potentiometers have a number of characteristics which differ from wirewound devices and may require special consideration for successful application.

Over the past several years, precision nonwirewound potentiometers have gradually replaced the wirewound device in many commercial, military, and aerospace applications. In most instances, the changeover was accomplished successfully. In other instances, problems arose that were ultimately solved. In the process, both users and manufacturers gained a better understanding of the performance characteristics of these devices.

Nonwirewound potentiometers, such as conductive plastic and cermet, differ in their nature from wirewounds. Conductive plastic provides two to four orders of magnitude improvement in rotational life. Many other equally important characteristics pertinent to potentiometer performance in the circuit, the system, and the environment are also different.

Basic nonwirewound element types and their construction are covered in detail in Chapter 7. The following sections discuss some important considerations for nonwirewound *precision* potentiometers such as contact resistance, output loading, effective electrical travel, and multiple taps. Included are some of the more subtle performance tradeoffs required for the optimum design. These subjects are treated in detail elsewhere in this book, and the index should be consulted for further study.

Contact Resistance. Contact resistance appears as a resistance between the wiper (contact) and the resistance element and may be shown schematically as in Fig. 6-17. Contact resistance may be thought of as the sum of fixed and variable components. The variable part is generally a fraction of the fixed component. The value of contact resistance is a function of the geometry of the resistive element, contact configuration, and the area of contact between the wiper and the resistance element. For a given geometry of element and contact, the contact resistance is proportional to the resistivity of the element ma-

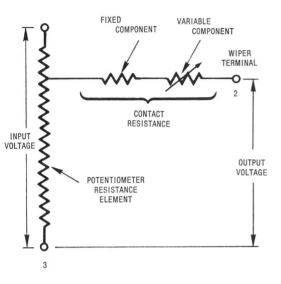

Fig. 6-17 Contact resistance appears as a resistance between the wiper (contact) and the resistance element

terial and may be expressed as a percentage of the total resistance. Contact resistance can be lower than 1% and can range upwards to 10% for certain elements.

In cases where the application requires the lowest possible value of contact resistance, use the largest possible potentiometer diameter and the longest electrical angle consistent with the available space. These measures will, in combination with the manufacturer's choice of resistive element geometry and contact design, act to minimize contact resistance. During the rotational life of the potentiometer contact resistance changes. The fixed component generally becoming smaller and the variable component larger. The significance of contact resistance is dependent on the circuit into which the potentiometer output operates. In potentiometers requiring tight conformity, the first question to consider is whether it is the deviation from the theoretical output that is important (such as a function generator) or whether the output fidelity can be more appropriately stated in terms of tracking accuracy of output smoothness between separate potentiometers (such as those used in remote position and follow servo systems).

Output Loading. In many applications the nonwirewound resistance element linearity must

be compensated with respect to output load. The output load ratio is defined as the nominal output load resistance divided by the element nominal total resistance (R_L/R_T). The load ratio should always be greater than ten to one. There are two important reasons for this limitation. If the load ratio becomes excessively low, the wiper current may become sufficient to seriously degrade the useful life of the resistance element. The other reason for the load ratio limitation is due to the manufacturing processes involved with nonwirewound resistance elements. It is difficult to achieve a total resistance tolerance less than approximately $\pm 5\%$. This situation creates no problem if the load ratio is 100:1 and the linearity tolerance is $\pm 1.0\%$. However, if the load ratio is 5:1 and the linearity tolerance is $\pm .05\%$ circuit analysis reveals the total resistance R_T and the load resistance R_L must be controlled to less than $\pm 1\%$. Fig. 6-18 shows the relationship between the load ratio, the tolerances of total resistance and load resistance, and associated linearity error.

SUM OF TOLERANCE ON TOTAL RESISTANCE AND LOAD RESISTANCE	$\dfrac{R_L}{R_T}$	5:1	10:1	20:1	50:1	100:1
±1.0%		0.028	0.015	0.0075	0.003	0.0015
±2.0%		0.055	0.030	0.015	0.006	0.003
±3.0%		0.082	0.045	0.023	0.009	0.0045
±5.0%		0.140	0.075	0.038	0.015	0.0075
±10.0%		0.280	0.150	0.075	0.030	0.015

TABLE 1

Fig. 6-18 Linearity error (%) for various load ratios

Electrical Travel. The electrical travel of non-wirewound potentiometers is built in during manufacture and cannot be modified later. The degree of accuracy achieved is a function of the manufacturing process and the size and the type of element. Nonwirewound resistance elements also have the unique characteristics that the output voltage does not change linearly in the immediate vicinity of the end terminations. For this reason the actual electrical travel as defined in wirewound technology has no meaning when applied to nonwirewound elements. A more meaningful specification is defined by the output voltage versus shaft position function and the required linearity or conformity. When these are properly defined a *theoretical electrical travel* and its associated angular tolerance can be specified.

Multiple Taps. When multiple taps are required on a nonwirewound potentiometer, special precautions must be taken. There are two types of taps available. The *current* tap consists of a conductive strip crossing the entire width of the resistance element perpendicular to the wiper path as shown in Fig. 6-19. The current tap acts as a miniature resistance short and disturbs the linearity of the element. The magnitude of this disturbance depends upon the relative size of the element and the manufacturer's process. The current tap can safely carry the same amount of current as the end terminations. The *voltage* or *zero width* tap consists of a conductor which barely touches the edge of the resistance track as shown in Feb. 6-20. A voltage tap has negligible effect on the linearity. Obviously the current carrying capability of this type of tap is limited.

When deciding which type of additional tap to specify for a given application it is unnecessary to consider the current carrying capabilities. It is more realistic to consider how the tap is used in the circuit. If a single voltage is applied to the tap a current tap is required. However, if an equal positive and negative voltage is applied between a center tap and the end terminals, a voltage tap may be used in the center. A voltage tap may be used when the tap is sensing voltage only and the measurement circuit has a high impedance. The method of specifying locations for the two types of taps differs. A voltage tap should always be located at some specified voltage with a tolerance. The angular location of a voltage tap is immeasurable due to the two-dimensional characteristic of the nonwirewound resistance element. The location of a current tap may be specified in terms of voltage or angular position. When angular position is specified, care must be exercised so that a measurement technique is defined and a realistic tolerance is assigned.

Temperature Coefficient of Resistance and Moisture Sensitivity. The total resistance of conductive plastic and cermet is known to be more sensitive to moisture and temperature than are wirewounds. The change in resistance occurring from exposure within the rated temperature range or from the extremes of room ambient *humidity* has little effect on other intrinsic characteristics. However, the changes in total resistance due to the temperature coefficient of resistance (TC) are sufficient to preclude the use of any external resistors as balancing resistors or as a voltage divider. To verify this, consider a typical 6% change in TR (for conductive plastic) from $-65°C$ to $+100°C$ (363 PPM/°C). To use a conductive plastic potentiometer as part of a divider network, the network should be built into the potentiometer resistance element. Series resistors made in this fashion will track the

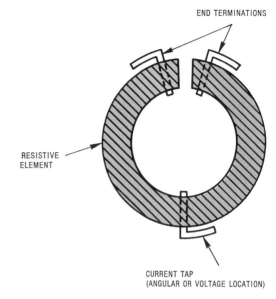

Fig. 6-19 Current tap for nonwirewound element

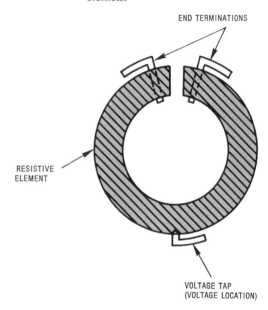

Fig. 6-20 Voltage tap for nonwirewound element

potentiometer section within the limits of normal linearity and conformity tolerances because the potentiometer and resistors are made of the same material and generally vary proportionately. The cost for these resistors (built into the potentiometer) is generally less than the cost of the resistors purchased separately combined with the labor to connect them externally. For optimum performance of potentiometers with built-in resistors, specify the actual voltage at which the potentiometer will operate. This will permit the matching of the series resistors to the potentiometer element at the actual working voltage.

In recent years, state of the art advances have reduced the change in total resistance resulting from exposure to humidity to approximately 5%. In most applications where a repeatable accurate TC is required, it is necessary to stabilize the moisture content of the resistance element with a few hours exposure at temperatures around 50 to 80°C.

LINEAR DISPLACEMENT TRANSDUCER

One of the basic precision potentiometer applications is that of converting mechanical linear motion to rotary motion and utilizing the resultant output. The linear displacement transducer assembly shown in Fig. 6-21 is a simple but very effective method. It uses a multiturn or single-turn servo mounted precision potentiometer, and can measure relatively great displacements compared to its size. This particular transducer uses

a spring return mechanism on the cable. The spring return is contained within the larger cylindrical body that accepts the operating shaft. The required effective electrical angle is based on the length of the cable required by the transducer application. This particular combination eliminates many of the mechanical interface problems inherent in gear and motor drive assemblies.

One application for this transducer assembly is in test aircraft. The assembly is mounted on the engine or in the cockpit or anywhere that a measure of linear motion (i.e., the throttle or a control surface) is required. The linear displacement of the mechanical linkage is transmitted to the potentiometer operating shaft with a 1 to 1 ratio through the cable. By applying a fixed voltage across the end terminal and the wiper, the mechanical motion is converted to an electrical signal which can be transmitted directly to a recording device or sent to a ground tracking station by external telemetry equipment. This is only one of many linear motion-to-electrical-signal transducer applications.

LOW TORQUE POTENTIOMETERS

Another application of precision potentiometers is one required by precision measuring instruments such as meteorological (weather) in-

Fig. 6-21 Cable type linear displacement transducer
(Space-Age Control, Inc.)

strumentation. A low torque potentiometer application generally requires torques less than 0.1 oz.-in., even with multiple sections. A servo mount ball bearing potentiometer of 1 inch diameter or less with an ⅛ inch diameter shaft is most suitable. The resistance element can be wirewound or conductive plastic. In the wirewound device, the key to low torque is the bridge between the ends of the element. Applications such as electronic weather vanes and anemometers require 360° of continuous mechanical rotation. Since most potentiometers have an effective electrical angle of 350°, the electrical angle must either be extended to as close to 360° as possible or the 10 degree travel area between the ends of the element must be designed with an extremely smooth transition surface.

In a wirewound device, it is best to keep the total resistance above 10K ohms for low torque requirements. This will force the diameter of the resistance wire to be small enough and the pitch for the winding on the resistance element close enough to provide a relatively smooth surface for the wiper to traverse. Generally, the torque will be highest over the bridged area between the ends of the resistance element. For example, if the application allows torques in the range of .05 oz.-in. with the wiper on the element, the torque over the bridge will generally be .07 to .10

oz.-in. Some variation of these values is possible by modifying the pressure of the wiper against the element. If the wiper pressure is decreased, the amplitude of the electrical noise is increased. There are ways of counteracting the increase of noise that involve the use of precious metal resistance wire which has an obvious effect on the cost of the unit as well as performance parameters.

This type of wirewound precision potentiometer usually requires connecting an external terminal to the resistance element with a small single wire tap. The circuit designer should design in a current-limiting device in the circuit with the potentiometer to control any surge current when the wiper is approaching the inactive bridge portion of the element or when the wiper is returning to the active area of the element.

COARSE/FINE DUAL CONTROL

One of the methods for multiple function front panel control is the dual concentric shaft precision potentiometer. This method is gaining popularity with precision instrument manufacturers. Fig. 6-22 shows the front panel of a spectrum analyzer. The inner and outer knobs in the frequency section of the front panel form the

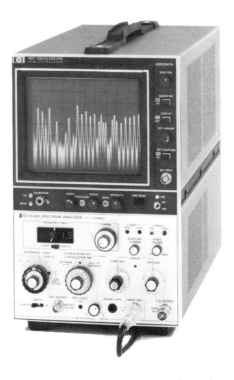

Fig. 6-22 Spectrum analyzer with coarse/fine dual concentric shaft control for frequency tuning (Hewlett-Packard)

frequency tuning control. This control consist of two parts, coarse tuning (larger knob) and fine tuning (smaller knob). The tuning control adjusts the center frequency or start frequency of the spectrum displayed.

The potentiometer is a multiturn, nonwire-wound device with two sections. Each section is operated independently by means of dual concentric shafts. Extremely fine resolution and precise tuning ability were the deciding factors in selecting a multiturn potentiometer with a conductive plastic resistive element in each section. The output of both the coarse and the fine tuning adjustment are fed into a summing operational amplifier. The voltages are then fed through a voltage-to-curent transfer circuit. The resultant current is applied to the field of a YIG oscillator where the frequency is determined.

POSITION INDICATION/TRANSMISSION

Fig. 6-23 shows a basic position transmission system arrangement using potentiometers for both transmitter and receiver (indicator). This simplified circuit shows how potentiometers can be used to transmit a relative mechanical position from one point to another.

The original mechanical motion may be the output of a driven system such as a servomechanism. It could also be an operator induced motion such as a front panel control of an instrument. The instrument might use a turns-counting dial to permit precise operator setting of certain critical adjustments. The reading on the turns-counting dial corresponds directly to the magnitude of some variable such as voltage level, temperature, frequency, or a time interval.

As long as the relative position of the wipers in Fig. 6-23 is the same, no input voltage will be applied to the servo amplifier and the motor will not turn. When the position of the transmitter R_1 is moved, an unbalance occurs and the differential input voltage to the amplifier produces an output drive signal to the motor. The motor turns in the proper direction to reduce the error (difference) signal to zero. The end result is the receiver potentiometer R_2 duplicates the position of the transmitter.

The precision with which the position is transmitted is dependent upon the accuracy of the two potentiometers and the gain of the servo amplifier system. If the gain is made too high, the receiver potentiometer will oscillate (hunt) as it attempts to find a position where the error voltage is zero. Resolution in the receiver potentiometer is very important as it determines the amount of gain which will be permitted without oscillation. If the resolution is too poor, then the error signal presented to the input of the amplifier may jump from a positive level to a negative level as the wiper moves from one turn to the next. Even if the oscillation — sometimes called *dither* — is not especially objectionable from an operations standpoint, it should not be permitted because of the resulting local wear on the element, the wiper, and the whole electromechanical system.

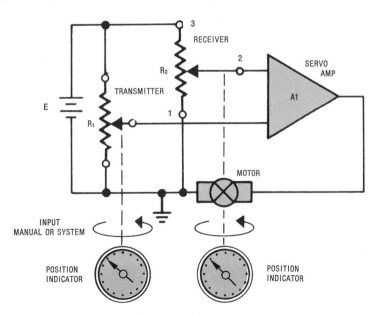

Fig. 6-23 A basic position transmission system using potentiometers for transmitter and receiver

THE X-22A, V/STOL AIRCRAFT

The aircraft pictured in Fig. 6-24 uses dual tandem ducted propellers to provide an aircraft for flight research and evaluation of this unique configuration. More importantly it provides a highly versatile aircraft capable of general research on vertical/short takeoff and landing (V/STOL) handling qualities using a variable stability system.

The variable stability system allows the pilot to change the dynamic characteristics of the aircraft in flight and to simulate the flying qualities of future aircraft that are on the drawing boards. The aircraft is in reality a flying simulator, allowing the pilot to feel the aircraft motions in a natural sense unlike fix based ground simulators.

Because of the uniqueness of the aircraft (it's the only one of its kind) constant monitoring is provided by an airborne telemetry system which transmits data to a mobile monitoring station. The data is observed on strip chart recorders and analog meters. A minicomputer constantly monitors safety of flight items and immediately warns of any aircraft parameter that departs from its normal range.

In the computer over 170 amplifiers and special function modules are wired to a patch board along with 100 wirewound, digital readout potentiometers. This patch board is shown in Fig. 6-25. Some of the potentiometers (3-10) are remotely mounted in the cockpit. Pilot comments have been favorable regarding reading and setting these devices in a crowded cockpit.

Fig. 6-24 An experimental aircraft (Bell Aerosystems/U.S. Navy)

Fig. 6-25 Patchboard used in aircraft of Fig. 6-24
(Flight Research Dept. of Calspan Corp.)

DENDROMETER

As a part of a project to investigate the relationship of tree growth to the factors of forest environment, scientists designed an electrical dendrometer band (a device that measures growth by measuring tree girth) utilizing a precision potentiometer.

As shown in Fig. 6-26 the potentiometer is mounted on a bracket at one end of a metal band that is passed around the circumference of the tree. A stainless steel wire, attached to the bracket via a spring, is wrapped several times around the potentiometer shaft and attached to the other end of the band. A simple reliable transducer is the result. Any growth in the tree diameter (girth) causes the band to pull the wire turning the shaft proportionately, thus, changing the resistance value of the potentiometer.

Two of the main advantages of the potentiometer dendrometer band are its relatively low cost and its ability to operate successfully in remote forested areas where reliable electrical power is difficult to obtain. These features enable tree growth measurements to be taken manually by simply reading the resistance of each poten-

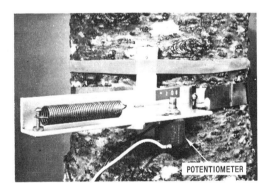

Fig. 6-26 Dendrometer for monitoring tree growth

tiometer/dendrometer band with a portable digital ohmmeter at regular intervals. No direct power supply of any sort is required, and bands can be located in upper areas of a tree with leads running down the trunk for easy monitoring.

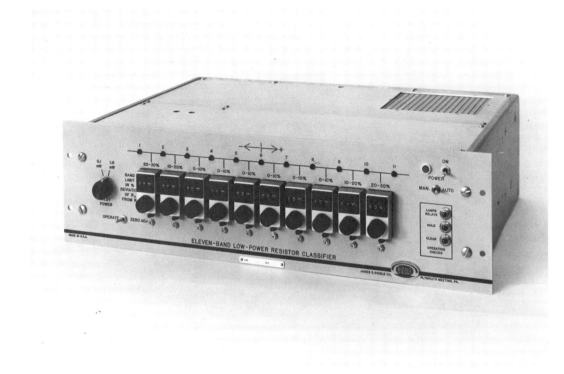

Fig. 6-27 A bridge used for high speed sorting of resistors and thermistors. (James G. Biddle Co.)

SORTING BRIDGE

The high speed sorting bridge in Fig. 6-27 is used with a suitable parts handler for accurately sorting resistors or thermistors into as many as ten different classifications. The unknown resistance is checked against ten individually set tolerance limit bands between ±0 and ±30%. In operation, the bridge selects the proper category and signals an automatic handler to feed the resistive component to the corresponding bin.

A power dissipation circuit senses the unknown resistance value and adjusts bridge voltage to maintain equal dissipation for various resistance values.

A unique dual-null multiband sorter permits comparison of the unknown resistance to several reference levels simultaneously rather than sequentially. This speeds the sorting process and eliminates error inducing switching.

The ten dials shown in the photo each operate a ten-turn precision wirewound potentiometer for tolerance setting control. The range of each dial is 10% and the potentiometer can be set anywhere in this band within 0.01%. Here is an example of precision devices being used in a control function as discussed in Chapter five.

MULTI-CHANNEL MAGNETIC TAPE RECORDER

A variety of adjustment, control and critical precision potentiometers are used in the multi-channel magnetic tape recorder pictured in Fig. 6-28.

Precision potentiometers are in the electronically controlled tape tensioning system which is part of the electronic control for the spooling motors. This system measures the actual tape tension on both the right and left sides. The tape tension sensor acts as tape storage and mechanical damping elements. The offset capstans shown in Fig. 6-29 cause the tension sensor to rotate in proportion to tape tension. Position of the sensor is converted into proportional voltage (actual value) by the directly driven high precision single turn potentiometer. The potentiometer is connected to the differential amplifier of the spooling motor control amplifier. The control voltage for the normal fast running mode or the manually controlled winding influence the reference input (set value) of the differential amplifier. With this system, the tape tension is electronically controlled even during the fast forward and rewind modes.

Fig. 6-28 Multichannel magnetic tape recorder. (Willi Studer, Switzerland)

During the braking procedure, the take-up spooling motor is electronically controlled until the tape comes to a complete standstill. Thus, in all modes, the tape tension is electronically controlled.

The single turn precision potentiometers on the right and left tape tension sensors are shown by arrows Fig. 6-30. This is a view of the underside of the tape drive system and related electronics. Potentiometers with conductive plastic elements are used in this application because long, reliable, noise-free life is required.

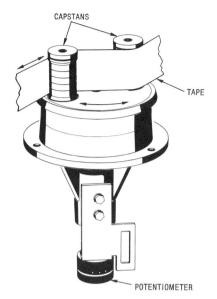

Fig. 6-29 A tape tension sensor
(Willi Studer, Switzerland)

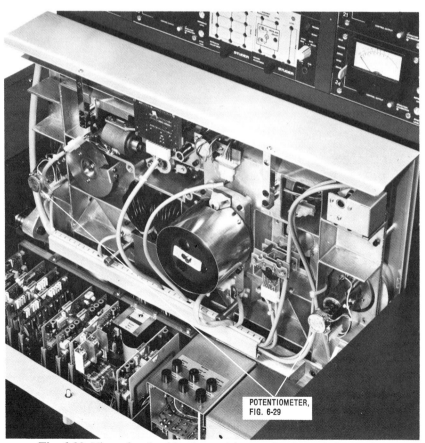

Fig. 6-30 View of underside of tape drive mechanics and electronics
(Willi Studer, Switzerland)

Notes

Notes

CONSTRUCTION DETAILS AND SELECTION GUIDELINES

Chapter 7

"*I am not building for a day. The trouble with some American manufacturers is just that very point. They cater to the passing whim. It pays to make things slowly, but to make them right. It is one of the fundamentals of business success — not measured by standards of today, but by those of a century hence. There are no seconds or thirds going out of my shops. Nothing but firsts — first, last and all the time.*"

T. A. Edison
Quoted in Popular Electricity Magazine
Vol. V, No. 7, November 1912

INTRODUCTION

There are five basic parts of any potentiometer:

    Resistive element
    Terminations
    Contact or wiper
    Actuator or shaft
    Case or housing

For each part there are several fundamental variations possible.

In this chapter the parts of a potentiometer are considered individually since each has special characteristics which offer an advantage or impose a limitation on the final assembly.

A careful study of the material presented will aid in the selection of the proper construction type for a particular application.

RESISTIVE ELEMENTS

The real heart of any potentiometer is the resistive element. It affects, to some degree, all potentiometer electrical parameters. There are two general classifications of resistive elements — *wirewound* and *nonwirewound*. The nonwirewound group can be further classified as cermet, carbon, metal film, or bulk metal. It is also possible to combine wirewound and conductive plastic (a special carbon composition) in one element to achieve improved performance of certain electrical parameters. In addition, cermet and conductive plastic have been combined by at least one manufacturer. Both of these combination elements are discussed under *Hybrid Elements*.

Wirewound Elements. Resistance wire can be used to form the resistive element in a potentiometer. Commonly used materials are one of three alloys:

Nickel-chromium
Copper-nickel
Gold-platinum

Nickel-chromium (75% Ni, 25% Cr) is the most common. Its temperature coefficient is typically less than $\pm$ 5ppm/°C. It has a resistivity of 800 ohms per circular mil foot. A circular mil foot (cmf) is a hypothetical quantity equivalent to one foot of wire that is one thousandth (.001) of an inch in diameter. Popular use of Ni-Cr resistance wire for resistive elements is largely due to its excellent TC and availability in many different diameters. The broad size range results in a wide selection of TR values with very low ENR ratings.

Copper-nickel (55% Cu, 45% Ni) wire has a resistivity of 300 ohms/cmf and a temperature coefficient of $\pm$ 20ppm/°C.

A less common material for resistive elements is a complex precious metal alloy of gold (Au) and platinum (Pt) together with small amounts of copper (Cu) and silver (Ag). The resulting resistivity is approximately 85 ohms/cmf with a high temperature coefficient of +650ppm/°C. This sacrifice in temperature coefficient results in an improvement in certain other parameters. For example, low resistivity and the ability to withstand harsh environments without oxidation of its surface. This helps keep wiper noise low even in severe environments.

Figure 7-1 lists the resistivities for various diameters of the three different resistance wire alloys. This is a partial listing. Other wire sizes are available.

DIAMETER (inches)	RESISTANCE (OHMS PER FOOT)		
	Ni-Cr	Cu-Ni	Au-Pt-Cu-Ag
0.008	12.50	4.69	1.33
0.005	32.00	12.00	3.40
0.002	200.0	75.00	21.25
0.001	800.0	300.0	85.00
0.0008	1250	468.8	132.8
0.0005	3200	1200	340.0
0.0004	5000	1875	531.3

Fig. 7-1 Relation of Resistance to Wire Diameter

Basically, the actual wire used depends upon the total resistance required, the resolution needed, and the space available. Smaller wire allows higher resistance in a given space and improved resolution. However, smaller wire is more fragile and therefore, difficult to wind. Power and current carrying requirements also influence the choice of resistance wire size.

Although it is possible to have a simple single straight wire element, such construction is impractical. As an illustration, assume the highest resistivity wire, listed in Figure 7-1, was used to construct a 5000 ohm resistive element. The finished potentiometer would be greater than one foot in length. The common construction technique for a resistance wire element requires many turns of resistance wire carefully wound on a carrier form or mandrel. This method allows a substantial resistance to be packaged in a small volume. A ceramic or plastic mandrel can be used. However, the most common type of mandrel is a length of insulated copper wire. After winding, this flexible carrier form can be coiled in a helical fashion to further compress the length required for the resistive element.

The copper wire mandrel has several practical advantages. It is available in precise round cross sections in long lengths and can be purchased with an insulating enamel already applied. These characteristics contribute toward high quality elements at relatively low manufacturing costs.

The characteristics of the mandrel are very important. Irregularities can make winding difficult and result in poor linearity and/or poor resolution. Mechanical instabilities can lead to undesirable stresses in the wire or loose windings which allow individual turns of wire to move. The advantages gained by using a resistance wire with a carefully controlled temperature coefficient will be lost if the mandrel expands and stretches the wire to produce substantial resistance changes with varying temperature. This is known as the *strain gage* effect. It occurs when the wire is stretched thus reducing its cross section and increasing its resistance. Any unbalance in coefficients of expansion between the resistance wire and mandrel can cause an intolerable variation in resistance due to temperature changes.

For accuracy and economy, high speed automatic machinery may be used to wind the element wire on long mandrels. These are cut to the proper lengths and installed in individual potentiometers at a later manufacturing phase. The photograph of Fig. 7-2 shows a machine which produces a continual helix of wound element. This element may be cut into individual rings for single turn potentiometers or helical elements for multiturn units. The very delicate resistance wire must be wound on the mandrel in a manner that produces uniformity and the right amount of total resistance in the exact length required. Irregularities in the winding process can result in a broken wire or overlapping turns. If the winding tension (the pull against the resistance wire as it is wound on a mandrel) is not just right, the turns of resistance wire will be loose

Fig. 7-2 An automatic machine used to produce a continual helix of resistance element (inset)

or over-stressed and the resulting element will exhibit a poor temperature coefficient.

Many factors are involved in the winding operation. These may not be obvious to the end user but are a major concern of every potentiometer manufacturer. The critical potentiometer buyer will be wise to compare the capability of various manufacturers before choosing a source.

The temperature coefficient of the finished potentiometer will be much poorer than that of the unwound, unstressed resistance wire. It is unfortunate for the circuit designer that data sheets for some wirewound potentiometers list temperature coefficient using values of the unwound resistance wire alone. If such a specification is listed, *never* make the assumption that this is the temperature coefficient of the completed potentiometer. Instead, check with your potentiometer source.

Potentiometers designed for high power, rheostat applications often use an insulated metal mandrel. This provides an excellent thermally conductive path for the heat generated within the resistance winding. The metal mandrel allows an increase in the power rating, compared to plastic or ceramic, which is especially significant in applications where power is dissipated by only a portion of the element.

In most cases, the unwound resistance wire is bare with no insulation. A slight amount of cement is used to bond the wire to the mandrel.

Too much cement will interfere with the wiper path; too little results in loose turns. Since the tolerance on the unformed resistance wire often approaches the total resistance tolerance for the completed unit, it is easily seen that quality is no accident.

Nonlinear Wirewound Elements. One or more of the following methods is used to achieve a nonlinear change in resistivity, ρ, with wiper travel.

1) A carefully shaped mandrel cross section to vary the resistance increment from one turn to the next.
2) Careful variation in the winding pitch to change the number of turns traversed for a given mechanical travel.
3) A change in the wire size and/or the wire material.
4) A combination of 2 and 3 above.
5) Carefully positioned taps to permit the addition to external connections.

Varying the mandrel shape. The drawing of Fig. 7-3 shows a variety of elements designed to produce different nonlinear functions. The winding mandrel is designed (shaped) to vary the resistivity in a nonlinear manner from turn to turn. The mandrel must be constructed such that the resistivity varies at the desired rate of change of $f(\theta)$. Mathematically:

$$\Delta\rho = \frac{d}{d\theta} f(\theta)$$

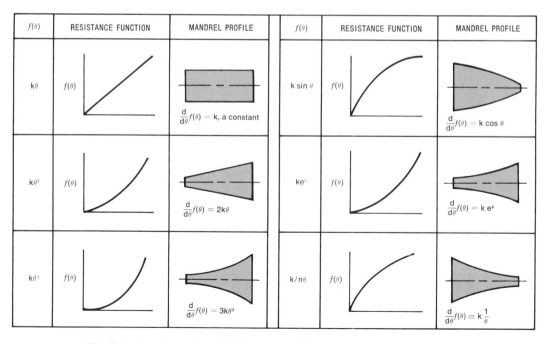

Fig. 7-3 A variety of mandrel shapes to achieve various output functions

The limits of the individual functions in Fig. 7-3 are determined by the steepness of mandrel slope and the ratio of maximum to minimum mandrel width. These ratios and thus mandrel slope ratios normally do not exceed 5:1. This is because of the practical maximum limitations of 20° mandrel slope and a 5:1 ratio of maximum to minimum mandrel width. Greater slope ratios are possible using other approaches described in the paragraphs that follow.

The slope ratio needed to yield a specified nonlinear function is not always the only indicator of the degree of difficulty of producing the function. For example, the two functions shown in

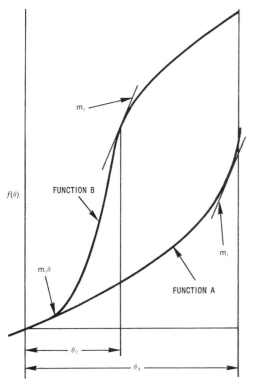

Fig. 7-4 Output functions with the same slope ratio may require different construction methods

Fig. 7-4 illustrate entirely different winding problems even though their slope ratios are identical. The two curves of Fig. 7-4 have equal slope ratios since the maximum slopes m_1 and m_2 are equal. Function B is more difficult to wind because the slope ratio must be achieved within the relatively small travel distance θ_2.

In Fig. 7-5 each straight line segment of the output acts as a linear potentiometer when the mandrel is stepped as shown. If the resistance

wire is uniformly spaced over each linear section the corresponding output of each will have a different slope. This is because the rate of change in $f(\theta)$ from turn to turn of resistance wire is greater if the mandrel is wider since the length of resistance wire is also greater.

The stepped mandrel is often a practical method to produce nonlinear wirewound resistance elements even though its straight sections may only approximate a nonlinear function. Closely spaced function changes are possible. Problems of securing wire to sloped mandrels are eliminated by use of stepped elements.

Varying the winding pitch. Changing the resistivity ρ in a nonlinear manner by altering the spacing between individual turns of the element is possible. Winding machines employing special servo techniques are available for controlling the

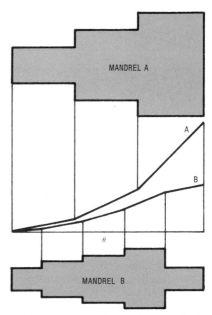

Fig. 7-5 Examples of stepping mandrel to change slope of output

winding spacing in a smooth fashion. This method is limited by the fineness of the wire that can be wound (the closeness of the turns at the *tight* end of the mandrel), and by the maximum wire spacing that can be tolerated (the resolution limit at the *loose* end of the mandrel). In practice, a 4:1 ratio in wire spacing is considered maximum.

Changing the wire. The range of achievable functions can be extended by changing wire size or material whenever one of the limitations is reached. In applications where the resolution must be essentially constant over the entire element length, multiple materials must be used.

In Fig. 7-6, the mandrel perimeter, winding space and wire size are held constant. By splicing different wires as shown and changing only the resistivity of the resistance wire, various slope

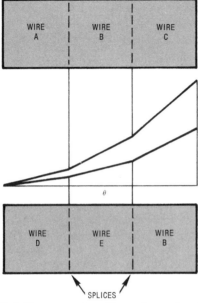

Fig. 7-6 Two examples of how change in wire resistivity changes slope with identical mandrel

ratios are possible. Thus, the slope of the straight-line output of each of the three sections of the element is directly proportional to the resistivity of the wire. The maximum practical slope ratio is 16. This is based on the ratio of maximum to minimum resistivity per unit length of winding using common resistance wire types. These range from 60 to 1000 ohms per circular mil foot depending on composition.

As discussed earlier in this chapter, choices of resistance wire compositions are limited by performance requirements including TC, life and noise. Most manufacturers use a limited number of wire types in a single wirewound element because of the cost of splicing. Instead, they stock a wide variety of wire sizes.

Slope ratio functions of 100:1 are possible by changing only wire size which ranges from less than .001 to .010 inches in diameter. As before, each change in wire size on an element requires a wire splice.

Varying winding pitch and wire. Control of resistivity can be maximized by varying wire spacing and wire size.

By using both .001″ and .010″ wire the maximum slope ratio possible with the same *winding factor* can be illustrated. Winding factor is the ratio of the number of turns per inch of mandrel that are actually wound to the maximum number that can theoretically be wound if turns are butted together. A maximum winding factor of 0.75 is normally practical for flat mandrels that are to be curved into a circle. This leaves enough space between turns so that turns of wire will not short together after the element is curved.

Assuming the winding factor is constant, ten times more .001″ diameter wire can be wound per inch of mandrel as .010″ diameter wire. Also .001″ diameter wire has 100 times the resistivity of .010″ diameter wire. By combining these extremes (10x100) a 1000:1 slope ratio is possible in a single resistive element. One shortcoming of this example is the relatively poor resolution of the .010″ wire compared to .001″ wire. When resolution is a factor then different wires should be considered to minimize the sacrifice in resolution with high slope ratios.

Tapping. Tapping is required when the function must go through an inversion as in the case of the sine function of Fig. 7-7. If the sine potentiometer were not tapped, the output function $\frac{E_0}{E_I}$ would follow its conventional curve over the first 180° of input rotation, but would continue upward as shown in Fig. 7-7A. The desired sine function is achieved by:

1) Providing a tap at the mid-point of the element and connecting this tap to one side of the excitation source, E_I.

2) Connecting the potentiometer end terminals to the other side of the excitation source.

The use of tapping to achieve difficult nonlinear functions together with other methods are coverered more extensively in Chapter 6.

Selection factors. Wirewound potentiometers offer very good stability of total resistance with time and temperature changes. Stability can be better than 0.01% in 1000 hours of operation. Additionally, these elements offer low noise in the static state, high power capabilities, and good operational life.

Wirewound elements do not offer as wide a selection of TR values as some other types; but, the range in trimmers is from 10 ohms to 20K ohms. Some manufacturers offer values as high

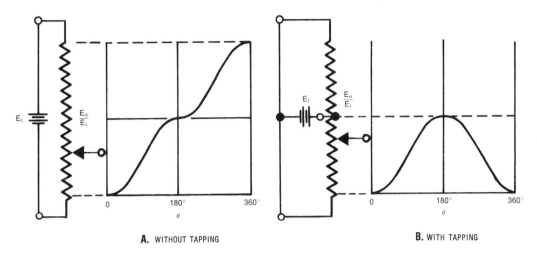

A. WITHOUT TAPPING   **B.** WITH TAPPING

Fig. 7-7 Use of tapping to achieve inversion of the sine-wave function

as 100K ohms at premium prices. Precisions are available with total resistances as low as 25 ohms and as high as 500K ohms.

One of the primary limitations of wirewound elements is the finite resolution steps which result from the wiper moving from turn to turn. These steps are distinct, sudden, repeatable changes in output. They may be as great as the resolution specification (% of total resistance) but are often less because of the bridging action of the wiper between turns. In systems that might be sensitive to such discrete steps, care should be taken to select a potentiometer with resolution good enough (small enough) to avoid difficulty.

The use of wirewound elements should be avoided in high frequency applications. The many turns of resistance wire exhibit an inductive reactance which increases directly with frequency. This effect is most noticeable in low total resistance potentiometers because the inductive reactance can be larger than the resistance, even at frequencies as low as 20kHz.

Performance of wirewound potentiometers is also affected by inherent capacitance. Capacitance exists from turn to turn and also between the winding and mandrel. Capacitance effects are most significant in high total resistance potentiometers that usually have more turns of wire.

Nonwirewound Elements. Variable resistive devices that are not made with resistance wire are categorized by the industry and military as *nonwirewound*. These element types are discussed in individual sections below.

Cermet resistive elements. One of the more recent materials developed combines very fine particles of ceramic or glass with those of precious metals to form a *ceramic metal* resistive material after firing in a kiln. *Cermet* is a term which may be applied to a wide range of materials as manufactured by different sources. Do not assume that all cermet potentiometers are the same. The comments included in the following paragraphs apply to cermet elements made by most manufacturers. Major manufacturers painstakingly compound, and carefully control, the composition of their cermet materials and processes. These are usually considered proprietary for competitive reasons so exact materials and details of manufacturing techniques are often closely guarded.

Cermet, also known as *thick film*, is defined as resistive and conductive films greater than .0001 inch thick, resulting from firing a paste or *ink* that has been deposited on a ceramic substrate. Similar materials and techniques are used to manufacture hybrid circuits and fixed resistor networks. For potentiometers, the condition of the *surface* of the film relative to wiper action (conductivity and abrasiveness) is a major concern.

The paste is applied to a flat ceramic substrate, usually alumina or steatite, by a *silk* screening operation. This is a mechanized precision stenciling process which uses screens of stainless steel or nylon. The ink is forced through the screen by a hard rubber squeegee. The drawing of Fig. 7-8 is an example of the screening process which applies the cermet ink to the ceramic substrate.

The shape of the element is controlled by

A.

B.

C.

D.

SQUEEGEE

SCREEN

SUBSTRATE

Fig. 7-8 Screening process for cermet element

small openings in the fine mesh screen that correspond to the desired pattern. The pattern of the screen openings is produced by a photographic process from a large scale artwork master. This process allows great versatility and provides high precision in screen production.

Composition of the resistive inks varies according to desired results. They all can be described as being composed of finely powdered inorganic solids (metals and metal oxides) mixed with a powdered glass binder (glass frit) and suspended in an organic vehicle (a resin mixture). Materials used include silver, palladium, platinum, ruthenium, rhodium, and gold.

Printing and firing of the inks is preferably done in a humidity and temperature controlled environment. This provides the best control of total resistance and temperature coefficient, and results in high yields and superior properties. A controlled temperature kiln with various temperature zones between 800°C and 1200°C is used to burn off the organic vehicle and causes a fusion of the glass particles with the ceramic substrate. The metallic particles provide a resistive film which is bonded to the substrate. Fig. 7-9 shows a kiln used to fire (glaze) the material.

A very wide range in resistance values can be achieved by varying:

1) The composition of the resistive ink.
2) The firing parameters (time and temperature).
3) The physical size of the element.

By using a substrate with good thermal characteristics, it is possible to get good power dissipation characteristics in a small space. In the single turn unit pictured in Fig. 7-10, the substrate is attached to the shaft in order to improve power dissipation. In this design a heat sink path is provided from the resistive element through the substrate, shaft, and bushing to the mounting panel.

The family of electronic ceramics includes steatite, forsterite, porcelain, zirconia, alumina, beryllia, and many other complex oxide ceramics. Of this family, the two most widely used materials are steatite and alumina.

Steatite is made from raw materials including talc (a mined inorganic material containing a major amount of magnesium oxide and silicon dioxides, plus a small amount of other oxide impurities), a stearate (lubricant), waxes (binder), and water. Alumina is made from high-purity aluminum oxide, talc, sterate, waxes, and water.

The two materials are processed from the raw

Fig. 7-9 The cermet ink is fired in a temperature controlled kiln

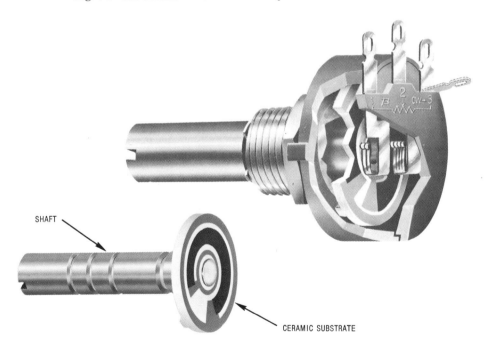

SHAFT

CERAMIC SUBSTRATE

Fig. 7-10 In this design, a ceramic substrate is attached directly to the shaft in order to increase power dissipation capability

151

state to a useable powder form in a similar manner. The process includes mixing, ball milling, sizing and drying, plus blending. Each step contributes to producing a dry powder with repeatable characteristics such as particle size distribution, bulk and tap densities—the latter being checked after the bulk material is vibrated and settles.

The ceramic substrates (or bases) can be formed by various methods. These include dry powder pressing, extruding, isostatic pressing (pressed from every side), casting (doctor-blade process), and injection molding. The most common process is dry powder pressing. A carefully controlled amount of the dry powder is placed in a steel and/or carbide die cavity and pressure is applied from either the top or the bottom or both. The pressure compacts the powder into a *green* (unfired) part. The green part is then fired in a high temperature kiln from 1300°C to 1760°C.

Firing the substrate causes a shrinkage which can range from as little as 8% to as much as 20%. This firing (actually a sintering) produces substrates with uniform densities and adequate dimensional tolerances.

Tight controls on batch processing, pressing parameters, and firing profiles (various temperature zones) produces substrates whose tolerances are held to thousandths of an inch. Fired surfaces can be improved as required by the processes of tumbling, lapping and/or grinding. Other forming methods such as extrusion, molding, and doctor-blade casting require a wet mix called a slip or slurry.

Many new materials and improvements in processing have been developed in the past ten years to allow production of economical, highly reliable substrates.

Selection factors. Potentiometers having a total resistance from 10 ohms to 10 megohms are practical. However, the entire resistance range is not available in all possible sizes and configurations.

Cermet elements offer very low (infinitisimal) resolution and good stability. Their noise performance is good in both the static and dynamic (CRV) condition.

Frequency response of cermet materials is very good and the practical application range extends well beyond 100MHz. The lower resistivity materials exhibit an equivalent series inductance, while the higher resistance cermets display an equivalent shunt capacitance.

Temperature coefficient for cermet potentiometers range from ± 50 ppm/°C to ± 150 ppm/°C, but average ± 100 ppm/°C or better depending on resistance range.

Operational life of cermet elements is excellent. The element surface is hard and very durable. Failures of cermet potentiometers, after extended mechanical operation, are more often wiper failures due to wear than problems in the element.

For trimming applications, cermet elements usually offer the best performance per dollar per unit space. Even though cermet elements are more abrasive than conductive plastic, wiper wear is low enough that mechanical life far exceeds trimmer requirements.

Carbon elements. Early carbon film potentiometers were usually made with a mixture of carbon powder and phenolic resin applied to a phenolic substrate and cured. Dramatic improvements in materials technology over the years have resulted in an upgrading of substrates and carbon-plastic resin compounds. One early improvement was the use of ceramic as a substrate for elements made of carbon and phenolic resin.

Depending on desired end results, a carbon composition element may be screened on (as with cermet), brushed on, sprayed on, applied with a transfer wheel, or dipped onto an insulative substrate. When sprayed onto a substrate, an automatically controlled spray gun is swept back and forth. Controlling the sweep speed determines the thickness of the resistive material which influences the resistivity of the element. A mask (stencil) may be used to control the resistive pattern.

The processing of all nonwirewound elements requires a manufacturing environment which is free from dust and other foreign particles. Foreign particles settling on a wet resistive film will interfere with stability, contact resistance variation and the element's total resistance.

After the resistive material is applied to the substrate, the resistance elements are transferred to an oven for curing. The procedure may be done by static oven batch curing or by infrared curing. During curing, the solvents are driven off and the organic resin cross links to form a durable plastic film. The resistivity increases with time or temperature. This increase in resistivity is predictable and may be compensated for during preparation of the carbon composition material. The finished element has characteristics similar to a carbon-film fixed resistor.

Various techniques are available for changing the resistivity of the elements. In addition to the amount of carbon, small quantities of powdered metals, such as silver, are sometimes used. The metals, being conductors, lower the resistivity and cause the temperature coefficients to become more positive. Altering the element geometry and placing shorting conductors under the element are two other methods which may be used to change the resistivity.

Potentiometers made with *molded carbon* are manufactured by molding a previously formed resistance element and other parts of the potentiometer together. These molded units are sometimes called hot molded carbon and are comparable to the carbon-pellet type of fixed resistors. The hot molded carbon element provides definite improvements in mechanical life and TC compared to ordinary carbon film elements.

Conductive plastic, the modern carbon film element, is made with one of the more recent plastic resins such as epoxy, polyesters, improved phenolics, or polyamides. These resins are blended with carefully processed carbon powder and applied to ceramic or greatly improved plastic substrates. The result is superior stability and performance. The importance of plastics technology to these improvements has probably been the reason for acceptance of the term *conductive plastic* or *plastic film* element.

Conductive plastic elements may vary considerably in temperature coefficient (TC). The resistivity range, ambient temperature range, materials preparation procedures, substrate material, and the curing techniques all influence the TC quality. TC values of -200 ppm/°C may be attained by optimizing the processing techniques of carbon or by incorporating metal powders or flakes into the system. Nickel, silver, and copper are most frequently used. However, these low TC's are usually found in the low resistivity ranges.

The substrates used may be either ceramic or plastic; however, modern plastic substrates result in better temperature coefficients due to the greater compatibility of the ink and the substrate.

For thinner films than those obtained with application methods mentioned above, carbon may be applied by vapor deposition. This method, while yielding an excellent fixed resistor, results in a film that is usually too thin to withstand wiper abrasion.

Conductive plastic material may also be deposited on an insulated metal mandrel and formed in a helix as shown in Fig. 7-11 for use in multi-turn potentiometers.

Selection factors. The carbon film potentiometer is usually the designer's first choice for an economical way to vary resistance in an electronic circuit. This is particularly true in commercial applications where specifications are less exacting and cost is a major concern. In addition to commercial applications, carbon film units with high quality elements and special construction techniques are also used in industrial and

Fig. 7-11 A conductive plastic film is applied to an insulated mandrel to provide an element for a multi-turn potentiometer

military equipment.

Advantages of carbon elements include low cost, relatively low noise during adjustment, and excellent high frequency performance. They also offer low inductive and distributed capacitive reactance. The operational life of carbon elements is very good and degradation characteristics are usually gradual rather than sudden catastrophic failures.

The resistive range of carbon elements extends as high as 20 megohms and as low as 10 ohms. Total resistance tolerance is typically ±10%.

The presence of substantial contact resistance in carbon elements limits applications where even moderate wiper current will be present. End resistance is usually high.

Carbon elements typically have poor moisture resistance and the load stability is not as good as cermet. Molded carbon elements can be expected to shift as much as 5% in a year. Although the carbon element has no resolution-type noise, the noise level at best can be quite high.

The outstanding characteristics of a conductive plastic element are low cost, low contact resistance variation, and extensive rotational life. The smooth surface produces extremely low resolution with virtually no friction or wear, even after a few million cycles of the wiper over the element. Linearities approaching those of wirewound elements are possible by blast trimming the elements with an abrasive after curing.

The temperature coefficient of carbon is negative. The magnitude of the TC differs for the various types of carbon potentiometers. Molded carbon units may exhibit a TC range of -2000 ppm/°C to -8000 ppm/°C whereas deposited carbon elements show TC's of about -1000 ppm/°C. Temperature coefficients as low as -200 ppm/°C are available in conductive plastic units.

The dynamic noise of conductive plastic potentiometers is quite low. This feature, coupled with the excellent resolution, permits the use of conductive plastic potentiometers in high-gain servo systems where other element materials would be unusable.

Conductive plastic elements offer good high frequency operation. No coils are present in the flat pattern design to produce inductive effects and the helical construction produces negligible inductive reactance. However, when the conductive plastic element is deposited on an insulated metal mandrel for multiturn potentiometers, some distributed capacitance is present between the element and the mandrel. This capacitance limits the high frequency performance of this construction very slightly.

Major limitations of conductive plastic

elements are low wiper current ratings, moderate temperature coefficient and low power capabilities.

Metal film elements. It is possible to vacuum deposit a very thin layer of metal alloy on a substrate to form a resistance element. Any metal which can be successfully evaporated or sputtered may be used, although only certain metals will yield the desirable characteristics of good temperature coefficient, useful resistivity, and a hard durable conductive surface. Typically, a member of the nickel-chromium alloy family is used to deposit a layer 100 to 2000 angstroms thick.

After deposition a very important part of the element processing is the stabilizing heat treatment. It is through precise control of this stage of manufacture that the complex strains inside the films are minimized. Carefully controlled processing makes it possible to achieve a temperature coefficient approaching wirewound elements. The uniformity of the process yields good linearity, extremely low resolution, and very low noise both at rest and during adjustment.

Selection factors. Due to their small size and construction, metal film elements are particularly low in reactive impedance. The housing and other packaging materials determine the effective parallel capacitance.

Metal film elements are practical only for lower resistance values. Total resistances are available from 10 ohms to 20K ohms. These elements are limited in power rating and have a rather short operational life. For these reasons, metal film elements are used primarily in those trimming applications where very low noise and good frequency characteristics are needed.

Bulk metal elements. Potentiometer elements may also be made with bulk or mass metal applied on a substrate in a much thicker layer than achieved by vapor deposition. One approach is a plating technique for a solid area of resistance metal, followed by precision photochemical etching of a zigzag pattern to increase the effective length of the element.

If the metal is carefully chosen to match properly with the substrate material the effective temperature characteristics of the two materials will compensate for each other. The result is an element with exceptionally low temperature coefficients.

Selection factors. Extremely low TC is the most significant advantage of bulk metal elements Less than 10 ppm/°C is possible.

Total resistances from 2 ohms to 20K ohms are obtainable in trimmer styles. For total resistances below 100 ohms a solid element may be used and the resolution is negligible. Larger resistance values require an etched pattern to increase the effective length of the element. This causes resolution to increase.

Contact resistance for bulk metal elements is very low but if an etched pattern is required, adjustment noise may be much higher.

Frequency response is excellent. The distributed capacitance is very low and inductance is negligible in either the etched or unetched pattern.

The limitations of bulk metal elements are cost, resolution in the higher TR values, and mechanical life. Their most frequent use is in trimmer applications where ambient temperature change is a critical factor.

Resistance taper. Resistance taper, as defined in Chapter 5, is the output curve of resistance measured between one end of the element and the wiper. It is expressed as a percentage of total resistance.

To achieve a given resistance taper, manufacturers vary the geometry of the element or the resistivity of the element material or both. This technique produces an element which is a linear approximation of the ideal theoretical taper and conforms to military and commercial specification tolerances. Fig. 7-12 shows three resistance tapers in MIL-R-94B. Fig. 7-13 shows a comparison of a linear approximation element and the ideal audio taper C of Fig. 7-12. Fig. 7-14 is a sampling of various film elements designed to provide a resistance taper.

A closer approximation to the ideal taper is possible with certain construction methods. An example is the molded carbon element which allows tight control of the element cross section. This can be made to conform to a given taper with a high degree of accuracy.

Selection factors. Taper A in Fig. 7-12 provides a rate of resistance change that is directly proportional to shaft rotation. Such tapers are often used for tone controls. Taper C is a left hand logarithmic curve which provides a small amount of resistance at the beginning of shaft rotation and a rapid increase at the end. This taper is most often applied as a volume (gain) control. Taper F, a right hand logarithmic, is the opposite of taper C. This taper is used for contrast controls in oscilloscopes and bias voltage adjustment.

The tolerance within which the resistance taper must conform to the nominal (ideal) taper is usually expresed only in terms of the resistance at 50% of full rotation. Military specifications require that the resistance taper *shall conform in general shape to the nominal curves* and that resistance value at 50% (±3%) of rotation shall be within ±20% (10% for cermet). For commercial controls this particular specification figure can be as high as ±40% tolerance at 50% rotation.

Hybrid elements. It is possible to combine a wirewound element with a conductive plastic coating to realize certain benefits. The hybrid element will exhibit the temperature coefficient and resistance stability of the wirewound element and the long operational life, low resolution and low noise of the conductive plastic element. Contact resistance will be about the same as with conductive plastic.

Wirewound plus conductive plastic increases the cost of hybrid elements significantly because of the extra processing involved. In a similar manner, conductive plastic may also be applied over cermet with similar advantages. This hybrid element approaches the TC and stability of

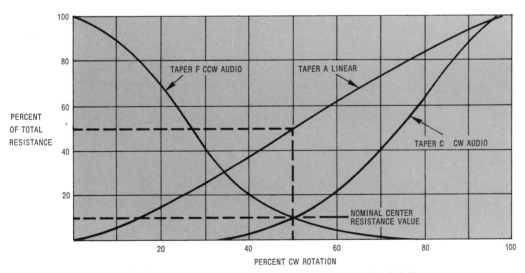

Fig. 7-12 Three resistance tapers taken from MIL-R-94B

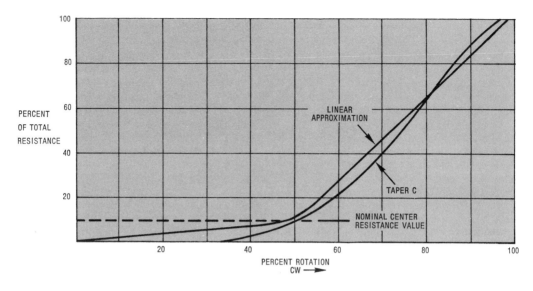

Fig. 7-13 A film element can be designed as a linear approximation to the ideal resistance taper

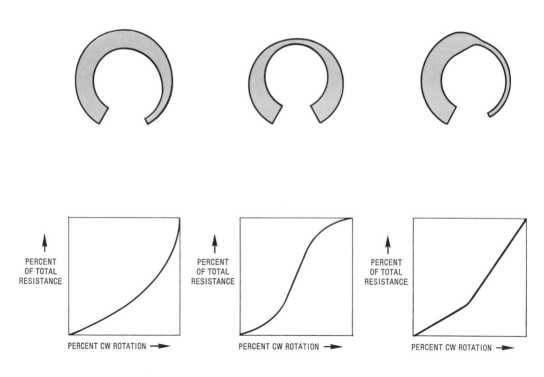

Fig. 7-14 Film element patterns and related graphs of resistance tapers

cermet.

Selection factors. The major application for wirewound-carbon elements is high-precision servo systems where the benefits in overall stability will justify substantially higher cost.

Nonlinear nonwirewound elements. The techniques used to create resistance tapers, described earlier, can be used to achieve a variety of nonlinear functions. A detailed presentation of the selection factors and applications for these devices is presented in Chapter 6. The following paragraphs explain the construction methods used to produce smooth nonlinear functions with conductive plastic elements.

The conductive plastic resistive track can be shaped to produce nonlinear functions. The slope of the output function is inversely proportional to the cross section of the resistive element. The minimum and maximum cross sections yield the highest and lowest resistivities respectively. This technique of cross section variation yields smooth output curves free from the scalloped output of tap and shunt techniques.

Sharp changes in the resistance element pattern do not produce corresponding sharp changes in the slope of the output function as is characteristic of the wirewound element. This differ-ence is due to the distributed vs. lumped resistive characteristics of the conductive plastic and wirewound respectively. The conductive plastic behaves as an electric field with the current flow distributed throughout the element cross section.

The flow lines in Fig. 7-15A illustrate the limitations on rate of change of slope without current collectors. Notice that the current lines are unaffected by the cross hatch region labeled *ineffective*. The current lines and output curves for this case would not change if the cross hatch section were removed. An identical shape is laid out in Fig. 7-15B with a conductive current collector (conductive termination material) is added at one edge of the ineffective area. Addition of the current collector has shaped the current lines and hence the output curve. By varying the cross section area and the resistivity of the element materials, a wide variety of functions is possible. Fig. 7-16 shows three examples of nonlinear conductive plastic elements.

Element Summary. Fig. 7-17 is a comparison of several common element types. The values are intended only as a guide. Look carefully at each individual specification when deciding which device to use for a specific application. The table should aid in narrowing possible choices.

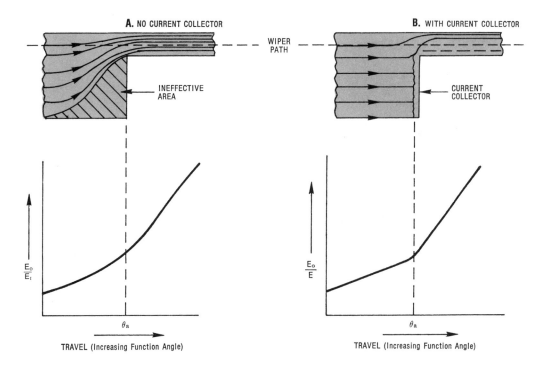

Fig. 7-15 Adding current collector at sharp change in resistive element pattern causes distinct changes in the slope ratio.

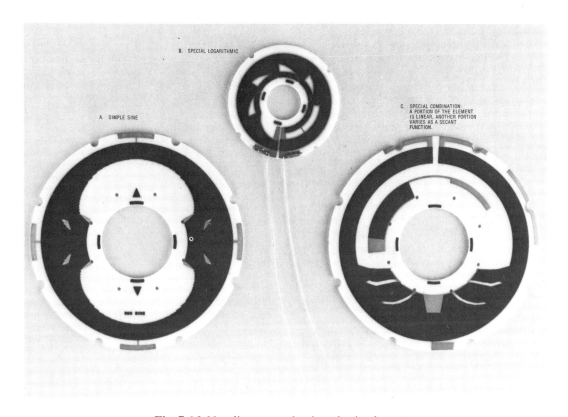

Fig. 7-16 Non-linear, conductive plastic elements

CHARACTERISTICS	WIREWOUND	CARBON		CERMET	METAL FILM	CONDUCTIVE PLASTIC
		MOLDED	DEPOSITED			
RESISTANCE RANGE	10Ω-100K	100Ω-10 MEG	100Ω-10 MEG	10Ω-5 MEG	50Ω-20K	100Ω-4 MEG
TEMP. COEFF.	±50 ppm/°C	±8000 ppm/°C	±1000 ppm/°C	±100 ppm/°C	±50 ppm/°C	< ±200 ppm/°C
RESOLUTION	0.1% to 1.0	—	—	< 0.05%	< 0.05%	< 0.05%
LINEARITY	0.1%	—	—	0.4%	0.2%	.05%
NOISE - STATIC	VERY LOW	HIGH	MODERATE	MODERATE	LOW	LOW
DYNAMIC	HIGH	MODERATE	MODERATE	MODERATE	LOW	LOW
ROTATIONAL LIFE	200,000 TO 1,000,000	5,000,000	1,000,000	500,000	100,000	4,000,000 REV.
FAILURE MODE	CATASTROPHIC	NOISY	NOISY	NOISY	RES. CHANGE OR CATASTROPHIC	NOISY
HIGH FREQUENCY PERFORMANCE	POOR	GOOD	GOOD	GOOD	EXCELLENT	GOOD

Fig. 7-17 Comparison of popular element types

TERMINATIONS

Obviously, there must be some means of connection to the element and wiper that is accessible to the user. These connections, called *terminations,* take many forms depending on specific needs and applications.

There are two basic requirements for termination. The first is making connection to the element. The second is providing some form of external access terminal. Element terminations are dependent on the type of element. Therefore, they will be discussed individually for the major element types, wirewound and cermet.

The external terminals are designed to be compatible with the popular mounting and wiring techniques used throughout the electronics industry. Fig. 7-18 illustrates the common forms. The general requirement is that a good solid electrical connection can be made without damage or stress to the potentiometers' interior or exterior.

Termination for Wirewound Potentiometers. There are five common methods of terminating wirewound potentiometer elements:

1) single-wire (pigtail)
2) silver-braze
3) pressure clips
4) solder
5) single-wire tap

In the single-wire form a portion of the element is left unwound. This pigtail is then routed to a terminal providing external access. The pigtail is attached to the terminal by soldering, brazing or spot welding. The length of the wire is kept short to produce very low end resistance. This method requires a high degree of assembler skill and is vulnerable to shock and vibration. Care must be taken not to induce stresses when the connection is made to the external terminal.

A preferable method of reliable connection to the element is to braze a small metal tab to a few turns of the resistance element. The advantage of this method is increased reliability since redundant connection is made to the resistance element. An additional benefit is excellent ability to withstand severe shock and vibration. A wire is welded or soldered to the tab and the external terminal. Sometimes the external terminal is connected directly to the tab.

Because the element wire is not discretely terminated at one point the silver-braze method of termination causes a slight increase in end resistance. Fig. 7-19 illustrates the brazing operation.

Pressure clips rely on mechanical connection between the clip and the element wire. The clip makes contact with one or more turns of resistance wire. Pressure clips are a potential source

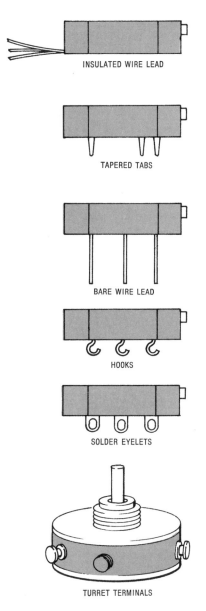

Fig. 7-18 A variety of typical external terminations

of problems because contaminations, such as solder flux, can lodge between the clip and the element. In addition, it is possible for a clip to change position slightly during temperature excursions. This can result in a variation in the number of wires being contacted and could cause noise or sudden output variations. Because of these undesirable possibilities the pressure clip method of termination is becoming obsolete in the potentiometer industry. Some manufacturers

159

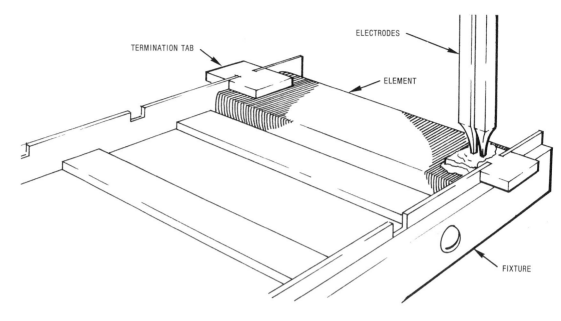

Fig. 7-19 Brazing operation for wirewound element termination

still use pressure clips in low cost potentiometers designed for non-critical applications.

A small wire may be soldered to the element to make connection to the external terminal. If properly accomplished with a high temperature solder a reliable connection will be made. Good assembly technique is necessary to insure that no stress is present at the element end of the wire.

Another single-wire technique which is commonly used to tap or connect to a single turn of resistance wire is one using percussion welding, This is primarily used to make precision non-linear elements by tapping and shunting as discussed earlier in this chapter. With this method a small diameter wire is connected to a percussion welder. The free end of the wire is positioned near a specific turn of resistance wire which has been selected by mechanical or electrical measurement. The end of the wire is placed on the target turn of resistance wire. A preset electrical charge is discharged through the junction of terminal wire and resistance wire resulting in a weld of the two molten surfaces that is as strong as the parent materials. Because fine wire and precise positioning is involved the operator usually works with a microscope. The opposite end of the termination wire is then attached to an external terminal.

Fig. 7-20 illustrates some popular element terminations used for wirewound potentiometers.

Termination for Cermet Potentiometers. Thick film conductive pads are used as a termination

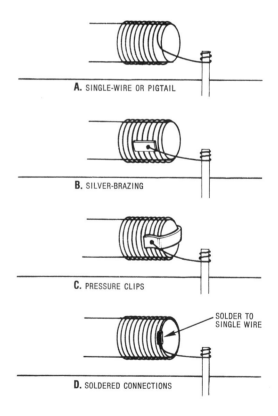

Fig. 7-20 Various methods of element termination for wirewound elements

160

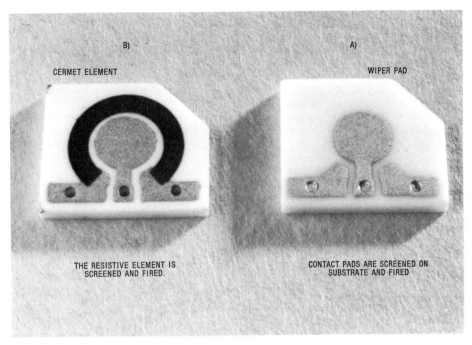

B)

CERMET ELEMENT

A)

WIPER PAD

THE RESISTIVE ELEMENT IS
SCREENED AND FIRED.

CONTACT PADS ARE SCREENED ON
SUBSTRATE AND FIRED

Fig. 7-21 Connection to the cermet element is made with conductive
pads under the ends of the element

means for cermet elements. There are two methods of application.

The most common method is to silkscreen the pads on the ceramic substrate using a conductive precious metal paste, such as paladium-silver. The substrate is then fired at a temperature of 950°C to remove all the solvent and binder. The resistance element is screened on the substrate with the ends overlapping the previously applied termination pads and the assembly is processed through a kiln a second time to fire the cermet. The result is a solid electrical bond between the termination pads and the resistive element. Figure 7-21 illustrates the two stages of this construction technique.

A second method is to fire the resistance and termination materials at the same time. For this process either the resistance or termination ink is applied first and usually dried. Then the other material is screened to the substrate and the two are co-fired.

Choice of these methods depends on the manufacturer's economic considerations and mechanization capability. From a performance standpoint, the end product is essentially identical.

Access to the outside world is made through the use of terminals attached to the substrate and element termination in one of several ways. In one, a tinned terminal with flared head is placed through a tapered hold with cermet termination material on its surface. The assembly is heated and the solder coating of the termination is re-flowed around the terminal pin. See Fig. 7-22A. The tapered configuration of the pin creates a physical interlock to improve resistance to pin pull-out.

A less secure method of termination, from a pull strength standpoint, uses a plain straight pin which is soldered into place using wave soldering. This technique, illustrated in Fig. 7-22B, depends entirely on the solder for mechanical strength and electrical connection between the pin and the termination material printed on the substrate.

Another termination method, Fig. 7-22C, is to install terminals by swaging (mechanically heading) them in place under very high pressure with an upset section on each side of the substrate. This provides a solderless bond directly between the cermet element termination and tin plated copper pins or leads. Residual stresses existing in the swaged pin insure that intimate contact is maintained through thermal cycling.

In this process wire is inserted through holes in an alumina substrate. The wires are clamped and mechanically upset filling the volume of the hole and forming a head on both sides of the element. An electrical and mechanical bond results with intimate contact between the wire and the termi-

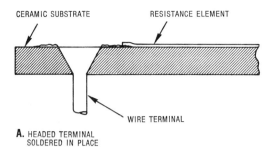

A. HEADED TERMINAL
SOLDERED IN PLACE

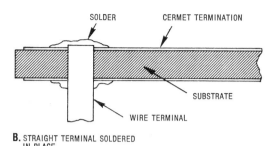

B. STRAIGHT TERMINAL SOLDERED
IN PLACE

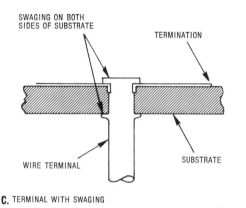

C. TERMINAL WITH SWAGING

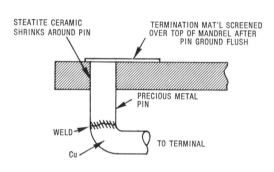

D. FIRED-IN
PIN TERMINATION

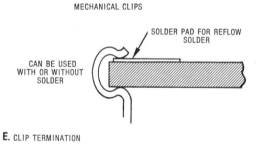

E. CLIP TERMINATION

Fig. 7-22 Terminations for Cermet Elements

nation material under the top head and along a portion of the shank because some cermet termination material is present in the holes.

Some cermet elements with steatite substrates have fired-in metal pins (usually precious metal to accommodate ceramic firing temperatures) as shown in Fig. 7-22D. These are installed and extend through and beyond the substrate before firing. When the assembly is fired the substrate shrinks tightly and holds the pins in place. They are then ground flush on one side and cermet termination material is printed over them. This results in connection between pin and element termination. To minimize the cost of the precious metal required, conventional copper leads are typically welded on to provide the external terminal.

Metal clip termination, Fig. 7-22E, eliminates the need for holes in the substrate but does require termination pads under the clips. After installation a solder reflow process is usually used to electrically and mechanically bond the clip to the cermet termination material.

Terminations for Other Nonwirewound Elements. Connection to other potentiometer ele-

ments generally use an underlying metal contact over which the element termination material is applied. The methods are completely analogous to those described for cermet. Another method, used very infrequently, involves conductive epoxy pastes to bridge between the element proper and the external access terminal.

CONTACTS

The wiper has a significant effect on many potentiometer parameters. Contact resistance, CRV, resolution, noise, power rating, operational life, and stability (with shock and vibration) are all influenced by the design of the movable contact.

Before discussing the factors related to contacts used in specific potentiometer designs consider what happens when two electrical members come together or make contact. The drawing of Fig. 7-23 shows two conducting members making contact and physically exerting pressure against one another.

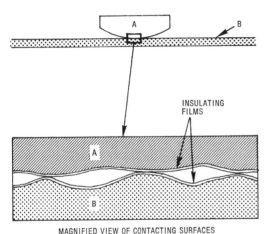

INSULATING FILMS

MAGNIFIED VIEW OF CONTACTING SURFACES

Fig. 7-23 A simple contact between two conducting members A and B

Although the surfaces may look smooth, an extreme enlargement of the interface reveals a non-uniform contact. The members touch only where the high points of member A meet the high points of member B. The area where the two members are actually touching is only a fraction of the apparent contact area.

Another important factor is illustrated in Fig. 7-23. All metals have films of oxides, sulfides, absorbed gases, moisture, or organic molecules on their surfaces. Even if thoroughly cleaned, the films will quickly reappear. On base metals such as copper and aluminum, these films can form to thicknesses of 50 angstroms in a few minutes. They continue to grow until they are several hundred angstroms thick.

Films also form on precious metal contacts, such as gold and platinum, but are much thinner and cause few problems in making proper electrical connection. For this reason, many potentiometer contacts are made of some precious metal alloy.

Certain factors can improve the electrical contact between the two members of Fig. 7-23. Friction, generated by the sliding action, can abrade a portion of the insulating film and smooth the surface to increase the effective contact area. Increased pressure between the members will also improve contact, although too much pressure will greatly increase wear and hence shorten the operational life of a potentiometer.

Lubricants are commonly used on metal and cermet elements to prevent oxidation so noise (CRV) will be low. Mechanical life is also increased which is especially important for precisions. The lube reduces wiper and element wear and acts as a vehicle to contain minute wear particles during extended mechanical cycling. Potentiometers used in servo systems are required to respond faithfully during the first sweep of the element even after sitting dormant in one spot for long periods of time. A lube will help assure this type of performance.

There is no one magic lubricant that can be used in potentiometers that will meet all requirements. The anticipated operational environment dictates the type of lubricant. For example:

1. Low temperature.
   Normally a type of light silicone *oil* lubricant can be used.
2. High Temperature.
   Here a silicone type *grease* can be used. Low viscosity oils may tend to migrate away from wear areas at high temperature and may not be acceptable.
3. Outer space vacuum applications.
   Dry lubricants are generally used. Molybdenum disulfide or niobium diselenide are very effective in a high vacuum as in outer space.

Many potentiometer manufacturers have their own special proprietary lubricants. These generally consist of various silicone fluids and greases prepared to their exacting specifications.

Movable contacts are generally made from a metal alloy which provides its own spring force thus simplifying the mechanical design of the assembly. In addition to aiding electrical conduction by helping to break down insulating films, spring pressure also aids in maintaining continuity during shock and vibration.

The physical form of the wiper takes many shapes as indicated in Fig. 7-24. The contact generally used with wirewound resistance elements is a single-fingered wiper similar to A in Fig. 7-24. It is formed in a manner that insures it will make physical contact with more than one turn of resistance wire. The contact material used is hard enough to minimize contact wear without having an abrasive quality which would shorten element life.

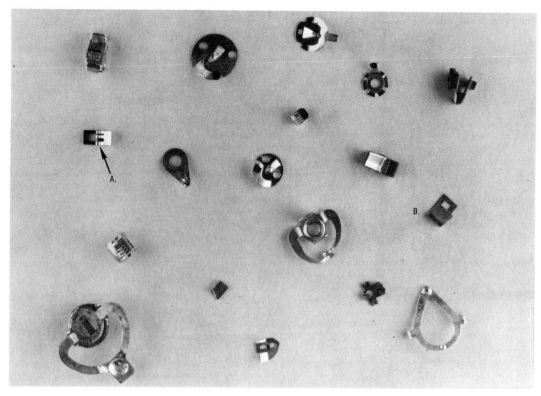

Fig. 7-24 Moveable contacts come in many different forms

The contact labeled B in Fig. 7-24 is typical of those used with nonwirewound elements. They are usually multi-fingered in order to decrease contact resistance; hence, the contact resistance variation, CRV. A careful study of Figures 7-25 through 7-28 will show why this is true.

In Fig. 7-25, 10 volts is applied across a resistive element. Equipotential lines are indicated by broken lines. Notice that they are all straight and equally spaced. Six different contact points, A through F, are indicated along one of the equipotential lines. If a very high resistance voltmeter were used to measure the contact voltage with respect to end terminal 1, each point would read 4 volts. This would seem to indicate that a single contact placed anywhere along the width of the element would yield a 4 volt poten-

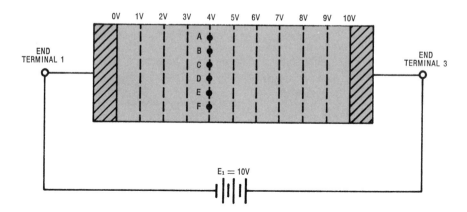

Fig. 7-25 Simple element showing equipotential lines

tial. If the wiper current is zero (i.e., an unloaded voltage divider) this is, in fact, true.

Look at what happens in Fig. 7-26 when the potentiometer is connected in a manner similar to the CRV and ENR demonstration circuits of Chapter 2. The equipotential lines start out fairly straight and evenly spaced at end terminal 1 but become much closer together and distorted in the vicinity of the single contact.

Consider the analogy of many people exiting a large theater with multiple aisles, only to find that all must pass through a single door in the lobby. The result is a crowding of the people. A similar effect happens with the electrons making up the current flow of the loaded voltage divider in Fig. 7-26. The result is current crowding around the contact.

Now look at what happens in Fig. 7-27 with the single contact placed along the edge of the element. The crowding effect is even worse than before. Any current, as in the case of a current rheostat or loaded voltage divider, passing through the very top of the element must not only make its way from left to right, but also must pass through more resistance material to get down to the contact. The end result is a higher contact resistance as evidenced by the higher meter reading in Fig. 7-27.

It is apparent that additional contacts would improve the situation. If multiple contacts are placed close together, however, current crowding will still result. If the theater in the analogy described above were redesigned by adding doors placed close together at one point in the

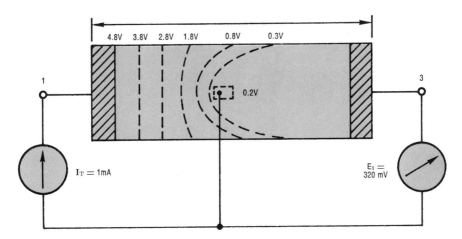

Fig. 7-26 Illustration of current crowding in single contact

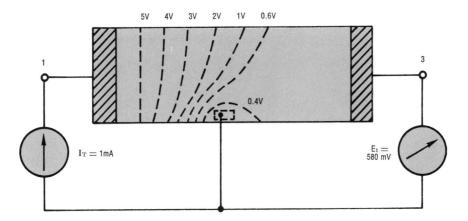

Fig. 7-27 A single-contact placed on edge causes increased contact resistance

lobby, crowding would be decreased. A better solution would be to separate the doors in an equidistant manner across the entire theater. The same is true for the contacts on the element of a potentiometer.

Fig. 7-28 illustrates the effect of placing five contacts, spaced equidistant, across the element. There is negligible current crowding and distortion in the equipotential lines. This results in the very low voltage at end terminal 3 which indicates a very low contact resistance.

The multi-wire wiper is an improvement over the sheet metal multi-finger wiper. In fact, the multi-wire wiper may be the single most important innovation in the past decade for nonwirewound potentiometers. It has greatly improved CRV performance and increased current carrying capacity of the wiper. The latter is of prime importance in current rheostats and loaded voltage dividers.

The effect of multiple contact wipers on CRV follows the same improvement pattern as for simple contact resistance. The worst possible case for CRV would be for a contact finger to lift completely, although this would rarely happen except possibly in cases of excess vibration or mechanical shock. It does aid in studying the effects of multiple contacts on CRV if complete loss of contact is assumed possible for a given finger.

When there is only one contact to begin with, the effect of losing a contact is disastrous. Even with two contact fingers, if any appreciable current is flowing through the wiper, the interruption of either will produce a substantial change in contact resistance with a corresponding change in output voltage. As the total number of contact fingers is increased, the effect of losing one of them becomes less significant.

In summary, both contact resistance and CRV are improved by using multiple-fingered wipers.

The optimum position of a single contact wiper is in the center of the width of the element. Multiple contacts should be distributed equidistantly across the width of the element.

One of the reasons nonwirewound potentiometers often outperform wirewound units under extreme shock or vibration is that multiple-fingered wipers are used with the quality nonwirewound units.

ACTUATORS

Many different variations of the mechanical means which moves the wiper across the resistive element are possible. The following paragraphs explain some of the common actuator types.

Rotary Shaft and Wiper. One of the most common configurations designed for frequent manual adjustments uses a shaft with a knob on one end and the wiper on the other end. This arrangement, less the knob, is also used when the potentiometer is in a servo system and driven by a motor, gear train, or any mechanical means.

In the simplest form, the shaft is passed through a friction or snap-in bushing (used for mechanical mounting purposes) and extended far enough to permit attachment of a knob. An arm is attached to the opposite end to transmit the rotary motion of the shaft to the wiper.

Generally, the wiper must be insulated from the shaft. Some reliable means must be provided to permit external electrical connection to the wiper. Where continuous rotation is not required, it is possible to use a length of very flexible wire to connect the wiper to an external terminal. This has some very severe limitations relative to the mechanical operating life of the potentiometer.

Another more reliable approach to making connection from the wiper to an external access terminal is by means of an additional sliding con-

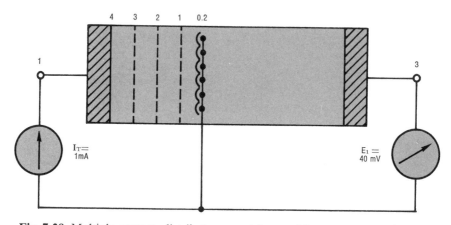

Fig. 7-28 Multiple contacts distribute current flow and lower contact resistance

tact. This sliding contact rides on a metal surface which is electrically connected to an external terminal. Some of the same considerations which applied to the element wiper will apply to this additional contact. The major differences is that both members may be of precious metal alloy and thus result in good contact with negligible contact resistance and contributed noise.

Fig. 7-29 illustrates several typical rotary shaft actuators. Careful design of the entire wiper assembly is necessary to yield proper performance with a reasonable manufacturing cost.

Many potentiometers are designed for a shaft rotation of less than one complete turn. This means that some form of mechanical stop must be used to limit the travel of the wiper to the element surface. For those cases where continuous rotation is required, the area between the ends of the element must be minimized but never shorted together. A smooth surface in this region is required for good performance as the wiper passes across it.

Rotary shaft actuators are also used in multiturn precision potentiometers, such as those

shown in Fig. 7-30. The mechanical problems are increased since the wiper must move along the length of the element as well as in a rotational manner to track on the helixed element path. Some mechanical means, a solid *stop,* must be used to prevent excessive rotation.

One of the general mechanical requirements for precision units is that the wiper assembly precisely follow the motion (mechanical input) of the actuator (shaft) system. This assures that each incremental motion applied to the external end of the shaft is faithfully transmitted into wiper travel.

Leadscrew Actuators. The mechanical drive requirements of adjustment potentiometers are quite different from those of frequently adjusted controls or precision devices. A reliable means of adjusting the position of the wiper is needed and some mechanical improvement in the ability of setting the wiper at exactly the right spot is necessary. Once adjusted, the position of the wiper should not change due to normal shock or vibration until manual adjustment is again desired. This is best done with a threaded shaft or

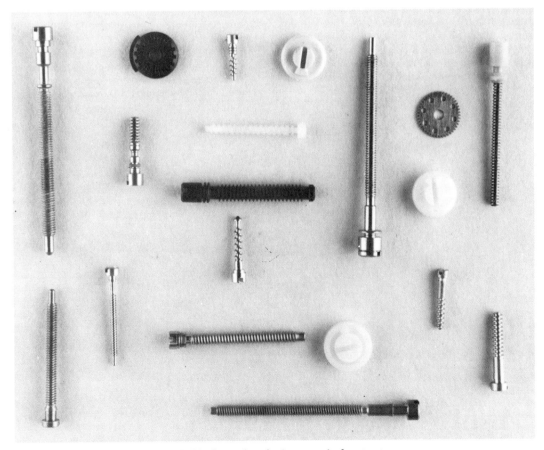

Fig.7-29 Several typical rotary shaft actuators

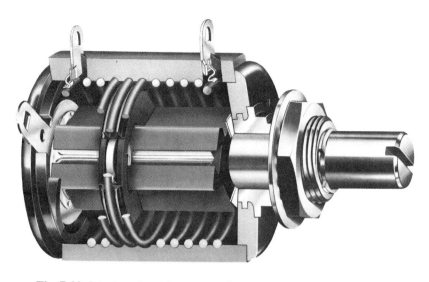

Fig. 7-30 Interior of multiturn potentiometers illustrate the mechanical detail necessary for reliable performance

leadscrew.

Fig. 7-31 illustrates a typical leadscrew actuated (rectangular) potentiometer. A simple thread along the body of the adjustment shaft engages grooves in the carriage that holds the wiper. This mechanism provides the translation from turns of input rotation to the required linear wiper travel along the straight element.

Notice the arrangement of Fig. 7-31. A shaft

seal retainer serves to prevent axial movement of the shaft and the entrance of moisture. This also keeps the leadscrew from turning during vibration or shock.

Most quality leadscrew potentiometers incorporate an automatic clutching action at the end of the travel. This prevents damage to the assembly due to overtravel. In some configurations, continued rotation of the screw causes a ratchet

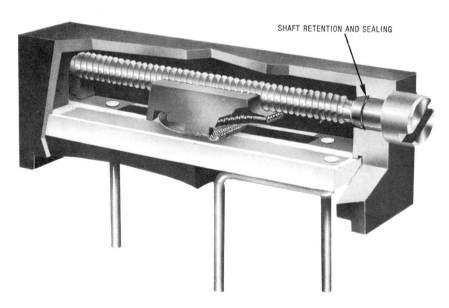

Fig. 7-31 The interior of a lead-screw actuated potentiometer illustrates that many turns of rotational motion are required to cause the wiper to traverse the entire element

action with a convenient audible click to tell the operator that the end of travel has been reached.

Wormgear Actuators. A greater length of resistance element can be included in a given linear dimension if the element is formed in a circular manner. This requires a different means of actuating the wiper. Fig. 7-32 shows a typical design for a worm gear actuated (square) potentiometer. The adjustment screw worm engages the teeth of a small plastic gear which is about the same diameter as the element.

The wiper assembly is placed between the plastic gear and the element, making contact to both. When the gear is rotated by turning the adjustment screw, the wiper moves along the element. At the end of the element, the wiper assembly encounters a mechanical stop to prevent further movement. Should the operator continue to turn the adjusting screw, the gear will turn and slide against the wiper in a clutch-like action without further motion or any

damage. If the direction of the adjustment screw rotation is reversed, the wiper correspondingly begins to move immediately.

Single Turn-Direct Drive. Another variation is somewhat like the basic wormgear design, but without the gear and worm. A simple rotor with a slot, and usually mechanical stops, permits rotary adjustment of the wiper position in a single turn unit. A typical example is shown in Fig. 7-33A. In this design, an o-ring seal prevents moisture entrance and also provides friction which serves as a mechanical restraint to prevent unwanted wiper movement. The design of Fig. 7-33B is a lower cost unit that does not provide the sealing feature.

Linear Actuators. For some servo applications, it is desirable to tie the wiper assembly directly to an external rod so that a linear motion causes a direct linear travel of the wiper. A typical example of this type of unit is shown in Fig. 7-34.

Some linear actuated potentiometers, used as

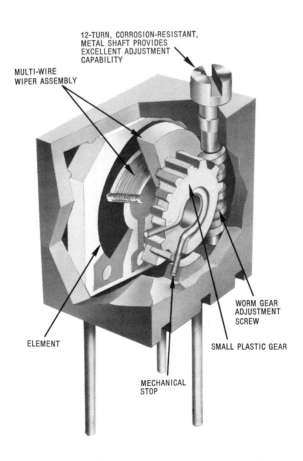

12-TURN, CORROSION-RESISTANT, METAL SHAFT PROVIDES EXCELLENT ADJUSTMENT CAPABILITY

MULTI-WIRE WIPER ASSEMBLY

WORM GEAR ADJUSTMENT SCREW

ELEMENT

SMALL PLASTIC GEAR

MECHANICAL STOP

Fig. 7-32 A typical worm-gear actuated potentiometer

A. SEALED CONSTRUCTION

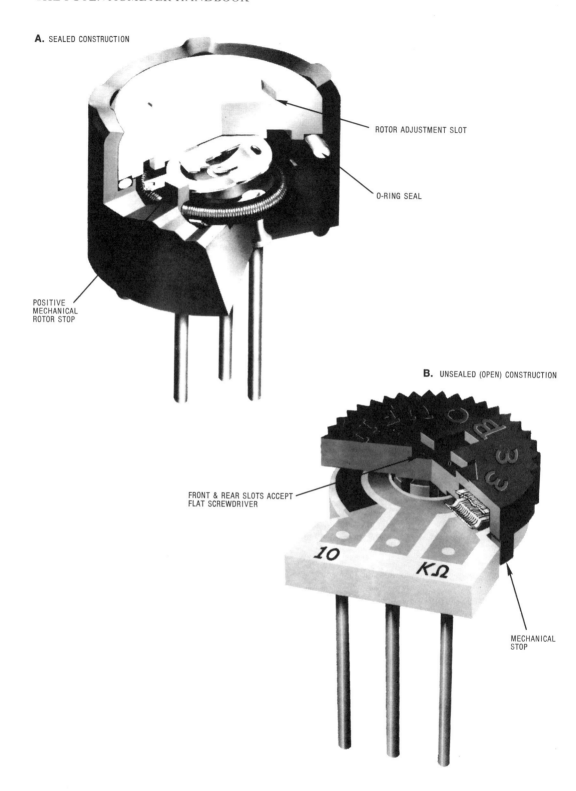

ROTOR ADJUSTMENT SLOT

O-RING SEAL

POSITIVE
MECHANICAL
ROTOR STOP

B. UNSEALED (OPEN) CONSTRUCTION

FRONT & REAR SLOTS ACCEPT
FLAT SCREWDRIVER

MECHANICAL
STOP

Fig. 7-33 Example of single-turn adjustment potentiometers

precision linear position feedback transducers, are several feet long. The wiper is tied directly to some moving member of the system.

The device shown in Fig. 7-35 is frequently used as an audio level control in the mixer panel used in a recording studio. It enables the operator to make rapid level changes and visually compare relative settings in an instant. Similar devices are also used in consumer music systems.

HOUSINGS

The housing of a potentiometer is very important. Much of the environmental qualities of the potentiometer are directly related to its enclosure. The degree of sealing achieved will dictate the ability of the unit to withstand moisture cycling. Although a quality potentiometer designed for harsh environments has an effective seal, it should never be considered hermetic (airtight).

The housing aids in stability and quiet performance by shielding the element and contact wiper surfaces from dust and dirt which will cause noisy operation. A good seal will aid in keeping out vapors which will cause oxide and film buildup on the element as well as on the wiper.

One of the major functions of any potentiometer housing is mechanical structure management. It is the framework which holds all of the other working members in their proper positions. This is especially true with precision potentiometers where the housing must keep the resistance element from changing shape or position with a changing outside environment.

The potentiometer housing must be properly designed to allow easy (and hence low cost) assembly. The inside of a well designed housing will have a variety of self jigging and holding features for various parts that make up the assembly. This provides a high quality level for a given cost.

The housing also provides the mechanical means for holding the leads or terminals securely in place. Mechanical stress and strain or elevated temperatures involved in soldering must be isolated from the internal members of the potentiometer. This prevents mechanical and electrical installation from affecting performance.

Economical manufacture of housing requires the use of complicated molds and molding presses of the type shown in Fig. 7-36.

SUMMARY

In order to select the most cost-effective potentiometer for a particular application, the user should be familiar with the many standard options available in potentiometer performance and construction. Application requirements can be matched against these options to make a final decision.

The tables in Fig. 7-37 are designed to aid this selection process in a general way. Where meaningful, specification and requirements are listed with those of lowest cost first in each category. These are marked with an asterisk (*). Element costs increase from left to right. By noting the application requirements of interest, the user can read the table to select standard resistive choices. Selection of a specific potentiometer design can then be decided based on severity of requirements vs. capability of the construction features as discussed in this chapter.

For unusually critical applications, the right side of the tables will alert the reader to important construction considerations relative to certain specifications that should be investigated with the potentiometer manufacturer before making a final product selection.

Methodical use of these tables will lead to the selection of construction features that best match a particular application. Data sheets from specific manufacturers can then be consulted to select a specific potentiometer model or type. Of course, packaging factors in Chapter Eight and other considerations discussed elsewhere in this book should influence the final choice of a potentiometer for the most cost-effective application.

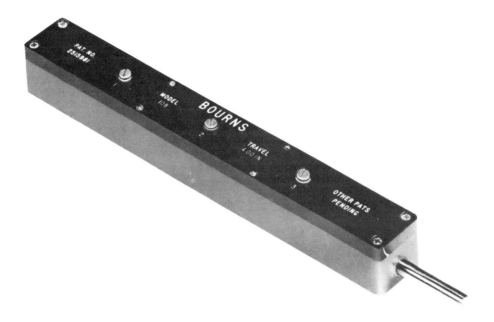

Fig. 7-34 This linear actuated unit is used to provide position feedback in a linear system

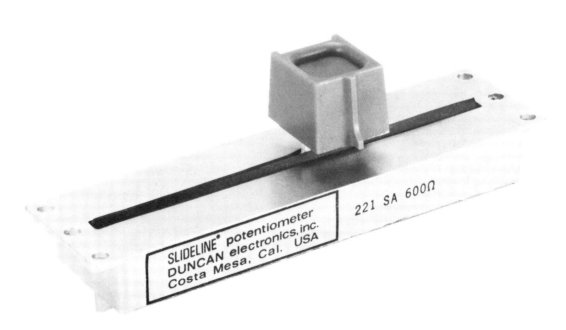

Fig. 7-35 Slider potentiometers such as this are frequently used in audio controls in studio mixer panels (Duncan Electronics, Inc. a subsidiary of Systron Donner)

Fig. 7-36 Precision potentiometer housings are molded by equipment such as this press

A. TRIMMER SELECTION GUIDE FOR COST-EFFECTIVE APPLICATIONS

APPLICATION REQUIREMENTS	STANDARD RESISTIVE ELEMENT CHOICES				CONSTRUCTION CONSIDERATIONS FOR CRITICAL APPLICATIONS (1)			
ELECTRICAL	Carbon	Cermet	Wirewound	Metal Film (2)	Wiper	Terminals	Actuator/Shaft Single-turn Multi-turn	Housing/Mounting
Total Resistance, ohms								
2 - 10				X				
10 - 20K		X	X	X				
20K - 50K	X	X	X					
50K - 2 meg	X	X						
2 meg - 5 meg	X							
*Total Resistance Tolerance, % max. (4)								
±20	X	X	X	X				
±10		X	X	X				
± 5			X	X				
Voltage Divider Adjustability % (3)								
.05 - .11		X		X	X		X	
.11 - .50		X	2k-50k	X				
.50 - 1.90	X	X	10-2k	X				
Current Rheostat Adjustability % (3)								
.05 - .11					X		X	
.11 - .50			2k-50k					
.50 - 1.90	X	X	10-2k	X				
*Power, watts max. at 70°C	(1)		(1)		X			X
.25	X	X	X	X				
.50		X	X	X				
1.00		X	X					
*Equivalent Noise Resistance, ohms max. (4)								
100			X	X	X			
10				X				
*Contact Resistance Variation, % max. (4)								
2.0	X	X			X			
1.0		X						
Circuit Frequency, Hertz								
DC — 1kHz	X		X	X			X	X
DC — 250kHz	X	X		X				
DC — 1MHz	X			X				
*TEMPERATURE								
Temperature Coefficient, PPM/°C, Max. (4)								
±400	X	X	X	X				X
±100		X	X	X				
± 50			X	X				
± 10				X				
*Operating Temperature °C (4)								
—55 to +105	X	X	X	X				X
—55 to +150		X	X	X				
—55 to +175		X	X					
MOISTURE RESISTANCE								
*Humidity (4)								
Poor	X							X
Good		X	X	X				
MECHANICAL								
*Rotational Life Cycles Maximum							X	
200 cycles	X	X	X	X				
over 200 cycles	X			X	X			
INSTALLATION						X		
Exposure to:								
Flux		X						X
Solder Temp	X	X	X	X				X
Solvents		X		X				X
*RELATIVE UNIT COST (5)								
a. (lowest cost)	X				Special materials. processes and testing may greatly increase product cost.			
b.		X						
c.			X					
d. (highest cost)				X				

Fig. 7-37 Selection guides for cost-effective applications

B. PRECISION POTENTIOMETER SELECTION GUIDE FOR COST-EFFECTIVE APPLICATIONS

APPLICATION REQUIREMENTS	STANDARD RESISTIVE ELEMENT CHOICES			CONSTRUCTION CONSIDERATIONS FOR CRITICAL APPLICATIONS (1)		
ELECTRICAL	Wirewound	Conductive Plastic	Hybrid W.W./C.P.	Wiper	Housing	Bearings
Total Resistance, ohms (1)						
25 - 100	X					
100 - 500K	X	X	X			
500K - 1 Meg		X				
*Total Resistance Tolerance, %						
±10 - ±20		X				
±10			X			
± 3 - ± 5	X					
*Resolution, % (3) (4)						
.009 - .090	X					
Essentially infinite		X	X			
*Linearity, independent Maximum, % (1)						
±0.50 - ±1.00		X				
±0.15 - ±0.50	X	X	X			
Equivalent Noise Resistance, Ohms (4)						
100	X			X		
Output Smoothness, % (4)				X		
0.1		X	X			
Power, Watts max. at 70°C (1) (or max. applied voltage)					X	
1.0 - 2.0	X	X	X			
2.0 - 5.0	X					
Circuit Frequency, Hertz						
DC - 1kHz	X	X	X			
DC - 1MHz		X				
TEMPERATURE						
*Temperature Coefficient, PPM/°C max. (3) (4)					X	
±200		X(2)				
±100			X			
± 50	X					
*Operating temperature, °C (4)					X	
—55° to 125°C	X	X	X			
—55° to 150°C	X					
MOISTURE RESISTANCE					X	
Fair		X				
Good	X		X			
MECHANICAL						
*Rotational Life, Shaft Revolutions, max.						X
2,000,000	X					
10,000,000			X			
25.000,000		X				
Torque	Depends on Specific design higher torque is least expensive					X
RELATIVE UNIT COST (5)						
1. (Lowest cost)	X					
2.		X				
3.			X		X	

NOTES TO SELECTION GUIDES

(1) Where certain specifications and performance are critical users may wish to discuss related key construction features of a particular potentiometer with the manufacturer.

(2) Includes "bulk metal".

(3) Specifications vary considerably depending on size and design of the resistive element, specific resistance and price. Check manufacturer's data sheet before final selection.

(4) When specifying potentiometers, care should be taken to match specifications to related type of element. As examples, when specifying conductive plastic precisions, do not call out wirewound specifications such as equivalent noise resistance or a TC of ±50 PPM. When specifying cermet trimmers do not call out ±10 PPM TC which is available only with metal film.

(5) Unit cost should be only one consideration for cost-effective application of potentiometers. See text for other factors.

(*) Specifications and requirements that generally result in a lower cost potentiometer are listed first. With other things being equal, first item will generally be most economical for that particular specification.

Notes

PACKAGING GUIDELINES

Chapter 8

In spite of the rapid technological advances and the growth of a discipline, electronic packaging is not easily defined in terms of some already existing technology. Rather, it is an overlapping of disciplines, which requires a breadth of knowledge rarely encountered before in emerging disciplines. While there may be no perfect definition of electronic packaging, it may, for practical purposes, be considered as the conversion of electronic or electrical functions into optimized, producible, electro-mechanical assemblies or packages. True, electromechanical assembly is not new, as electrical equipment has always required some mechanical form factor; however, the binding of sound interdisciplinary principles into a single discipline of electronic packaging is new.

Charles A. Harper
Handbook of Electronic Packaging, 1969

INTRODUCTION

Information in earlier chapters dealt primarily with the electrical characteristics and proper application of potentiometers. Another important consideration when selecting potentiometers is the physical manner in which the unit will be packaged into the overall system. Chapter six includes special mounting requirements for precision potentiometers and should be studied if precisions are of interest.

The purpose of this chapter is to remind the user of those common sense facts of electronic packaging applicable to potentiometers. The material presented, while second nature to the experienced designer, provides an excellent guide to anyone involved with variable resistive components. Circuit designers and packaging engineers are urged to give *early* consideration to the

ultimate need for and placement of potentiometers (especially trimmers) to avoid future frustration of a technician as in Fig. 8-1.

PLAN PACKAGING EARLY

All too often, insufficient planning and care results in a potentiometer being placed in an inaccessible location where the maintenance or calibration person cannot get to the potentiometer at all or must practically disassemble the instrument before the adjustment can be made. This is particularly true of trimmers and especially those that are added at the end of the design check-out stage when it is finally realized that an adjustment *will* be needed in the circuit.

Trimmers. Fig. 8-2 and 8-3 illustrate some ways to make trimmers accessible.

Fig. 8-1 Planning ahead can avoid frustrations of the calibrations personnel

DETERMINE
ACCESSIBILITY NEEDS

General. As an initial step, think about who is going to be making the adjustments, how often, and under what conditions. Try to visualize the entire adjustment process.

If the adjustment is to be made fairly often, such as once or twice a day for calibration, then the potentiometer shaft or actuator should be accessible without having to take the instrument or equipment apart. On the other hand, it might be that a particular adjustment should be made *only* by very skilled technicians with elaborate test equipment. To make this adjustment too easily available, say from the front panel, would be to invite disaster! This type of adjustment is best hidden by a service panel or placed at the back of equipment depending on accessibility of the latter.

Some adjustments are made only when the instrument or system has experienced a critical component failure, and compensation for replacement component variations is required. In this case it is probable that the top cover of the equipment will already have been removed for maintenance, so placing the potentiometer such that it is accessible under these conditions is adequate and probably wise.

Watch out for subassemblies! If the initial adjustment is made in subassembly state, consider whether that same adjustment might have to be varied in the field after all panels, switches, controls, etc., are in place. Maybe a special extender card or cable harness is used by factory assembly personnel is making initial calibrations; will a maintenance man in the field have those same accessories? If not, then make sure he can gain proper access to all the adjustments he will be required to make.

It is usually easy to make all the necessary adjustments on an instrument when it is sitting alone on a bench. Think what happens when it is installed in a big rack cabinet with other equipment above, below, and even behind it. Will it be necessary to pull the instrument out of the rack in order to make minor adjustments? Try to eliminate or minimize this type of maintenance.

On one line of instruments now on the market, a minor operational amplifier drift adjustment requires setting a particular trimmer. Unfortunately, three circuit cards have to be removed before the potentiometer can be reached for adjustment. The fact that all the cards have to be in place during the adjustment made life most difficult for service personnel.

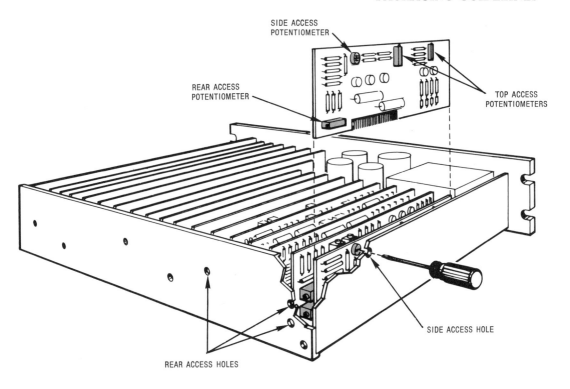

Fig. 8-2 Access to trimmers on PC cards mounted in a cage can be provided by careful placement of potentiometers near the top or rear

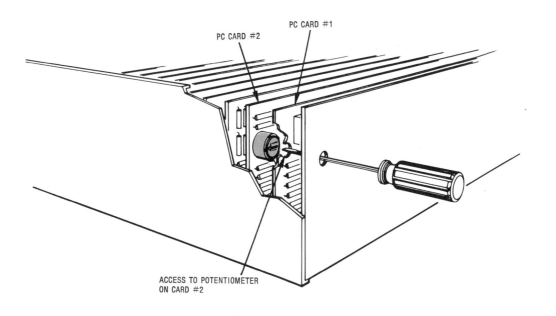

Fig. 8-3 Access hole provided through one PC card permits adjustment of potentiometer on another card

CONSIDER OTHER PACKAGING RESTRICTIONS

General. There are other restrictions which may limit the choices in designing with potentiometers. For example, front panel space may not be available to include all of the required controls and leave room for adjustment access. Harsh environment during operation may make it necessary to provide moisture seals or other protection over access holes and around adjustment shafts.

Potentiometers may be used in critical circuits that will not allow long leads or printed circuit runs because of unwanted stray capacitance or possible noise pickup. This means that the entire critical circuit must be fabricated with all the components, including the potentiometer, in proximity. If the circuit must be packaged in an inaccessible area of the assembly, an extension can be attached to the adjustment shaft. Another consideration is undesirable capacitance or noise introduced by an ordinary screwdriver during the adjustment process. It might be necessary to use only a nonconducting screwdriver. Information of this type should be included in the service manual.

Trimmers. Location of a particular trimmer on a printed circuit card may be influenced by other components. For example, it is not wise to place a critical trimming potentiometer adjacent to a high-power resistor that radiates a significant amount of heat.

Practical packaging techniques can also dictate potentiometer mounting positions. Various trimmers may be scattered over a given printed circuit card, but only one access direction for adjustment is available. The layout designer must carefully arrange all of the circuit components so that all trimmers can be adjusted from the same direction without obstruction by adjacent components.

The metal adjustment screw in the typical trimmer potentiometer is not electrically connected to any terminal. In a rectangular or square multiturn potentiometer, there is a certain amount of distributed capacitance between the metal shaft and the element. If the potentiometer is at a low level, high impedance point in a circuit, it may be necessary to use a metal bushing mounted potentiometer which allows the shaft to be grounded.

CHOOSE THE PROPER PHYSICAL FORM

General. Potentiometers are available in many different mechanical variations. Fig. 8-4 illustrates some of the possible choices for trimmer potentiometers. Often, equivalent electrical performance may be obtained with different physical shapes and sizes. The mounting means may be chosen from a variety of possibilities. Terminal choices include wire leads, printed circuit pins, and solder lugs.

For some applications, physical size limitations will be one of the determining factors. In other cases, the necessary direction of adjustment access will be more important.

Application in precision servo systems may require a servo mounted potentiometer (Chapter 6) or other factors may make a bushing mount more practical. Proper choice of a potentiometer for any application requires consideration of all physical and electrical requirements. For specifications such as power, size is often the controlling factor and adequate mounting space must be allowed.

The Potentiometer Packaging Guides, Fig. 8-4 thru Fig. 8-7, shows some common mounting areas and space occupied above or behind the mounting area for adjustments, controls, and precisions. These are intended as *general guidelines* to help in planning adequate space for mounting variable resistive components..

Before making a final selection, check the specific product data sheet and confirm that the potentiometer size vs. specifications and performance (Chapter 7) are within possible circuit design requirements. To be conservative, leave room for the next larger sized potentiometer. This could simplify a future circuit redesign that may not be contemplated at this early stage.

MOUNTING METHODS

Trimmers. Potentiometers may be installed or physically mounted in many different ways. In simple applications, printed circuit pin trimmers may be inserted directly into the printed circuit card and wave soldered. This gives both electrical connection and mechanical mounting. It is wise to use large enough lands (printed circuit conductor around hole) on printed circuit boards to prevent pulling off the copper circuitry during adjustment. In designs which must be capable of withstanding severe mechanical vibration and shock, it would be wiser to use a more substantial mounting means. Consider screw mounted potentiometers or conformal coatings in the circuit board assembly. Reinforcing a pin mounted unit with cement is another possibility.

Trimmers with mounting holes may be grouped together with screws or threaded rod, and angle brackets to mount the complete assembly to a panel or circuit board.

Mounting requirements are often determined by access demands as discussed earlier. Consider the drawing of Fig. 8-8. A printed circuit mounted potentiometer is placed right at the

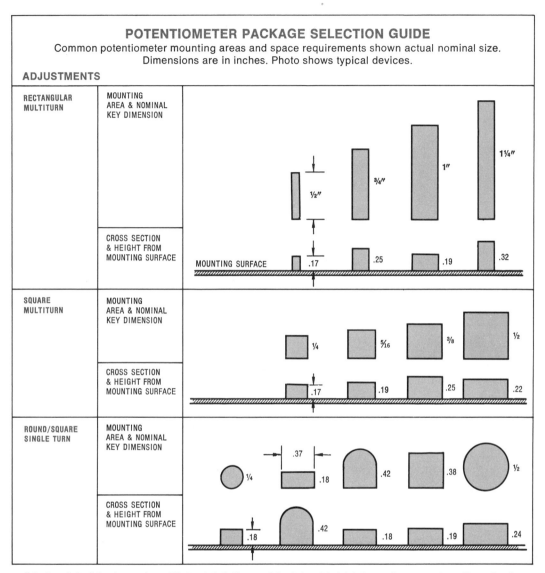

POTENTIOMETER PACKAGE SELECTION GUIDE

Common potentiometer mounting areas and space requirements shown actual nominal size.
Dimensions are in inches. Photo shows typical devices.

ADJUSTMENTS

RECTANGULAR MULTITURN	MOUNTING AREA & NOMINAL KEY DIMENSION	
SQUARE MULTITURN		
ROUND/SQUARE SINGLE TURN		

Note: Normal circuit board areas shown. Adjustment shaft, if any, and terminals not included. Some units may mount on side or edge. For final selection, see manufacturer's data sheets for options including terminal and actuator location.

Fig. 8-4 Adjustment potentiometer package selection guide

POTENTIOMETER PACKAGE SELECTION GUIDE

Common potentiometer mounting areas and space requirements shown actual nominal size.
Dimensions are in inches. Photo shows typical devices.

CONTROLS

MOUNTING AREA AND NOMINAL DIAMETER	

½   ⅝   ⅝ SQUARE

SHAFT CENTERLINE

| SPACE OCCUPIED BEHIND MOUNTING SURFACE EXCLUDING TERMINALS | |

$^{29}\!/_{64}$ $^{25}\!/_{64}$ ⅝

MOUNTING SURACE

| MOUNTING AREA AND NOMINAL DIAMETER | |

¾ $^{15}\!/_{16}$ $1\ ^{5}\!/_{32}$

| SPACE OCCUPIED BEHIND MOUNTING SURFACE EXCLUDING TERMINALS | |

¼ $^{29}\!/_{64}$ ⅝

MOUNTING SURFACE

Fig. 8-5 Control potentiometer package selection guide

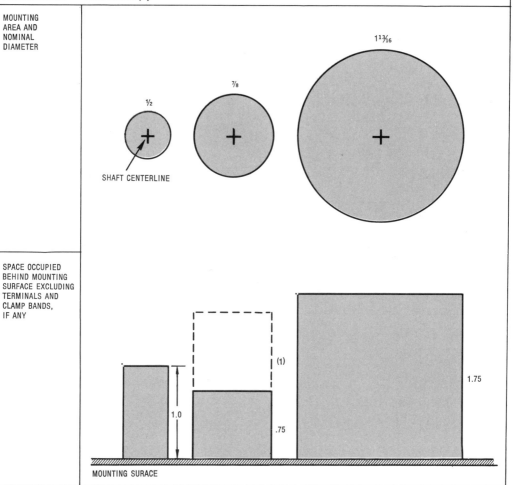

POTENTIOMETER PACKAGE SELECTION GUIDE

Common potentiometer mounting areas and space requirements shown actual nominal size.
Dimensions are in inches. Photo shows typical devices.

PRECISIONS — TEN TURN (1)

MOUNTING AREA AND NOMINAL DIAMETER

$1\frac{13}{16}$

$\frac{7}{8}$

$\frac{1}{2}$

SHAFT CENTERLINE

SPACE OCCUPIED BEHIND MOUNTING SURFACE EXCLUDING TERMINALS AND CLAMP BANDS, IF ANY

(1)

1.0

.75

1.75

MOUNTING SURACE

NOTES — Continued
(1) Multiturn precisions vary greatly in length depending on number of turns, specifications and specific design. See manufacturer's data sheets.

Fig. 8-6 Precision ten turn potentiometer package selection guide

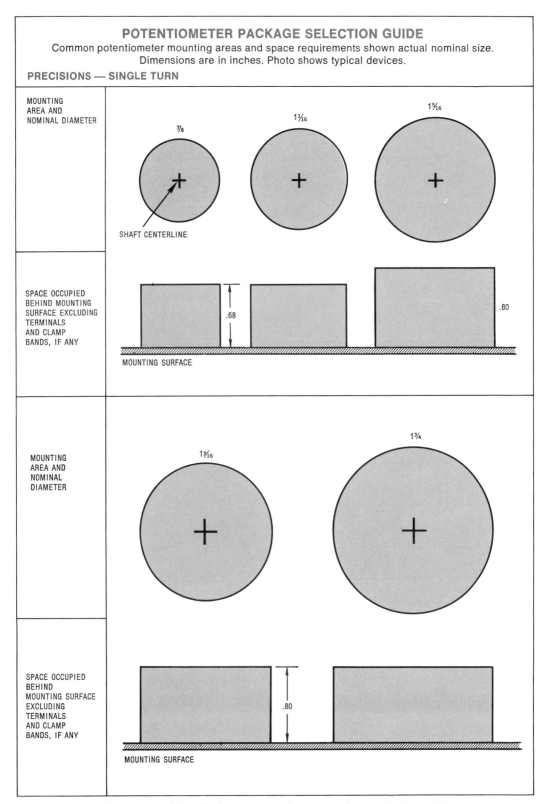

Fig. 8-7 Precision single turn potentiometer package selection guide

POTENTIOMETER PACKAGE SELECTION GUIDE

Common potentiometer mounting areas and space requirements shown actual nominal size.
Dimensions are in inches. Photo shows typical devices.

PRECISIONS — SINGLE TURN

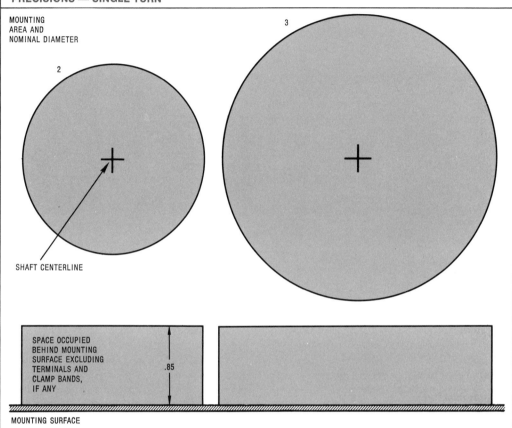

MOUNTING
AREA AND
NOMINAL DIAMETER

2

3

SHAFT CENTERLINE

SPACE OCCUPIED
BEHIND MOUNTING
SURFACE EXCLUDING
TERMINALS AND
CLAMP BANDS,
IF ANY

.85

MOUNTING SURFACE

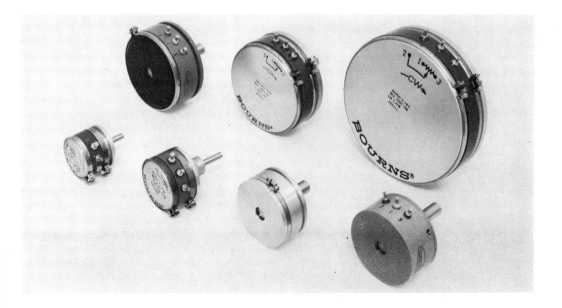

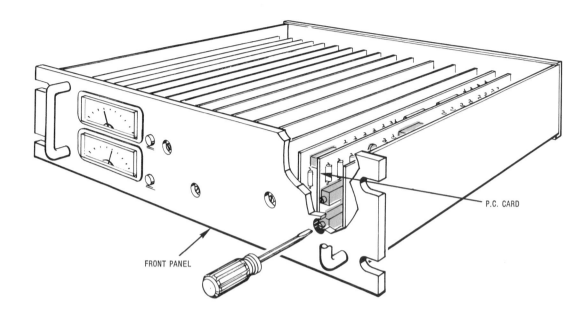

Fig. 8-8 Trimmer potentiometer access through front panel

edge of the card in such a manner that the small adjusting head may be inserted into an access hole in the front panel. The hole provides a limited access filter in that a rather small screwdriver is required in order to make an adjustment. Where the head is actually in the hole in the panel, the hole acts as a guide for the adjusting tool. The major limitation to this type of approach is the necessity of pulling the printed circuit card directly away from the panel. Where a plug-in card is used, it may be that the potentiometer must be pulled back from the edge of the card. It will usually be necessary to provide a larger access hole for this arrangement to allow for mechanical tolerances so the screwdriver and adjustment shaft will line up.

Fig. 8-9 illustrates a mounting variation in which the entire body of the potentiometer is inserted in a hole in the printed circuit card. This reduces the maximum seated height by the thickness of the card, and thus permits closer board-to-board spacing.

In some applications, adjustment access from the circuit side (side opposite components) of the printed circuit card is necessary. This can be done in the manner shown in Fig. 8-10. The wire pins are carefully bent over, as described in the next section, to permit inverted installation of the potentiometer. A small amount of cement to solidly secure the trimmer in place is good insurance against possible damage during adjustment. If this approach is used where the printed circuit cards are wave soldered, the entire hole should be masked off to prevent possible damage to the potentiometer.

Controls. Control type potentiometers are normally bushing mounted directly to an externally accessible panel, although limited access controls might be mounted on a bracket inside the assembly and adjustment accomplished with a screwdriver inserted through an access hole.

Applications such as those discussed in Chapter 5 use a mounting style convenient for manual adjustment. The style shown in Fig. 8-11 is known as a bushing mount. Note the threaded bushing on the front of the unit and the operating shaft which extends beyond the end of the bushing. Some bushings incorporate a locking feature. The operating shaft generally extends ½ inch beyond the end of the bushing. This length of exposed shaft is adequate for the attachment of a knob or turns counting dial. The shaft is generally available with a plain, slotted, or flatted end.

It may be that an insulated shaft extension is

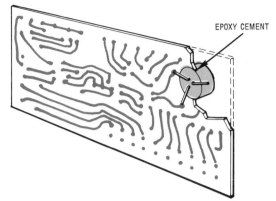

EPOXY CEMENT

Fig. 8-9 Low-profile mounting

necessary to provide isolation in critical low-level circuits, applications where distributed capacitance may be important, or where the operating voltage is high.

Snap-in mounting is convenient, economical and especially satisfactory where shock and vibration are not a major concern. These designs reduce the installation labor costs involved by eliminating the threaded bushing, nut, and lockwasher. The potentiometer is simply pressed into a drilled or punched hole until the locking fingers snap into place. Care should be taken in selecting mounting hole size and panel thickness when packaging snap-in type potentiometers. Appearance from the front of the panel is cleaner looking than nuts, washers, and other retainers. Fig. 8-12 shows products with snap-in mounting including a rectangular push-button potentiometer. The bezel on the latter model provides a neat appearance without extra mounting hardware.

A rigid printed circuit board placed behind and parallel to a front panel is a good way to mount and interconnect components. Controls that are

to be adjusted from the front of the panel are mounted on the PC board with adjustment shafts extending through the panel. Mounting and wiring of front panel components and related devices is simplified by eliminating individual wires. Fig. 8-13 illustrates this packaging technique.

In cases where the printed circuit board is at right angles to a panel, mounting such as that shown in Fig. 8-14 may be used. In this arrangement, the terminals of the potentiometer are soldered to the printed circuit board and a supporting bracket is soldered or mechanically clamped in place. This eliminates the need for a more complex bracket that might require screws for attachment and a nut and washer on the potentiometer. The adjustment devices can be mounted at varying distances behind the panel for greater front panel packaging density.

A wide variety of mounting hardware is available to make component packaging easier as illustrated in Fig. 8-15. These accessories are available directly from the potentiometer distributor or manufacturer.

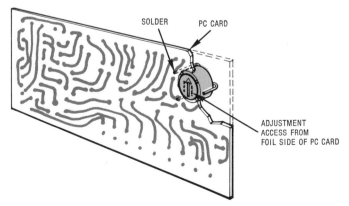

SOLDER   PC CARD

ADJUSTMENT
ACCESS FROM
FOIL SIDE OF PC CARD

Fig. 8-10 Potentiometer mounted inverted to permit foil-side adjustment access

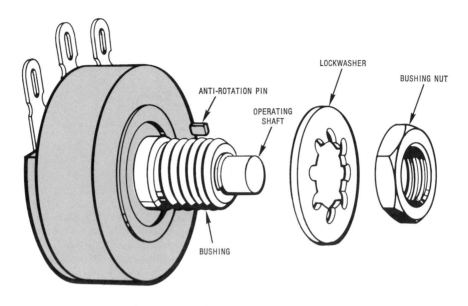

Fig. 8-11 Bushing mount potentiometer

Fig. 8-12 Snap in mounting potentiometer

STRESSES AND STRAINS

General. A few simple precautions in handling and mounting potentiometers can prevent component damage and avoid system problems. Actually, most potentiometers are rugged and reliable, but they do have their limits.

On units with insulated wire leads, hold onto the leads when stripping the ends of the wire. Do not pull directly against the potentiometer terminal and body.

Trimmers. The terminal pins may need to be bent at an angle for some particular mounting scheme. Don't just force the lead over by bending it at the potentiometer body. Use a pair of long nosed pliers as shown in Fig. 8-16 to relieve the stresses which might otherwise be induced into the potentiometer.

If a lead is bent in the wrong direction, straighten it out with the pliers and re-bend it correctly. Never twist the pins or solid leads as that might rupture the connection to the element terminations. Twisting also can provide a leakage path in an otherwise sealed unit by breaking the bond between the package material and the wire lead.

Avoid pulling on the leads. Forcing a package to lay down after the pins have been soldered in place can result in an open circuit or an intermittent connection.

Precision and Controls. Many precision potentiometers have a small anti-rotation pin on the front surface of the bushing mount package. If the anti-rotation feature is not used, then either remove the pin by clipping or grinding it off, or use a small washer between the potentiometer and the mounting panel. Otherwise, when the nut is tightened down on the bushing, the anti-rotation pin is forced between the potentiometer

and panel with unwanted stresses again introduced.

If wire cables are used to connect to the terminals of the potentiometer, use some form of strain relief to prevent a possible pull on the terminals. A small plastic tie is a good investment.

SOLDERING PRECAUTIONS

Potentiometers, like most electronic components, are subject to damage if excessive heat is applied to their terminals or housings during installation. When soldering by hand, use only enough heat to properly flow the solder and make a good electrical joint. Continued application of heat from an iron can soften the case material surrounding the terminals. This can make the potentiometer more susceptible to future stresses even though it may not cause immediate failure.

Soldering should be done in a physical attitude such that gravity will help keep any excess solder or flux out of the interior of the potentiometer. Some low-cost potentiometers are susceptible to flux entering along the leads. If it gets on the wiper or element, it may cause an open circuit or at least a very erratic output during adjustment.

Properly performed, wave soldering is usually more gentle than soldering by hand. Improper control of the time and temperature can result in a damaged component as well as a poor solder job. One area which needs special attention is the application of flux. Applied too generously, the flux can enter an unsealed potentiometer with unfortunate results as discussed previously.

In some instances, it may be wise to delay installation of the potentiometers until after wave soldering has been completed. Small solder masks or round toothpicks can be used to keep the circuit board holes open for later installation of potentiometers.

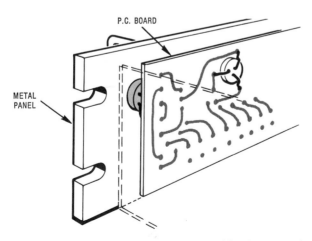

Fig. 8-13 Printed circuit board simplifies front panel wiring

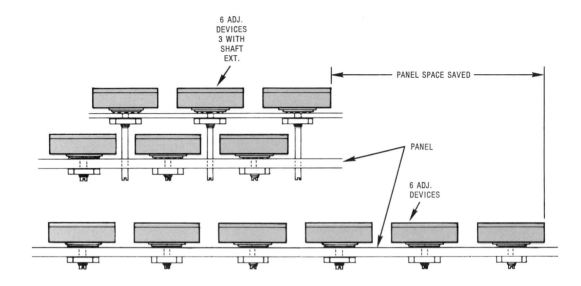

Fig. 8-14 Shaft extensions allow versatile placement of potentiometers and increased panel packaging density

Fig. 8-15 Adjustment potentiometer mounting hardware

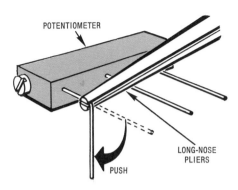

POTENTIOMETER

LONG-NOSE
PLIERS

PUSH

Fig. 8-16 To bend leads, hold pliers to
avoid stress on the potentiometer

SOLVENTS

Solvents are frequently used to remove flux
residue from printed circuit cards. If the circuit
card contains potentiometers, careful selection
of the solvent is necessary because certain compounds can be harmful to potentiometers.

Some solvents will attack the plastic housing
material or the adhesive used to assemble the
unit. Before using any cleaning techniques, consider the potential incompatibilities between the
solvent used and *all* components to be subjected
to the process.

Where the entire printed circuit card assembly
is to be totally immersed in a cleaning solvent,
it is possible that some solvent may enter the
potentiometer's package. If the solvent contains
some dissolved flux (and that's its primary function), then it is possible that flux residue will remain on the element surface after the solvent
evaporates. This is sure to cause erratic wiper
behavior and noise.

The severity of a solvent on a material is a
function of the temperature of the solvent,
exposure time, and agitation of the fluid. A great
variety of solvents are used in the electronics
industry under a multitude of trade names. Experience has shown that for consistent results in
cleaning circuit board assemblies, it is wiser to
buy solvents by their chemical name. Fig. 8-17
lists several common solvents that are acceptable
(or unacceptable) for use with potentiometers.
Appropriate precautions, such as venting, should
always be observed when handling hazardous
fluids.

One solvent that is not recommended for use
with potentiometers or other electronic components is the azeotrope of trichlorotrifluoroethane
with methylene chloride. Although there are un-
doubtedly other solvents that might have a bad

effect on potentiometers, this is one of the most
common.

If the solvent problems cannot be corrected by
using a more gentle procedure or changing fluids,
then it may be wise to delay installation of the
potentiometers until after the main cleaning is
done.

ENCAPSULATION

In most applications where the potentiometer's
function is to provide occasional control or adjustment, it is usually positioned outside any en-
capsulated section. However, if the potentiometer
is only used for adjustment of circuit performance during assembly, then the entire circuit,
including the potentiometer, may be potted.

Typical encapsulation procedures use either
pressurized encapsulation material or the application of a vacuum to aid in the removal of air
bubbles which might produce voids in the coating. It is safer to check with the potentiometer
manufacturer for this application. Unless the
potentiometer is properly sealed, it is possible for
some of the encapsulant to enter the package and
cause problems. Although readjusting of the
potentiometer is not required (hence, there is
no worry about material getting on the element
away from the present position of the wiper), it
is possible for the potting material to actually lift
the wiper off the element. This results in an open
circuit for the potentiometer and a rejected
circuit board assembly.

One successful technique for avoiding this
problem is to apply a somewhat generous coating
of a cement to all probable entry points on the
exterior of the potentiometer. If the cement is
allowed to cure at room conditions, then it can
act as a barrier to the normal encapsulant.

COMMON ACCEPTABLE SOLVENTS
1. Trichlorotrifluoroethane
2. Trichlorotrifluoroethane and isopropyl
   alcohol
3. Trichlorotrifluoroethane and ethyl alcohol
4. Trichlorotrifluoroethane and acetone
5. Trichloroethylene
6. Perchloroethylene
7. Chloroform
8. 1, 1, 1 trichloroethane
9. Methyl chloroform

SOLVENTS TO AVOID
1. Azeotrope of trichlorotrifluoroethane
   with methylene chloride

Fig. 8-17 Some common solvents

Notes

TO KILL A POTENTIOMETER

Previous chapters have presented all of the positive things about potentiomters and how to develop cost-effective designs for optimum performance and reliability. This book would be incomplete without introducing KUR KILLAPOT, that mischievous, misdirected character who seems to leave his mark wherever pots are used. As you can tell from his outfit, he's been around at *least* since pots were found under a dinosaur. One historian claims that after the invention of the wheel, KUR was the first to run over somebody! Some long-timers will remember his efforts as a gremlin in World War II. His more successful phenomenons are still unexplained.

KUR *intends* to be helpful but when it comes to using pots he's apt to cause problems. For example, he means well but sometimes gets a little heavy handed . . . uses a boulder when a pebble would do . . . or his Cro-Magnon brain doesn't quite grasp some fundamentals like Ohm's law . . . so he tries to make pots do things they can't or shouldn't. He's always around when pots are designed in, specified, installed and in service.

Actually KUR isn't bad but it does seem like sooner or later he screws something up . . . always with the *best* of intentions.

For those who revere pots and always want to use them properly to gain the ultimate in performance, reliability, and cost-effectiveness from these rugged and reliable components, read on: By following KUR's antics, perhaps you'll avoid ways of damaging a potentiometer or a circuit. We ask you to take heed by learning from KUR's mistakes. A summary of how to do this is at the end of the chapter.

All of his twenty-three adventurous mishaps that follow are arranged under these headings:

MAYHEM
MECHANICAL ABUSE AND MISUSE

SLAUGHDERING
SLAUGHTER BY SOLDERING

ZAP!
ELECTRICAL ABUSE AND MISUSE

When KUR takes the direct approach and does his dastardly deed early in the game, he uses the mechanical crunch. Several possibilities will be discussed. Even if he avoids them all, his carelessness is sure to trigger other sure fire ways of inflicting pain, suffering, or death upon potentiometers by MAYHEM — mechanical abuse and misuse.

A small projection, normally called an anti-rotation pin, is provided on the face of bushing mounted precision potentiometers and on many industrial types of controls. Its *intended* function is to keep the potentiometer body from turning after installation on a panel or bracket. This requires an extra hole be drilled or punched in the panel to receive it. If KUR forgets to drill this mating hole and jams the pin right against the panel he can produce the proper conditions for a strain in the case. After KUR's JAM SESSION, the adjustment shaft will be out of square with the panel so it will look a little strange by today's standards. This may not produce an immediate fracture of the potentiometer body. The unequal stress generated will be sure to cause some effect in eons to come, even if it is only a loss in linearity or increased rotational torque (better known as a bind). If KUR is not careful, a delayed failure will occur in the field. This will get him out of his cave and among the saber-toothed tigers which may be hazardous to his health.

This situation can be made even worse. Greater strain and degradation may occur in time when the pot is mismounted near some source of heat, for example, the pot placed close to a really hot power resistor or lossy (technical term meaning disipating more heat than usual) transformer. Excessive internal power dissipation can also give the same bad results.

JAM SESSION

KUR has found that unless the potentiometer case is completely destroyed, it will still continue to function as a clamp — long after it stops functioning as a potentiometer.

Versatile Cable Clamp

KUR tries to save money and be creative. He has discovered that a rectangular trimmer mounted to a panel or printed circuit card by means of screws make a VERSATILE CABLE CLAMP. Unfortunately, this provides an almost fail-safe means of failure with several disasterous modes.

Worse results are obtained when the cable is bunched up near the center of the pot to give maximum stress. KUR learned he can get instant failure or a more subtle long-term effect depending on the torque on the screws.

VERSATILE CABLE CLAMP

Another ineptitude which will produce almost the same effect is the use of the potentiometer to clamp a bracket or a subpanel to SAVE A NUT AND BOLT. If the assembly is thick enough so the threaded bushing barely extends beyond it, KUR gets instant failure by tightening the bushing nuts down very tight on a thread or two. On the other hand, a delayed failure is most likely if he uses only light tightening. In this way things will loosen later and result in a few surprises . . . such as loose panels and components.

THE SQUEEZE PLAY

When stacking several trimmer potentiometers together with mounting hardware, KUR wants to be sure they are *really* tight. In his over-zealous urge to do things right he sometimes does 'em wrong. This is one of those times! Using the SQUEEZE PLAY he overtightened the screws, thereby squeezing the potentiometers to death. Won't he ever learn?

Another way KUR inflicts punishment on a pot is by pulling hard on its solid terminals or flexible wire leads in a TUG O' WAR during installation or after wiring is complete. It takes *cave man* forces on some terminals to pull them out by the roots. KUR didn't uproot them, but did cause a simple internal open connection that was not obvious even to the experienced eye. Excessive pull on the terminals can cause an intermittent connection, which may not show up until after the system is shipped out of the cave and into the field.

TUG·O·WAR

DOING THE TWIST

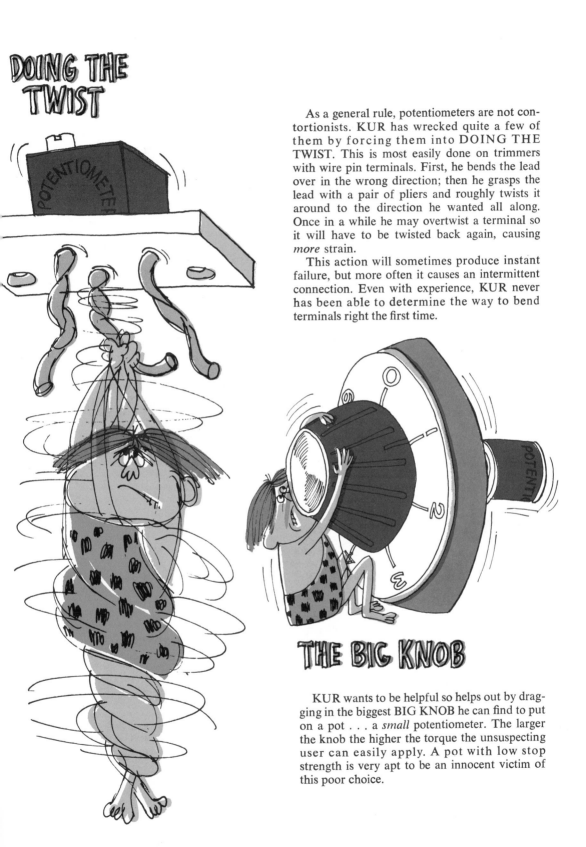

As a general rule, potentiometers are not contortionists. KUR has wrecked quite a few of them by forcing them into DOING THE TWIST. This is most easily done on trimmers with wire pin terminals. First, he bends the lead over in the wrong direction; then he grasps the lead with a pair of pliers and roughly twists it around to the direction he wanted all along. Once in a while he may overtwist a terminal so it will have to be twisted back again, causing *more* strain.

This action will sometimes produce instant failure, but more often it causes an intermittent connection. Even with experience, KUR never has been able to determine the way to bend terminals right the first time.

THE BIG KNOB

KUR wants to be helpful so helps out by dragging in the biggest BIG KNOB he can find to put on a pot . . . a *small* potentiometer. The larger the knob the higher the torque the unsuspecting user can easily apply. A pot with low stop strength is very apt to be an innocent victim of this poor choice.

Another clever idea from KUR's caveman brain that he applies in his eagerness to finish an assembly or get it repaired quickly is to use a panel mounted precision potentiometer as a HANDY HANDLE to pick up an entire assembly. Wrestling with a heavy chassis in this way can damage the potentiometer by bending the bushing or breaking or cracking the housing. This rough treatment might even bend a panel, so hopefully he'll remember potentiometers are *not* handy handles.

HANDY HANDLE

TIPOVER TERROR

Another disaster KUR learned the hard way is TIPOVER TERROR . . . accomplished by laying a chassis or cabinet on its face with the extending potentiometer shafts supporting most of the weight. This puts an axial or radial load on the shafts that may exceed the limits, especially if the chassis is dropped against a workbench or stepped on by a mastodon. The result may be a bent shaft, damaged bearing, or loosening of the rear cover of the potentiometer.

SLAUGHDERING
SLAUGHTER BY SOLDERING

Since KUR just discovered fire, he's still in the dark about sophisticated soldering techniques and gets burned with his SLAUGHDERING — slaughter by soldering.

HOTTER' N'HOT

Until he learned better KUR always liked to solder with a big soldering iron or turn up the bonfire on his solder flow pot to get things HOTTER N' HOT. He thought this would improve his solder joints and speed up the process. Instead, excess heat can soften material surrounding the terminals so that it will be more vulnerable to damage with a quick pull or twist. In extreme cases, the heat may penetrate the potentiometer and cause an internal joint failure or mechanical problem.

IT TAKES TIME

KUR is conscientious and wants to be thoroughly thorough and believes IT TAKES TIME to do it right. This is great in most cases but not when he does it with soldering. For example, a terminal may be heated adequately for soldering almost instantly but he takes his time. Heat gets inside the potentiometer where it can damage the terminations or cause a shift in the resistance value. He even slowed down the travel rate through his wave soldering pot until he got wise.

When it comes to preparing for soldering, KUR wanted to be ready. He figured if a drop of flux is good a flood would be better. FLUX IT AGAIN was his motto. He was an expert at flux flooding until he found that noisy contacts or even intermittent opens can be caused by too much flux. Actually it's hardly ever a problem on sealed potentiometers but KUR still hasn't learned to economize by using less flux. He would find this takes less clean up time too.

SOLVENT SOAK

FLUX IT AGAIN

KUR is the biggest SOLVENT SOAK around especially when he uses too much flux. He doesn't choose his solvent with care to be sure it's adequate but not too severe. If he only read Chapter 8 to pick up techniques and solvents to do his particular job, he could get rid of his old solvent soak image.

ZAP!
ELECTRICAL ABUSE AND MISUSE

KUR likes to sneak into engineering and play with the circuit designs on the drawing boards. You can't imagine some of the creative opportunities for electrocution—ZAP!—electrical abuse and misuse—he's left in his wake. He even designed in (unknowingly, of course) conditions that could result in possible damage to the potentiometer by an unsuspecting technician making a necessary adjustment.

Thinking he was planning ahead, he set the stage for multiple component failure or domino effect where the death of some other component takes the potentiometer with it.

KUR found too late that he could zap a pot faster than saying saber-toothed tiger with one of three simple methods: Exceed the power rating of the element, cause excessive wiper current flow, or operate the unit at a very high voltage which can cause voltage breakdown between the element and a grounded case or bushing. Any of these methods can be quite disasterous but KUR hasn't figured them out yet. He wants to warn you with a few basic illustrations that KUR still can't fully comprehend.

MORE POWER TO THE POT

Causing the potentiometer to dissipate power in excess of its rating will produce internal temperatures that will really warm things up! In fact with **MORE POWER TO THE POT** it may get that warm glow deep inside. This can produce a direct failure of varnishes and other insulating materials, or might be enough heat to cause a deformation of the element or surrounding parts.

Heat combined with some form of mechanical strain as discussed earlier in this chapter can cause trouble. High enough heat can soften the plastic used for the body and housing allowing movement of various parts that release the strain.

Remember, from Chapter 2, the power rating of a potentiometer is somewhat dependent upon the manner in which it is used. Thus, excessive current in only a portion of the element might easily exceed the power rating.

Damage due to excess concentrated power is likely when several potentiometers each dissipating full rated power are mounted close together. KUR watches for this condition but forgets to derate the power accordingly.

KUR's earliest killing of a potentiometer, the CHECK-OUT BURN-OUT, was at incoming inspection when he used a common VOM multimeter to measure end resistance or just to look at the output of the potentiometer. He set the meter to the Xl resistance range with one lead connected to the wiper and the other to one end of the element; then, he turned the shaft until the meter read minimum resistance. As luck would have it, the power source in the VOM caused a wiper current of 300 to 400 milliamperes! This either destroyed the potentiometer right on the spot or at least burned some rough spots on the element. KUR still doesn't know that a VOM plus a POT equals a NO-NO. Of course, he hasn't learned to use a digital ohmmeter either.

CHECK-OUT BURN-OUT

UP THE WIPER CURRENT

Of all the electrical techniques for potentiometer execution, KUR has been burned on UP THE WIPER CURRENT most often. Maybe it's because it's too subtle for his stone-age mind. Cermet and plastic film potentiometers are especially easy to damage with too much wiper current. Excessive noise and rough adjustment can be current induced even though the unit doesn't fail completely. Still KUR keeps designing in excessive current loads that wipe out wipers and fry elements in a flash. In fact, some of his prehistoric circuits are still in use today.

ZAP IT LATER

KUR carelessly designed the ZAP IT LATER, Method A Circuit in which execution of a potentiometer might be performed during final checkout or even some time after the equipment has been in service.

This circuit is one for delivering a constant current to the load. A VR diode is used to establish a constant voltage from the base of the transistor to the positive supply. The emitter current, and hence the collector current flowing through the load, will be determined by the difference in the VR diode voltage and the base-emitter voltage of the transistor in conjunction with the value of the potentiometer resistance.

KUR's instructions for initial adjustment (chiseled in stone, of course) had the potentiometer set to maximum resistance. Then, while monitoring the current level, he adjusted the potentiometer for the proper output current. This will defer execution until a technician in the field unsuspectingly turns the potentiometer too far clockwise and exceeds the current rating of the potentiometer. The excess current might also zap the transistor and even damage the load too, which results in a difficult repair job.

KUR should learn from this experience that by placing a fixed resistor in series with the potentiometer to limit the minimum total emitter resistance he can avoid this problem entirely.

METHOD A

KUR's ZAP IT LATER circuit, Method B, designed in 1,000,000 B.C., causes delayed execution. It provides an indication of charge completion by monitoring the voltage across the battery. Potentiometer R_1 permits an adjustment of the threshold voltage at which the lamp is turned on.

If the wiper of the potentiometer is turned full clockwise, a large current will flow through the VR diode, the wiper, and into the base of the transistor. If KUR's design is really poor he can be sure of slaying the potentiometer before the VR diode or transistor open up. On the other hand, if the VR diode shorts out, then he can almost be assured that the pot and transistor will be destroyed too. Once again, although KUR provides careful instructions to the person making the initial adjustments, an unsuspecting technician in the field was stuck with the dirty work and zapped the potentiometer.

Still another of KUR's infamous circuits, ZAP IT LATER, Method C, has a potential for maiming. This is a very simple unijunction oscillator in which the potentiometer is used to vary the charging rate of the timing capacitor and hence the operating frequency. The lower the resistance, the faster the charging rate and the higher the frequency.

KUR provided a control knob adjustment like this and didn't have to wait long for some one to turn it all the way. Suddenly the oscillator stopped! A conflict occurred between the potentiometer with minimum resistance trying to charge the capacitor and the unijunction transistor trying to discharge it. High currents resulted, and as luck would have it, the potentiometer, the unijunction, and the output pulse transformer were wiped out. One, two, three! Zap! Zap! Zap!

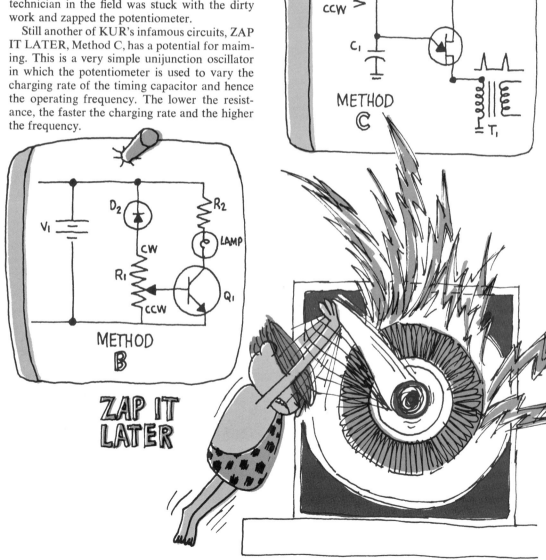

METHOD C

METHOD B

ZAP IT LATER

In the previous circuit arrangements, failure was induced by adjustment. In some applications, KUR caused a failure in a delayed manner during normal operation without the need for *any* adjustment. CIRCUIT SURPRISE is a good example. Here the potentiometer is used to generate a sawtooth output voltage waveform, or perhaps the setup is used to produce an output voltage indicative of the shaft position. A capacitor reduces the noise. Note that the output voltage must change from a zero value to a maximum as the wiper reaches the counterclockwise end of the element. This sudden change in voltage causes a high pulse current through the capacitor. After a while, it is quite probable that either a portion of the element will be eroded away or the wiper will become damaged by the high pulse currents. Erratic output will soon be followed by complete failure.

In some of the previous circuits, you saw how KUR managed to set up the massacre of several components with one stroke of the pencil on the drawing board. This is often called the domino effect. It is possible to design a circuit which will perform well as long as all components are good, then set up the domino game after one part fails due to some other cause.

Study the ZAPPO circuit above for a simple voltage regulator. As long as all of the parts are good, the position of the potentiometer can be varied over its entire range without causing damage. Minimum output voltage is achieved with the wiper at the counterclockwise position, and maximum voltage (with no regulation either) results when the wiper is moved to the extreme clockwise position.

Consider what might happen if transister Q_2 were to fail by shorting. The output voltage would jump to the maximum. KUR, noting this, might first try to reduce the output voltage by adjusting the potentiometer. He turns the potentiometer counterclockwise; the wiper reaches the end, and zappo! He wipes out the potentiometer, transistor Q_1, and the VR diode.

Another of KUR's circuit arrangements which looks perfectly acceptable chiseled in stone but relies on the laws of probability, is called SHORT STUFF. Here a control voltage is developed by the potentiometer used as a variable voltage divider. The voltage is then transmitted over a cable to some remote point where the current load may be very insignificant. So far, no problem.

Ah, but where you have an external cable leading from one area to another, you have the opportunity for a short. Consider what might happen if the wires in the cable become shorted to each other or even if the "hot" line gets shorted to ground. KUR cranks the potentiometer control knob clockwise trying to get more output. Once again, another pot fatality.

SHORT STUFF

HIGH VOLTAGE SURPRISE

Occasionally, KUR has an opportunity to use a potentiometer in a circuit which operates at a high voltage with respect to ground and results in a HIGH VOLTAGE SURPRISE. By using a potentiometer with a grounded metal bushing, the full voltage is applied between the element and the bushing frame. This might lead to direct voltage breakdown if the voltage difference is great enough or erratic behavior and possible long-term failure as arcing eats away at the element.

Once (just once!) KUR insulated a bushing mounted potentiometer properly, then an unsuspecting technician came along to make an adjustment using the bare metal shaft . . . high voltage surprise through the technician between the metal shaft and ground! Who was the most surprised? KUR, because he was the technician!

Seriously now, we hope KUR's adventures, while damaging or completely destroying several potentiometers and a few circuits, have been constructive. A summary of how to avoid these problems is on the next two pages. This story was told so you potentiometer users will be more aware of some of the problems that can be induced by misapplication or carelessness.

Potentiometers are inherently very rugged and reliable. With reasonable care in installation and use they effectively perform their function. When correctly applied in a circuit, they are one of industry's most cost-effective components.

TO GET BEST RESULTS
AND MAKE POTENTIOMETERS IMMORTAL

APPLIES TO	PAGE	TO AVOID	DO
ALL POTENTIOMETERS	195	Inoperative or damaged potentiometers.	Mount on flat surfaces. Tighten screws or nuts with reasonable torque.
	196	Open or intermittent terminations and damaged terminals.	Never pull on terminals and leads with excessive force.
	200	Possible damage or degradation from excess and severe solvent.	Use sparingly and never soak components any longer than necessary.
	202	Failure of varnishes and other insulating materials, deformation of element or surrounding parts, softening of plastic and possible shifting of various parts to release internal strain.	Operate potentiometer within power ratings. Watch for overpowering a portion of element.
	203	Noise and rough adjustment, with possible damage to the element.	Operate wiper at current levels within specified values.
	202	Burned out or damaged element due to excess current.	Never use a common VOM (multimeter) to measure end resistance or monitor potentiometer output. Use a digital ohmeter instead.
	203	Excess current in the wiper circuit.	Place a fixed resistor in series with the potentiometer to limit current.
	204	Damage to potentiometers and other components.	Never design circuits that allow excess power or current to flow through the potentiometer element or wiper circuit.
	204	Burning out potentiometers, unijunctions, output pulse transformers and other components.	Never design circuits that allow potentiometer adjustment settings that will cause excess current through any part of the circuit.
	205	High pulse currents through the wiper.	Design circuits that limit current through the system.
	205	Damage to potentiometer and other components.	Design circuits that prevent failure of other components when any component fails.
RECTANGULAR OR SQUARE, SCREW MOUNTED TRIMMERS	195	Inoperative or damaged potentiometers.	Use cable clamps instead of trimmers to secure cables.

APPLIES TO	PAGE	TO AVOID	DO
POTENTIOMETERS WITH PIN TERMINALS	197	Damaged terminals and potential open connection.	Reform pins with care. Never overbend and reform or twist any terminal excessively.
POTENTIOMETERS WITH SOLDERED TERMINALS	199	Softening material around terminals and failure or damage to internal joints or mechanical problems.	Apply only enough heat to make good solder joint. Use travel rate through wave soldering to accomplish good solder joint and avoid excessive heat.
	200	Noisy contacts or intermittent contacts.	Economize by using only sufficient flux to make a good solder joint.
PANEL MOUNTED POTENTIOMETERS WITH THREADED BUSHINGS		Damaged housing, increased rotational torque, loss of linearity, shaft out of square, general degradation because of anti-rotation pin without matching hole in panel.	Provide matching hole in panel for anti-rotational pin, or cut pin off.
		Stripped threads, loose potentiometers and assemblies.	Use threaded bushing long enough for full thread engagement in nut. Use nuts and bolts to fasten other parts together.
		Damage to the mechanical stop of a potentiometer with knob.	Use a reasonable size knob to operate the potentiometers.
		Bent bushings, broken or cracked housings, and bent panels and bent shafts.	Never lay chassis or panel on its face with potentiometer shafts supporting the weight of the system. Block up the panel to protect extended shafts. Never use potentiometer to pick up chassis or circuit board.
POTENTIOMETERS IN CIRCUITS WITH REMOTE CABLE OR LEADS	206	Damage to circuit or components.	Design circuit to limit current and voltage if cable or lead shorts to ground.

APPENDICES

Appendix I.
Standards of the Variable Resistive Components Institute

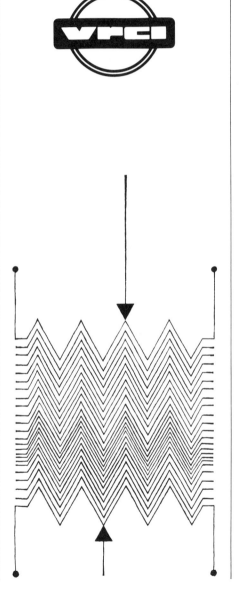

vrci-p-100a
terms
and definitions

This document provides, in two parts (VRCI-P-100, VRCI-P-200), a standard for nomenclature and preferred inspection and test procedures for wirewound and nonwirewound precision potentiometers. It combines all industry-approved terms, definitions and test procedures for precision potentiometers.

The inherent construction difference between wirewound and non-wirewound resistance elements, with respect to both geometry and methods of termination, results in somewhat different output characteristics. Those same construction characteristics that provide for substantially infinite resolution in nonwirewound pots also eliminate distinct step-off voltages at the ends of electrical travel, a reference point for some wirewound potentiometer definitions and tests. For this reason, the VRCI has, where necessary, provided alternate definitions and test procedures for both types of pots. This approach provides the user with a better understanding of each type of unit and provides a uniform source of communication throughout the industry.

In addition to the many technical papers on precision potentiometer characteristics and performance, that aided in the preparation of this document, many of the procedures, terms and definitions were adopted or modified from the following Military and Industry Specifications: MIL-R 12934; MIL-R 39023; MIL-STD-202; NBS Handbook 77-Volume 1, "Electricity and Electronics"; and EIA Subcommittee SQ 6.3 Report.

VARIABLE RESISTIVE COMPONENTS INSTITUTE

1 symbols and general terms

1.1 ■ LIST OF SYMBOLS

C = CONFORMITY

CT = CENTER TAP

CW = CLOCKWISE

CCW = COUNTERCLOCKWISE

E = TOTAL APPLIED VOLTAGE

e = OUTPUT VOLTAGE

e_i = INPHASE OUTPUT VOLTAGE,

e_q = QUADRATURE VOLTAGE

e/E = OUTPUT RATIO (Output Voltage Ratio)

V_e = END VOLTAGE

R_T = TOTAL RESISTANCE

R_L = LOAD RESISTANCE

R_e = END RESISTANCE

TC = TEMPERATURE COEFFICIENT OF RESISTANCE

RTC = RESISTANCE-TEMPERATURE CHARACTERISTIC

A = OUTPUT SLOPE

G = GRADIENT

θ = SHAFT POSITION

ϕ = PHASE SHIFT

θ_T = THEORETICAL ELECTRICAL TRAVEL

θ_A = ACTUAL ELECTRICAL TRAVEL

LEGEND:

□ WIREWOUND

■ WIREWOUND AND NONWIREWOUND

■ NONWIREWOUND

1.2 ■ GENERAL TERMS

1.2.1 ■ PRECISION POTENTIOMETER A mechanical-electrical transducer dependent upon the relative position of a moving contact (wiper) and a resistance element for its operation. It delivers to a high degree of accuracy a voltage output that is some specified function of applied voltage and shaft position.

1.2.1.1 □ WIREWOUND PRECISION POTENTIOMETER A precision potentiometer characterized by a resistance element made up of turns of wire on which the wiper contacts only a small portion of each turn.

1.2.1.2 ■ NONWIREWOUND PRECISION POTENTIOMETER A precision potentiometer characterized by the continuous nature of the resistance element in the direction of wiper travel.

1.2.3 ■ CUP A single mechanical section of a potentiometer which may contain one or more electrical resistance elements.

1.2.4 ■ GANG An assembly of two or more cups on a common operating shaft.

1.2.5 ■ SHAFT The mechanical input element of the potentiometer.

1.2.6 ■ SHAFT POSITION An indication of the position of the wiper relative to a reference point.

1.2.7 ■ TERMINAL An external member that provides electrical access to the potentiometer resistance element and wiper.

1.2.8 ■ INTEGRAL RESISTOR An internal or external resistor preconnected to the electrical element and forming an integral part of the cup assembly to provide a desired electrical characteristic. The resistor may be a separate entity, a part of the wirewound or nonwirewound resistance element, or a layer type resistor formed on the same insulating substrate as the resistance element.

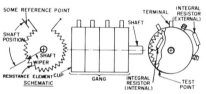

FIGURE 1.2.8 Precision potentiometer

1.2.9 ■ TEST POINT An additional terminal used only to facilitate measurements.

1.2.10 ■ TAP

1.2.10.1 ■ CURRENT TAP: An electrical connection fixed to the resistance element which is capable of carrying *rated* element current and may distort the output characteristic. **Note: Current taps on non-wirewound units commonly have significant width, but low resistance. See paragraph 3.13.**

1.2.10.2 ■ VOLTAGE TAP: An electrical connection fixed to the resistance element which introduces no significant distortion in the output characteristic. A voltage tap usually has significant tap resistance and *may* not be capable of carrying rated element current. **Note: The distinction between current and voltage taps basically applies to taps on non-wirewound units. Most taps on wirewound potentiometers are attached to one turn of wire and can carry rated element current. They do not usually have an effect on resolution or output characteristics.**

2 input and output terms

2.1 ■ INPUT TERMS

2.1.1 ■ TOTAL APPLIED VOLTAGE (E) The total voltage applied between the designated input terminals.

Note: When plus ($+$) and minus ($-$) voltages are applied to the potentiometer, the Total Applied Voltage (commonly called peak-to-peak applied voltage) is equal to the sum of the two voltages. Each individual voltage is referred to as zero-to-peak applied voltage.

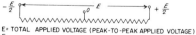

E = TOTAL APPLIED VOLTAGE (PEAK-TO-PEAK APPLIED VOLTAGE)

$\frac{E}{2}$ = ZERO-TO-PEAK APPLIED VOLTAGE.

FIGURE 2.1.1 Total applied voltage

2.2 ■ OUTPUT TERMS

2.2.1 ■ OUTPUT VOLTAGE The voltage between the wiper and the designated reference point. Unless otherwise specified, the designated reference point is the CCW terminal.

2.2.2 ■ OUTPUT RATIO The ratio of the Output Voltage to the designated input reference voltage. Unless otherwise specified the reference voltage is the Total Applied Voltage (see 2.1.1).

2.2.3 ■ TOTAL VARIABLE OUTPUT The difference between the maximum and minimum Output Ratios. These ratios correspond to the Minimum Voltages at each input terminal.

2.2.4 ☐ ■ END VOLTAGE

2.2.4.1 ☐ END VOLTAGE — WIREWOUND The voltage between the wiper terminal and an end terminal when the shaft is positioned at the corresponding End Point. End Voltage is expressed as a percent of the Total Applied Voltage.

2.2.4.2 ■ END VOLTAGE — NONWIREWOUND The voltage between the wiper terminal and an end terminal when the shaft is positioned at the corresponding Theoretical End Point. End Voltage is expressed as a percent of the Total Applied Voltage.

2.2.5 ■ MINIMUM VOLTAGE The smallest or lowest voltage between the wiper terminal and an end terminal when the shaft is positioned near the corresponding end of Electrical Continuity Travel. Minimum Voltage is expressed as a percent of the Total Applied Voltage.

2.2.6 ☐ JUMP-OFF VOLTAGE (WIREWOUND POTENTIOMETERS ONLY) The magnitude of the first measurable voltage change as the wiper moves from the overtravel region onto the Actual Electrical Travel. It is expressed as a percent of the Total Applied Voltage.

2.2.7 ■ SHORTED SEGMENT A portion of the resistance element over which the Output Ratio remains constant within specified limits as the wiper traverses the segment with a specified Load Resistance.

2.2.8 ■ OUTPUT SLOPE The ratio between the rate of change of Output Ratio and the rate of change of shaft travel.

MATHEMATICALLY: $A = \dfrac{\triangle \dfrac{e}{E}}{\triangle \dfrac{\theta}{\theta_T}}$

θ_A may be substituted for θ_T where applicable

Note: The theoretical output slope is the first derivative of the normalized Theoretical Function Characteristic.

MATHEMATICALLY: $A = \dfrac{d\,f\,(\theta/\theta_T)}{d(\theta/\theta_T)} = \dfrac{d(e/E)}{d(\theta/\theta_T)}$

2.2.9 ■ SLOPE RATIO The ratio of the largest to the smallest Output Slopes of a monotonic Theoretical Function Characteristic.

2.2.10 ■ GRADIENT The rate of change of Output Ratio relative to shaft travel

MATHEMATICALLY: $G = \dfrac{d(e/E)}{d\theta}$

2.3 ■ LOAD TERMS

2.3.1 ■ LOAD RESISTANCE (R_L) The external resistance as seen by the Output Voltage; (connected between the wiper and the designated reference point).

Note: No load means an infinite Load Resistance.

2.3.2 ☐ LOADING ERROR The difference between the Output Ratio with an infinite Load Resistance and the Output Ratio with a specified finite Load Resistance, at the same shaft position.

Note: Elimination of Loading Error, by compensating the resistance element to give the desired output with a specified Load Resistance, is referred to as "Load Compensation."

3   rotation and translation

3.1 ■ DIRECTION OF TRAVEL For rotary potentiometers, clockwise (CW) or counterclockwise (CCW) when viewing the specified mounting end of the potentiometer. The designation of terminals in the figure corresponds to the direction of shaft travel..

For translatory potentiometers, "extending" or "retracting" when viewing the specified end of the potentiometer.

The Output Ratio and shaft position increase with clockwise (or extending) direction of travel unless otherwise specified.

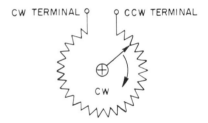

FIGURE 3.1 View of shaft and element from specified mounting end

3.2 ■ TOTAL MECHANICAL TRAVEL The total travel of the shaft between integral stops, under specified stop load. In potentiometers without stops, the mechanical travel is continuous.

3.3 ☐ ■ MECHANICAL OVERTRAVEL

3.3.1 ☐ MECHANICAL OVERTRAVEL — WIREWOUND The shaft travel between each End Point (or Theoretical End Point for Absolute Conformity or Linearity units) and its adjacent corresponding limit of Total Mechanical Travel.

3.3.2 ■ MECHANICAL OVERTRAVEL — NONWIREWOUND The shaft travel between each Theoretical End Point and its adjacent corresponding limit of Total Mechanical Travel.

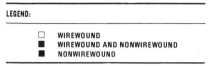

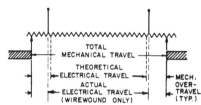

FIGURE 3.3.2 Mechanical overtravel

Note: The relationship of the electrical travels to each other and to the input terminals shown above is given for illustration only and may vary from one potentiometer to another.

3.4 ■ BACKLASH The maximum difference in shaft position that occurs when the shaft is moved to the same actual Output Ratio point from opposite directions.

3.5 ☐ END POINT (WIREWOUND POTENTIOMETERS ONLY) The shaft positions immediately before the first and after the last measurable change(s) in Output Ratio, after wiper continuity has been established, as the shaft moves in a specified direction.

3.6 ■ THEORETICAL END POINT The shaft positions corresponding to the ends of the Theoretical Electrical Travel as determined from the Index Point.

3.7 ■ INDEX POINT A point of reference fixing the relationship between a specified shaft position and the Output Ratio. It is used to establish a shaft position reference.

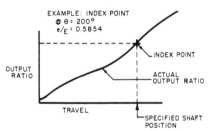

FIGURE 3.7 Index point

3.8 ☐ ACTUAL ELECTRICAL TRAVEL (WIREWOUND POTENTIO-METERS ONLY) The total travel of the shaft between End Points.

3.9 ■ THEORETICAL ELECTRICAL TRAVEL The specified shaft travel over which the theoretical function characteristic extends between defined Output Ratio limits, as determined from the Index Point.

3.10 ☐ ■ ELECTRICAL OVERTRAVEL

3.10.1 ☐ ELECTRICAL OVERTRAVEL — WIREWOUND The shaft travel over which there is continuity between the wiper terminal and the resistance element beyond each end of the Actual Electrical Travel. (Theoretical Electrical Travel is substituted for Actual Electrical Travel in Absolute Conformity or Linearity units.)

3.10.2 ■ ELECTRICAL OVERTRAVEL — NONWIREWOUND The shaft travel over which there is continuity between the wiper terminal and the resistance element beyond each end of the Theoretical Electrical Travel.

3.11 ■ ELECTRICAL CONTINUITY TRAVEL The total travel of the shaft over which electrical continuity is maintained between the wiper and the resistance element.

3.12 ■ TAP LOCATION The position of a tap relative to some reference. This is commonly expressed in terms of an Output Ratio and/or a shaft position. When a shaft position is specified, the Tap Location is the center of the Effective Tap Width.

3.13 ■ EFFECTIVE TAP WIDTH The travel of the shaft during which the voltage at the wiper terminal and the tap terminal are the same, as the wiper is moved past the tap in one direction.

Note: In some instances, particularly nonwirewound pots, the tap width may be essentially zero (i.e., no flat zone) but the tap may have a significant effect on conformity. In these cases the term "Effective Tap Width" should not be applied. Instead, the effect of the tap on the output characteristics should be considered in terms of conformity.

3.14 ☐ ■ PHASING POINT — WHEN INDEX POINT (3.7) IS NOT EMPLOYED

3.14.1 ☐ PHASING POINT — WIREWOUND A reference point on a cup of a gang, usually an Output Ratio, an End Point, or an intermediate tap.

3.14.2 ■ PHASING POINT — NONWIREWOUND A reference point on a cup of a gang, usually an Output Ratio or an intermediate tap (not an end tap).

3.15 ■ PHASING (SEE ALSO SIMULTANEOUS CONFORMITY PHASING PARA. 5.10) The relative alignment of the Phasing Points of each cup of a gang potentiometer.

Note: Unless otherwise specified, phasing requirements apply to a single specified Phasing Point in each cup and all cups are aligned to the Phasing Point of the first cup.

4 *resistance*

4.1 ■ TOTAL RESISTANCE (DC INPUT IMPEDANCE) The DC resistance between the input terminals with the shaft positioned so as to give a maximum resistance value.

4.2 ■ DC OUTPUT IMPEDANCE The maximum DC resistance between the wiper and either end terminal with the input shorted.

4.3 ☐ ■ MINIMUM RESISTANCE

4.3.1 ☐ MINIMUM RESISTANCE — WIREWOUND The resistance measured between the wiper terminal and any terminal with the shaft positioned to give a minimum value.

4.3.2 ■ MINIMUM RESISTANCE — NONWIREWOUND Refer to Tap Resistance (4.5) or Minimum Voltage (2.2.5) for applicable definition.

4.4 ☐ ■ END RESISTANCE

4.4.1 ☐ END RESISTANCE — WIREWOUND The resistance measured between the wiper terminal and an end terminal with the shaft positioned at the corresponding End Point.

4.4.2 ■ END RESISTANCE — NONWIREWOUND Refer to End Voltage (2.2.4.2) for applicable definition.

4.5 ■ TAP RESISTANCE (NONWIREWOUND POTENTIOMETERS ONLY) The minimum resistance obtainable between a tap terminal and a wiper position on the resistance element, measured without drawing wiper current.

Note: This definition applies only to intermediate taps. For End Terminations refer to End Voltage (2.2.4.2)

4.6 ■ APPARENT CONTACT RESISTANCE (NONWIREWOUND PO-TENTIOMETERS ONLY) Refer to Output Smoothness (6.2)

4.7 ☐ ■ EQUIVALENT NOISE RESISTANCE (ENR)

4.7.1 ☐ EQUIVALENT NOISE RESISTANCE — WIREWOUND Refer to Noise (6.1).

4.7.2 ■ EQUIVALENT NOISE RESISTANCE — NONWIREWOUND Refer to Output Smoothness (6.2).

4.8 ☐ TEMPERATURE COEFFICIENT OF RESISTANCE (WIRE-WOUND POTENTIOMETERS ONLY) The unit change in resistance per degree celsius change from a reference temperature, expressed in parts per million per degree Celsius as follows:

$$T. C. = \frac{R_2 - R_1}{R_1(T_2 - T_1)} \times 10^6$$

Where:
R_1 = Resistance at reference temperature in ohms.
R_2 = Resistance at test temperature in ohms.
T_1 = Reference temperature in degrees celsius.
T_2 = Test temperature in degrees celsius.

4.9 ■ RESISTANCE — TEMPERATURE CHARACTERISTIC (NON-WIREWOUND POTENTIOMETERS ONLY) The change in Total Resistance over a specified temperature range expressed as a percent of the Total Resistance at a specified reference temperature.

$$RTC = \frac{R_2 - R_1}{R_1} \times 100$$

Where:
R_1 = Resistance at reference temperature in ohms.
R_2 = Maximum or minimum resistance at any of the test temperatures, in ohms.

Note: Although Temperature Coefficient of Resistance can be applied to Nonwirewounds, the Tempco of many Nonwirewounds is not linear over the normal use temperature range and this can be misleading.

5   *conformity and linearity*

5.1 ■ FUNCTION CHARACTERISTIC The relationship between the Output Ratio and the shaft position.

MATHEMATICALLY: $\dfrac{e}{E} = f(\theta)$

5.2 ■ CONFORMITY The fidelity of the relationship between the actual function characteristic and the theoretical function characteristic.

MATHEMATICALLY: $\dfrac{e}{E} = f(\theta) \pm C$

5.3 ■ ABSOLUTE CONFORMITY The maximum deviation of the actual function characteristic from a fully defined theoretical function characteristic. It is expressed as a percentage of the Total Applied Voltage and measured over the Theoretical Electrical Travel. An Index Point on the actual output is required.

MATHEMATICALLY: $\dfrac{e}{E} = f(\theta/\theta_T) \pm C$; $0 \le \theta \le \theta_T$

Note: The theoretical function characteristic is assumed to be a smooth curve when it can be described by a mathematical expression. When empirical data are provided, the points are assumed to be joined by straight line segments.

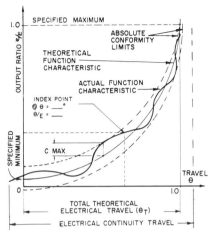

FIGURE 5.3 Absolute conformity

5.4 ■ LINEARITY A specific type of conformity where the theoretical function characteristic is a straight line.

MATHEMATICALLY: $\dfrac{e}{E} = f(\theta) \pm C = A(\theta) + B \pm C$

Where:
A is given slope; B is given intercept at $\theta = 0$.

5.5 ■ ABSOLUTE LINEARITY The maximum deviation of the actual function characteristic from a fully defined straight reference line. It is expressed as a percentage of the Total Applied Voltage and measured over the Theoretical Electrical Travel. An Index Point on the actual output is required.

The straight reference line may be fully defined by specifying the low and high theoretical end Output Ratios separated by the Theoretical Electrical Travel. Unless otherwise specified, these end Output Ratios are 0.0 and 1.0, respectively.

MATHEMATICALLY: $\dfrac{e}{E} = A(\theta/\theta_T) + B \pm C$

Where:
A is given slope; B is given intercept at $\theta = 0$.
Unless otherwise specified:
$A = 1$; $B = 0$.

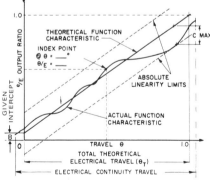

FIGURE 5.5 Absolute linearity

5.6 □ TERMINAL BASED LINEARITY (WIREWOUND POTENTIO-METERS ONLY) The maximum deviation, expressed as a percent of the Total Applied Voltage, of the actual function characteristic from a straight reference line drawn through the specified minimum and maximum Output Ratios which are separated by the Actual Electrical Travel. Unless otherwise specified, minimum and maximum Output Ratios are, respectively, zero and 100% of Total Applied Voltage.

MATHEMATICALLY: $\dfrac{e}{E} = A(\theta/\theta_A) + B \pm C$

Where:

A is given slope; B is given intercept at $\theta = 0$.
Unless otherwise specified:
$A = 1; B = 0$.

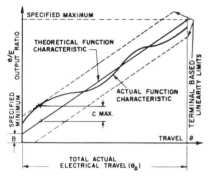

FIGURE 5.6 Terminal based linearity — wirewound

5.7 □ ZERO BASED LINEARITY (WIREWOUND POTENTIOMETERS ONLY) The maximum deviation, expressed as a percent of Total Applied Voltage, of the actual function characteristic from a straight reference line drawn through the specified minimum Output Ratio, extended over the Actual Electrical Travel, with its slope chosen to minimize the maximum deviations. Any specified End Voltage requirement may limit the slope of the reference line. Unless otherwise specified, the specified minimum Output Ratio will be zero.

MATHEMATICALLY: $\dfrac{e}{E} = P(\theta/\theta_A) + B \pm C$

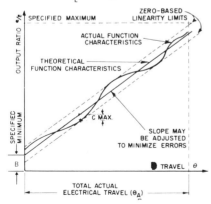

FIGURE 5.7 Zero based linearity — wirewound

Where:
P is unspecified slope limited by the End Voltage requirements, at the maximum output ratio end.
Unless otherwise specified:
$B = 0$.

5.8 □ ■ INDEPENDENT LINEARITY (BEST STRAIGHT LINE)

5.8.1 □ INDEPENDENT LINEARITY — WIREWOUND The maximum deviation, expressed as a percent of the Total Applied Voltage, of the actual function characteristic from a straight reference line with its slope and position chosen to minimize deviations over the Actual Electrical Travel, or any specified portion thereof.

Note: End Voltage requirements, when specified, will limit the slope and position of the reference line.

MATHEMATICALLY: $\dfrac{e}{E} = P(\theta/\theta_A) + Q \pm C$

Where:
P is unspecified slope; Q is unspecified intercept at $\theta = 0$. And both are chosen to minimize C but are limited by the End Voltage requirements.

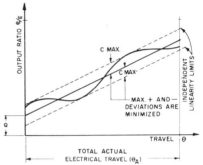

FIGURE 5.8.1 Independent linearity — wirewound

5.8.2 ■ INDEPENDENT LINEARITY — NONWIREWOUND The maximum deviation of the actual function characteristics from a straight reference line with its slope and position chosen to minimize the maximum deviations. It is expressed as a percentage of the Total Applied Voltage and is measured over the specified Theoretical Electrical Travel. The slope of the reference line, if limited, must be separately specified. An

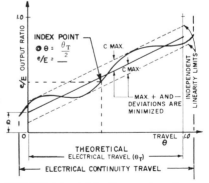

FIGURE 5.8.2 Independent linearity — nonwirewound

217

Index Point on the actual output is required. Unless otherwise specified, the Index Point will be at $\theta = \dfrac{\theta_T}{2}$.

MATHEMATICALLY: $\dfrac{e}{E} = P(\theta/\theta_T) + Q \pm C$

Where:

P is unspecified slope; Q is unspecified intercept at $\theta = 0$. And both are chosen to minimize C but are limited by the End Voltage requirements.

5.9 ■ TOLERANCE LIMITS

5.9.1 ■ CONSTANT LIMITS Permissible Conformity deviations specified as a percentage of the Total Applied Voltage.

Note: Unless otherwise specified, all definitions in this document employ Constant Limits.

5.9.1.1 ■ ZERO-TO-PEAK CONSTANT LIMITS Permissible Conformity deviations specified as a percentage of Zero-To-Peak Applied Voltage.

Note: The numerical value of zero-to-peak errors is double that of equal peak-to-peak errors, because the reference zero-to-peak applied voltage is one-half of the Total (peak-to-peak) Applied Voltage (see 2.1.1).

5.9.2 ■ PROPORTIONAL LIMITS Permissible Conformity deviations specified as a percentage of the theoretical Output Ratio at the point of measurement.

Note: Proportional limits may become impossibly restrictive in the vicinity of zero theoretical output and should be modified to provide a practical tolerance in that region, if the theoretical Output Ratio approaches zero.

5.9.3 ■ MODIFIED PROPORTIONAL LIMITS Any combination of Constant and Proportional Limits.

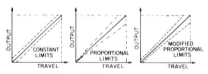

FIGURE 5.9.3 Tolerance limits

5.10 ■ SIMULTANEOUS CONFORMITY PHASING The relative alignment of the cups of a gang potentiometer, from a common index point, such that the Output Ratios of all cups fall within their respective Conformity limits over the Theoretical Electrical Travel.

5.11 ■ VOLTAGE TRACKING ERROR The difference, at any shaft position, between the Output Ratios of any two commonly actuated similar electrical elements, expressed as a percentage of the single total voltage applied to them.

6 general electrical characteristics

6.1 ☐ NOISE (WIREWOUND POTENTIOMETERS ONLY) Any spurious variation in the electrical output not present in the input, defined quantitatively in terms of an equivalent parasitic, transient resistance in ohms, appearing between the contact and the resistance element when the shaft is rotated or translated. The Equivalent Noise Resistance is defined in-

dependently of the resolution, the functional characteristics, and the total travel. The magnitude of the Equivalent Noise Resistance is the maximum departure from a specified reference line. The wiper of the potentiometer is required to be excited by a specified current and moved at a specified speed.

6.2 ■ OUTPUT SMOOTHNESS (NONWIREWOUND POTENTIOMETERS ONLY) Output Smoothness is a measurement of any spurious variation in the electrical output not present in the input. It is expressed as a percentage of the Total Applied Voltage and measured for specified travel increments over the Theoretical Electrical Travel. Output Smoothness includes effects of contact resistance variations, resolution, and other micro-nonlinearities in the output.

6.3 ■ RESOLUTION A measure of the sensitivity to which the Output Ratio of the potentiometer may be set.

6.4 ☐ THEORETICAL RESOLUTION (LINEAR WIREWOUND POTENTIOMETERS ONLY) The reciprocal of the number of turns of wire in resistance winding in the Actual Electrical Travel, expressed as a percentage.

N = Total number of resistance wire turns.

$\dfrac{1}{N}$ x 100 = Theoretical Resolution percent.

6.5 ☐ TRAVEL RESOLUTION (WIREWOUND POTENTIOMETERS ONLY) The maximum value of shaft travel in one direction per incremental voltage step in any specified portion of the resistance element.

6.6 ■ VOLTAGE RESOLUTION The maximum incremental change in Output Ratio with shaft travel in one direction in any specified portion of the resistance element.

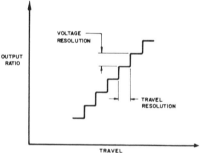

FIGURE 6.6 Wirewound resolution

Note: The illustration above is valid only for wirewound potentiometers because of the "stepped" nature of the output function. For determination of the effects of resolution in a nonwirewound potentiometer, refer to Output Smoothness (6.2).

6.7 ■ DIELECTRIC WITHSTANDING VOLTAGE Ability to withstand under prescribed conditions, a specified potential of a given characteristic between the terminals of each cup and the exposed conducting surfaces of the potentiometer, or between the terminals of each cup and the terminals of every other cup in the gang without exceeding a specified leakage current value.

6.8 ■ INSULATION RESISTANCE The resistance to a specified impressed DC voltage between the terminals of each cup and the exposed conducting surfaces of the potentiometer, or between the terminals of each cup and the terminals of every other cup in the gang, under prescribed conditions.

6.9 ■ POWER RATING The maximum power that a potentiometer can dissipate under specified conditions while meeting specified performance requirements.

6.9.1 ■ POWER DERATING The modification of the nominal power rating for various considerations such as Load Resistance, Output Slopes, Ganging, nonstandard environmental conditions and other factors.

6.10 ■ LIFE The number of shaft revolutions or translations obtainable under specific operating conditions and within specified allowable degradations of specific characteristics.

7 *ac characteristics*

7.1 ■ TOTAL INPUT IMPEDANCE The impedance between the two input terminals with open circuit between output terminals, and measured at a specified voltage and frequency with the shaft positioned to give a maximum value.

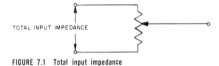

TOTAL INPUT IMPEDANCE

FIGURE 7.1 Total input impedance

7.2. ■ OUTPUT IMPEDANCE Maximum impedance between slider and either end terminal with the input shorted, and measured at a specified voltage and frequency.

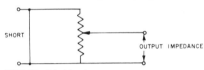

SHORT

OUTPUT IMPEDANCE

FIGURE 7.2 Output impedance

7.3 ■ QUADRATURE VOLTAGE The maximum value of that portion of the output voltage which is $\pm 90°$ out of time phase with the input voltage, expressed as volts per volt applied, measured at a specified input voltage and frequency.

7.4 ■ PHASE SHIFT The phase difference, expressed in degrees, between the sinusoidal input and output voltages measured at a specified input voltage and frequency with the shaft at a specified position.

MATHEMATICALLY: $\phi = \sin^{-1}(e_q/e) = \tan^{-1}(e_q/e_i)$

Where:

ϕ = phase shift in degrees
e_q = quadrature voltage
e_i = inphase output voltage
e = output voltage

8 *mechanical characteristics*

8.1 ■ SHAFT RUNOUT The eccentricity of the shaft diameter with respect to the rotational axis of the shaft, measured at a specified distance from the end of the shaft. The body of the potentiometer is held fixed and the shaft is rotated with a specified load applied radially to the shaft. The eccentricity is expressed in inches, TIR.

8.2 ■ LATERAL RUNOUT The perpendicularity of the mounting surface with respect to the rotational axis of the shaft, measured on the mounting surface at a specified distance from the outside edge of the mounting surface. The shaft is held fixed and the body of the potentiometer is rotated with specified loads applied radially and axially to the body of the pot. The Lateral Runout is expressed in inches, TIR.

8.3 ■ PILOT DIAMETER RUNOUT The eccentricity of the pilot diameter with respect to the rotational axis of the shaft, measured on the pilot diameter. The shaft is held fixed and the body of the potentiometer is rotated with a specified load applied radially to the body of the pot. The eccentricity is expressed in inches, TIR.

8.4 ■ SHAFT RADIAL PLAY The total radial excursion of the shaft, measured at a specified distance from the front surface of the unit. A specified radial load is applied alternately in opposite directions at a specified point. Shaft Radial Play is expressed in inches.

8.5 ■ SHAFT END PLAY The total axial excursion of the shaft, measured at the end of the shaft with a specified axial load supplied alternately in opposite directions. Shaft End Play is expressed in inches.

8.6 ■ STARTING TORQUE The maximum moment in the clockwise and counterclockwise directions required to initiate shaft rotation anywhere in the Total Mechanical Travel.

8.7 ■ RUNNING TORQUE The maximum moment in the clockwise and counterclockwise directions required to sustain uniform shaft rotation at a specified speed throughout the Total Mechanical Travel.

8.8 ■ MOMENT OF INERTIA The mass moment of inertia of the rotating elements of the potentiometer about their rotational axis.

8.9 ■ STOP STRENGTH

8.9.1 ■ STATIC STOP STRENGTH The maximum static load that can be applied to the shaft at each mechanical stop for a specified period of time without permanent change of the stop positions greater than specified.

8.9.2 ■ DYNAMIC STOP STRENGTH The inertia load, at a specified shaft velocity and a specified number of impacts, that can be applied to the shaft at each stop without a permanent change of the stop position greater than specified.

VARIABLE RESISTIVE COMPONENTS INSTITUTE

INDUSTRY STANDARD

WIREWOUND
AND NON-WIREWOUND
TRIMMING
POTENTIOMETERS

This standard, presented in two parts (VRCI-T-110 and VRCI-T-215) provides industry-approved terms, definitions and preferred inspection and test procedures for wirewound and non-wirewound trimming potentiometers.

The inherent construction difference between wirewound and non-wirewound resistance elements, with respect to both geometry and methods of termination, results in somewhat different output characteristics. Those same construction characteristics that provide for substantially infinite resolution in non-wirewound trimming potentiometers also eliminate distinct changes in output ratios at the ends of adjustment travel. However, the intended applications of trimming potentiometers are such that the location of these distinct points is unimportant. For this reason, the institute provided unified definitions and test procedures for wirewound and non-wirewound trimming potentiometers. This approach simplifies both the specification and testing of trimming potentiometers, inasmuch as the difference between wirewound and non-wirewound trimming potentiometers (except for resolution) are not relevant to intended trimming potentiometer applications.

In addition to the many technical papers on trimming potentiometer characteristics and performance that aided in the preparation of this document, many of the procedures, terms and definitions were adopted or modified from the military specifications: MIL-R 22097, MIL-R 27208, MIL-R 39015, and MIL-STD 202, and MIL-R-39035.

VARIABLE RESISTIVE COMPONENTS INSTITUTE

trimming potentiometer standard

VRCI-T-110
REVISION A
TERMS AND DEFINITIONS
APPROVED BY
VARIABLE RESISTIVE COMPONENTS INSTITUTE
February, 1974

INTRODUCTION

This document provides a standard for nomenclature for trimming potentiometers. It covers industry-approved terms and definitions for wirewound and non-wirewound trimming potentiometers.

The purpose of this standard is to establish improved communications between the manufacturer and the user. The VRCI Standards and Nomenclature Committee believes that this standardization will create a better mutual understanding of intended specifications and maintain a perspective on the intended application of the trimming potentiometer.

TABLE OF CONTENTS

1.1   LIST OF SYMBOLS

1.1.1 ELECTRICAL

E	–	TOTAL APPLIED VOLTAGE
e	–	OUTPUT VOLTAGE
e/E	–	OUTPUT RATIO
ENR	–	EQUIVALENT NOISE RESISTANCE
V_m	–	MINIMUM VOLTAGE
R_t	–	TOTAL RESISTANCE
R_e	–	END RESISTANCE
R_m	–	ABSOLUTE MINIMUM RESISTANCE
R_L	–	LOAD RESISTANCE
TC	–	TEMPERATURE COEFFICIENT OF RESISTANCE
RTC	–	RESISTANCE TEMPERATURE CHARACTERISTIC
CRV	–	CONTACT RESISTANCE VARIATION
CT	–	CENTER TAP
SL	–	WIPER (SLIDER)
P	–	POWER HANDLING CAPABILITY IN WATTS

1.1.2 MECHANICAL

CW	–	CLOCKWISE ROTATION
CCW	–	COUNTERCLOCKWISE ROTATION
ST	–	SINGLE TURN TRIMMING POTENTIOMETER
MT	–	MULTITURN, SCREW ACTUATED TRIMMING POTENTIOMETER
SS	–	STOPS, SOLID
C	–	CONTINUOUS ROTATION
SC	–	STOPS, CLUTCH ACTION
L	–	INSULATED WIRE LEAD TERMINALS
P	–	PIN TERMINALS, FLAT BASE MOUNT NORMALLY FOR PRINTED CIRCUIT APPLICATION.
W	–	EDGE MOUNTED TERMINALS, ADJUSTMENT SHAFT 180° FROM TERMINALS.
X	–	EDGE MOUNTED TERMINALS, ADJUSTMENT SHAFT 90° FROM TERMIANLS.
S	–	SOLDER LUG

1.2 GENERAL TERMS

1.2.1 TRIMMING POTENTIOMETER: An electrical mechanical device with three terminals. Two terminals are connected to the ends of a resistive element and one terminal is connected to a movable conductive contact which slides over the element, thus allowing the input voltage to be divided as a function of the mechanical input. It can function as either a voltage divider or rheostat.

1.2.1.1 WIREWOUND TRIMMING POTENTIOMETER: A trimming potentiometer characterized by a resistance element made up of turns of wire on which the wiper contacts only a small portion of each turn.

1.2.1.2 NON-WIREWOUND TRIMMING POTENTIOMETER: A trimming potentiometer characterized by the continuous nature of the surface area of the resistance element to be contacted. Contact is maintained over a continuous, unbroken path. The resistance is achieved by using material compositions other than wire such as carbon, conductive plastic, metal film and cermet.

1.2.2 RESISTANCE ELEMENT: A continuous, unbroken length of resistive material without joints, bonds or welds except at the junction of the element and the electrical terminals connected to each end of the element or at an intermediate point such as a center tap.

1.2.3 ADJUSTMENT SHAFT: The mechanical input member of a trimming potentiometer which when actuated causes the wiper to traverse the resistance element resulting in a change in output voltage or resistance.

1.2.3.1 SINGLE TURN ADJUSTMENT: Requires 360° or less mechanical input to cause the wiper to traverse the total resistance element.

1.2.3.2 MULTITURN ADJUSTMENT: Requires more than 360° mechanical adjustment to cause the wiper to traverse the total resistance element.

1.2.4 TERMINAL: An external member that provides electrical access to the resistance element and wiper.

1.2.4.1 LEADWIRE TYPE: (L) Flexible insulated conductor.

1.2.4.2 PRINTED CIRCUIT TERMINAL: (P, W&X) Rigid uninsulated electrical conductor so arranged, suitable for printed circuit board plug-in.

1.2.4.3 SOLDER LUG TERMINAL: (S) Rigid uninsulated electrical conductor so arranged, suitable for external lead attachment.

1.2.5 WIPER: (SL) The wiper is the member in contact with the resistive element that allows the output to be varied with the mechanical member adjustment.

1.2.6 STOP-CLUTCH: (SC) A device which allows the wiper to idle at the ends of the resistive element without damage as the adjustment shaft continues to be actuated in the same direction.

1.2.7 STOP – SOLID: (SS) A positive limit to mechanical and/or electrical adjustment.

1.2.8 STACKING: The mounting of one trimming potentiometer adjacent to or on top of another utilizing the same mounting hardware.

1.2.9 THEORETICAL RESOLUTION: (Wirewound only) The theoretical measurement of sensitivity to which the output ratio may be adjusted and is the reciprocal of the number of turns of wire in resistance winding expressed as a percentage.

N = Total number of resistance wire turns.

$\frac{1}{N}$ X 100 = Theoretical resolution percent.

2 INPUT AND OUTPUT TERMS

2.1 INPUT TERMS

2.1.1 TOTAL APPLIED VOLTAGE: (E) The total voltage applied between the designated input terminals.

2.2 OUTPUT TERMS

2.2.1 OUTPUT VOLTAGE: (e) The voltage between the wiper terminal and the designated reference point. Unless otherwise specified, the designated reference point is the CCW terminal (See 3.1).

2.2.2 OUTPUT RATIO (OUTPUT VOLTAGE RATIO): (e/E) The ratio of the output voltage to the designated input reference voltage. Unless otherwise specified, the reference voltage is the total applied voltage.

2.3 LOAD TERMS

2.3.1 LOAD RESISTANCE: (R_L) An external resistance as seen by the Output Voltage (connected between the wiper terminal and the designated reference point).

---

3 ROTATION AND TRANSLATION

3.1 DIRECTION OF TRAVEL: Clockwise (CW) or counterclockwise (CCW) rotation when viewing the adjustment shaft end of the potentiometer. The designation of terminals in the figure corresponds to the direction of wiper travel.

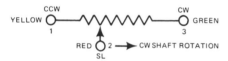

Figure 3.1 Circuit and Travel Diagram

3.2 MECHANICAL TRAVEL:

3.2.1 MECHANICAL TRAVEL—SOLID STOPS: (SS) The total travel of the adjustment shaft between integral stops. Continuity must be maintained throughout the travel.

3.2.2 MECHANICAL TRAVEL—CLUTCHING ACTION: (SC) The total travel of the adjustment shaft between the points where clutch actuation begins. Continuity must be maintained throughout the travel and during clutch actuation.

3.2.3 MECHANICAL TRAVEL—CONTINUOUS ROTATION: (C) The total travel of the adjustment shaft when the wiper movement is unrestricted at either end of the resistive element as the adjustment shaft continues to be actuated.

3.3 ADJUSTMENT TRAVEL (ELECTRICAL): The total travel of the adjustment shaft between minimum and maximum output voltages.

3.4 CONTINUITY TRAVEL: The total travel of the shaft over which electrical continuity is maintained between the wiper and the resistance element.

---

4 GENERAL ELECTRICAL CHARACTERISTICS

4.1 TOTAL RESISTANCE: (R_t) The dc resistance between the input terminals with the wiper positioned to either end stop, or in dead band for continuous rotation potentiometers.

4.2 ABSOLUTE MINIMUM RESISTANCE: (R_m) The resistance measured between the wiper terminal and each end terminal with the wiper positioned to give a minimum value.

4.3 END RESISTANCE: (R_e) The resistance measured between the wiper terminal and an end terminal when the wiper is positioned at the corresponding end of mechanical travel. Absolute minimum resistance and end resistance are synonymous for continuous rotation trimmers.

4.4 TEMPERATURE COEFFICIENT OF RESISTANCE: (TC) The unit change in resistance per degree celsius change from a reference temperature, expressed in parts per million per degree celsius as follows:

$$TC = \frac{R_2 - R_1}{R_1 (T_2 - T_1)} \times 10^6$$

Where:

R_1 = Resistance at reference temperature in ohms.
R_2 = Resistance at test temperature in ohms.
T_1 = Reference temperature in degrees celsius.
T_2 = Test temperature in degrees celsius.

4.5 RESISTANCE-TEMPERATURE CHARACTERISTIC: (RTC) The difference between the total resistance values measured at a reference temperature of $25°C$ and the specified test temperature expressed as a percent of the Total Resistance.

$$RTC = \frac{R_2 - R_1}{R_1} \times 100$$

Where:

R_1 = Resistance at reference temperature ($25°C$) in ohms.
R_2 = Resistance at the test temperature in ohms.

4.6 CONTACT RESISTANCE VARIATION: (CRV) The apparent resistance seen between the wiper and the resistance element when the wiper is energized with a specified current and moved over the adjustment travel in either direction at a constant speed. The output variations are measured over a specified frequency bandwidth, exclusive of the effects due to roll-on or roll-off of the terminations and is expressed in ohms or % of R_t.

4.7 EQUIVALENT NOISE RESISTANCE: (ENR) Wirewound only. Any spurious variation in the electrical output not present in the input, defined quantitatively in terms of an equivalent parasitic, transient resistance in ohms, appearing between the contact and the resistive element when the shaft is rotated or translated. The Equivalent Noise Resistance is defined independently of the resolution, functional characteristics and the total travel. The magnitude of the Equivalent Noise Resistance is the maximum departure from a specific reference line. The wiper of the potentiometer is required to be excited by a specific current and moved at a specific speed.

4.8 CONTINUITY: Continuity is the maintenance of continuous electrical contact between the wiper and both end terminals of the resistive element.

4.9 SETTING STABILITY: The amount of change in the output voltage, without readjustment, expressed as a percentage of the total applied voltage.

4.10 DIELECTRIC STRENGTH: The ability to withstand the application of a specified potential of a given characteristic, between the terminals and all other external conducting member such as shaft, housing and mounting hardware without exceeding a specified leakage current value.

4.11 INSULATION RESISTANCE: The resistance to a specified dc voltage impressed between the terminals and all other external conducting members such as shaft, housing and mounting hardware.

4.12 POWER RATING: The maximum power that a trimming potentiometer can dissipate across the total resistive element under specified conditions while meeting specified performance requirements.

4.13 LIFE:

4.13.1 ROTATIONAL LIFE: The number of cycles obtainable under specific operating conditions while remaining within specified allowable degradation. A cycle is defined as one complete traversal of the wiper over the resistive element in both directions.

4.13.2 LOAD LIFE: The number of hours at which a device may dissipate rated power under specified operating conditions while remaining within specified allowable degradations.

4.14 ADJUSTABILITY: Defines the precision with which the output of a device can be set to the desired value.

4.14.1 ADJUSTABILITY (OUTPUT RESISTANCE): The precision with which the output resistance of a device can be set to the desired value.

4.14.2 ADJUSTABILITY (OUTPUT VOLTAGE RATIO): The precision with which the output voltage ratio of a device can be set to the desired value.

---

5 GENERAL MECHANICAL CHARACTERISTICS

5.1 TORQUE

5.1.1 STARTING (OPERATING) TORQUE: The maximum moment in the clockwise and counterclockwise directions required to initiate shaft adjustment anywhere in the mechanical travel.

5.1.2 STOP TORQUE: The maximum static moment that can be applied to adjustment shaft at each mechanical stop for a specified period of time without loss of continuity or mechanical damage affecting operational characteristics.

5.2 SOLDERABILITY: The ability of the terminals to accept a uniform coating of solder under specified conditions.

5.3 WELDABILITY: The ability of materials to be welded together under specified conditions.

5.4 TERMINAL STRENGTH: The ability of the terminals to withstand specified mechanical stresses without sustaining damage that would affect utility of the terminals or operation of the trimming potentiometer.

5.5 IMMERSION SEALED: The ability of the unit to withstand submersion in acceptable cleaning solutions used in normal soldering processes without performance degradation under specific environmental conditions.

vrci-p-200a inspection and test procedures

Test procedures reported in this revised standard cover all characteristics that can be measured without seriously affecting the remaining life in a potentiometer. These procedures specifically exclude environmental exposure tests, life tests and power ratings. The purpose of VRCI-P-200 is to assist you in obtaining better correlation of inspection results between the users facility and the manufacturer's plant. Although the VRCI realizes that there are alternate methods to those proposed here, it believes that standardization can better achieve a common basis for evaluation. If you prefer to use an alternate procedure, you should realize the responsibility associated with such substitution.

The test procedures are to be performed at, or corrected to standard conditions as follows, unless otherwise specified: 25° Celsius (formerly Centigrade), 760 mm of HG and 50% relative humidity. Where measurements are not corrected to these standard conditions, the burden of proof of equivalency lies with the individual tester.

Many of these procedures call for specifying important test parameters with numerical values; for example, shaft loads, operating speeds, applied voltages and currents, etc. Various factors governed their selection — standardization, availability of equipment, ease of test and, above all, the prevention of damage to the potentiometer. It should be noted, however, that these are only recommended typical values and are subject to modification in individual cases depending upon specific requirements of end use.

TABLE OF CONTENTS

8	Mechanical Characteristics
8.1	Shaft Runout
8.2	Lateral Runout
8.3	Pilot Diameter Runout
8.4	Shaft Radial Play
8.5	Shaft End Play
8.6	Starting Torque
8.7	Running Torque
8.8	Moment of Inertia
8.9	Stop Strength

LEGEND:

☐ WIREWOUND
■ WIREWOUND AND NONWIREWOUND
■ NONWIREWOUND

1 *equipment description*

1.1 ■ MECHANICAL EQUIPMENT

1.1.1 ■ POTENTIOMETER MOUNTING FIXTURE A fixture to rigidly hold the test specimen by the normal mounting means leaving the shaft free to move. When utilized for Dynamic Stop Strength measurements, the mounting should be of massive construction to prevent absorption of energy during impact.

1.1.2 ■ SHAFT POSITIONING DEVICE A device to provide a means for moving the shaft to any position relative to the potentiometer body and maintaining a stable setting during electrical measurements.

The shaft positioning device must not apply any axial or radial loads on the shaft of a rotary potentiometer.

1.1.3 ■ TRAVEL MEASURING DEVICE A device composed of a shaft position indicator and a potentiometer mounting fixture, which will precisely indicate shaft position relative to the potentiometer body. The device has an over-all accuracy of at least 0.1°, or such other value required to make total measuring errors for any test less than 1/10th the specified tolerance. THE TRAVEL MEASURING DEVICE MOUNTING PROVISION SHOULD IN NO WAY DISTORT OR MAR THE FINISH OF THE PART NOR EXERT ANY RADIAL OR AXIAL LOADS.

1.1.4 ■ LOAD DEVICE A device that provides a means for applying a force or torque of known magnitude to a pot shaft. The device, such as a spring scale, weight at a known radius, torque wrench, etc., should have an accuracy of 1/10th the specified value.

1.1.5 ■ DIAL INDICATOR The minimum dial division must be equal to or less than 1/10th the specified tolerance and must have a readability to 0.00005 inches for measurements below 0.001 inches. The indicator should not be used over a range of more than 1/3 of the total travel of the probe without error correction. If a dial indicator is used with a pivot-type pointer, the longitudinal C/L of the stylus must remain normal to the workpiece during measurement or the applicable correction factor must be used.

1.1.6 ■ DIAL INDICATOR HOLDING FIXTURE The dial indicator holding fixture must hold the dial indicator rigidly, maintaining its proper attitude to the workpiece during measurement. It should provide for fine adustment of the indicator position such that the dial indicator is not damaged as the probe approaches the workpiece.

1.1.7 ■ POTENTIOMETER SHAFT HOLDING FIXTURE This fixture must hold the test specimen by the shaft in either a horizontal or vertical position leaving the potentiometer body free to move.

1.1.8 ■ CYLINDRICAL SHAFT ADAPTOR An adaptor with a smooth cylindrical surface and which, when mounted on the pot shaft, adds eccentricity no greater than 1/10th the specified runout tolerance.

1.1.9 ■ DEAD WEIGHT LOAD The load may be applied using a spring scale, weights or equivalent, accurate to 1/10th the specified value.

1.1.10 ■ SHAFT LOAD ADAPTOR When it is necessary to transmit loads larger than can be accommodated with frictional clamping devices and without relative movement of more than 1/10th the specified allowable change in stop position, a suitable permanent attachment to the shaft, such as a pinned bushing, is used.

1.1.11 ■ INERTIA LOAD An inertia load of known magnitude and capable of being driven in opposite directions at a specified constant velocity should be used. Input energy should be removed rapidly so that the inertia load is the only source of energy during impact. The inertia load is the sum total of Moments of Inertia of all moving components attached to it. The potentiometer shaft and ALL COUPLINGS and INERTIA LOAD MOUNTINGS MUST BE OF HEAVY DUTY CONSTRUCTION TO MINIMIZE LOSS OF ENERGY DURING IMPACT, THEREBY TRANSMITTING THE FULL INERTIA LOAD TO THE SHAFT.

1.1.12 ■ CONSTANT SPEED DRIVE The drive must be able to operate at a constant velocity of 4 ± 1 RPM or $18 \pm 3''/$ minute. For Output Smoothness measurements, the short term speed variation must not exceed 1/10th the specified Output Smoothness. The device should have a slip clutch provision to prevent damage to the mechanical stops in the unit, where applicable.

1.1.13 ■ MOMENT OF INERTIA ADAPTOR An adaptor of known inertia to couple the suspended steel wire to the rotating elements of the potentiometer. It must attach to each at the centerline of its mass.

1.1.14 ■ MASS OF KNOWN INERTIA A mass calculated to have a known moment of inertia, similar in dimension and weight to the unknown inertia to be evaluated.

1.1.15 ■ TIMING CLOCK A timing clock with readability and repeatability to 0.1 second.

1.1.16 ■ TEMPERATURE TEST CHAMBER The temperature of the chamber should be adjustable within $\pm 3°C$ of the test temperature. The chamber should be stable within $\pm 0.5°C$ at any given point in the proximity of the test specimens. The temperature gradient in this area should not exceed $\pm 1°C$. If larger gradients exist, the temperature must be monitored with a thermocouple immediately adjacent to the test specimen. Air flow around the area of the test specimens should be at least 60 CFM.

1.1.17 ■ VARIABLE SPEED DRIVE The drive must be capable of variability from 1/4 to 20 RPM, or mean speeds of 1.0 to 80 inches/minute. Provisions should be made, where applicable, for a slip clutch to prevent damage to the potentiometer mechanical stops.

1.2 ■ ELECTRICAL EQUIPMENT

1.2.1 ■ VOLTAGE RATIO EQUIPMENT The Kelvin-Varley Voltage Divider or a modification of it is recommended for measurement of the output voltage ratio of precision potentiometers. The voltage dividers for precision potentiometers are usually of two types.

(1) Decade Voltage Dividers (4 or 5 decades)

(2) Digital Ratiometers (4 or 5 places)

The Decade Voltage Divider is used in conjunction with a null detector described in Paragraph 1.2.19. The Digital Ratiometers are generally self-nulling with direct numerical readouts. The equipment accuracy, resolution and repeatability must

LEGEND:

☐ WIREWOUND
■ WIREWOUND AND NONWIREWOUND
■ NONWIREWOUND

equal or be less than 1/10th of the specified tolerance. The voltage applied should never exceed the voltage/power rating of the test unit.

1.2.2 ■ RESISTANCE MEASURING DEVICE CARE MUST BE TAKEN WHEN USING ANY RESISTANCE MEASURING DEVICE THAT THE CURRENT DRAWN DOES NOT EXCEED THE CURRENT CARRYING CAPACITY OR RATING OF THE UNIT.

1.2.2.1 ■ OHMMETER (HAND SET VOLT-OHMMETER) This type ohmmeter generally is not sufficiently accurate for quantitative measurements and, when used on its lowest scale, applies a voltage without internal impedance limits. Its use should be avoided.

1.2.2.2 ■ WHEATSTONE BRIDGE When resistance tolerances are less than 10% and the resistance values are above 10 ohms, use a Wheatstone bridge with an accuracy of 1/10 the tolerance to be measured. For resistance values above 1 megohm it is necessary to use a guarded Wheatstone bridge. (Commercially available Wheatstone bridges have an accuracy of 0.01% to 10 megohms and 0.5% to 1000 megohms.) A null detector recommended for use in the bridge circuit is described in Paragraph 1.2.19.

1.2.2.3 ■ KELVIN BRIDGE For resistance values less than 10 ohms a Kelvin (Thomson) bridge is used. The accuracy of the bridge required is 1/10 the specified tolerance to be measured. (Commercially available Kelvin bridges have an accuracy of 0.25% from 0.0005 ohms and 0.5% from 0.0001 to 0.0005 ohms.)

A null detector recommended for use in the bridge circuit is described in Paragraph 1.2.19.

1.2.2.4 ■ DIGITAL OHMMETER Digital ohmmeters are self balancing Wheatstone bridges and can be used in place of Wheatstone or Kelvin bridges.

1.2.3 ■ POWER SUPPLIES

Note: A DC voltmeter should be used with all power supplies to determine actual voltage to permit proper evaluation of traces in percent total applied voltage.

1.2.3.1 ■ FOR OUTPUT RATIO MEASUREMENT 10 ±3 volts DC with no limitations on voltage stability, current regulation or line regulation. If its capacity is sufficient for the current drawn by potentiometer, a battery may be used.

1.2.3.2 ■ BALANCED POWER SUPPLY FOR OUTPUT RATIO MEASUREMENT 10 ±3 volts each side of center tap. Halves balance ±0.01%; halves balance stability ±0.01% per hour. No requirements for voltage or current regulation, line regulation or total voltage, as long as balance is maintained.

1.2.3.3 ■ POWER SUPPLY FOR ERROR TRACE OR OUTPUT SMOOTHNESS RECORDING Same as 1.2.3.1 or 1.2.3.2 with additional requirement for voltage ripple not to exceed 1/10 value of Conformity or Output Smoothness limit.

1.2.4 ■ RECORDER A continuous paper chart recorder with a flat frequency response within 3 db from DC to a minimum of 100 Hz, and a maximum drift less than 1%. The recorder is combined with a high gain DC amplifier with sufficient power output to drive the chart recorder and, if desired, a null detector. The amplifier zero line stability should be better than 1% per hour.

1.2.5 ■ CONSTANT CURRENT SOURCE The source should produce a 1.0 ±5% milliamperes DC current under load. A

suggested circuit for a constant current generator is shown in Figure 1.2.5A.

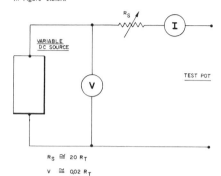

$R_S \cong 20\ R_T$

$V \cong 0.02\ R_T$

WHERE R_T = TEST POT TOTAL RESISTANCE

FIGURE 1.2.5A Constant current source

1.2.6 ■ OSCILLOSCOPE A high gain DC oscilloscope should have a high persistence screen to retain the image for a minimum of 1/2 second. It should have a minimum input impedance of one megohm and a flat frequency response from DC to a minimum of 50 KHz.

1.2.7 ■ LOW PASS FILTER The filter has the following characteristics:

Band Pass Frequency:	0-1000 Hz
Insertion Loss:	2 db
Attenuation Outside Band Pass:	6 db/octave
Input Impedance:	2 Megohms
Output Impedance:	2 Megohms
Maximum Current:	5 Milliamperes

A suggested filter circuit is presented in Figure 1.2.7A.

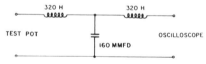

FIGURE 1.2.7A Low pass filter

1.2.8 ■ OUTPUT SMOOTHNESS FILTER The standard filter below is basically designed to measure output discontinuities (sudden output changes) occurring over 0.5° travel or less at 4 r.p.m. The low pass time constant of this filter is 20 milliseconds. At 4 r.p.m., this value corresponds to a shaft travel of 0.5°.

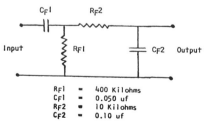

R_{F1}	=	400 Kilohms
C_{F1}	=	0.050 uf
R_{F2}	=	10 Kilohms
C_{F2}	=	0.10 uf

FIGURE 1.2.8A Output smoothness filter

1.2.9 ■ VARIABLE RESISTOR A variable resistor with a range to 0.5 megohms in minimum steps of 1 ohm and a minimum current rating of 10 ma.

1.2.10 ■ HIGH VOLTAGE SOURCE A source variable from zero to the maximum specified VRMS @ 60 Hz of sine waveform with 5% maximum distortion. A current limiter to limit the leakage current to 150% of the maximum specified value should be provided.

1.2.11 ■ AC VOLTMETER The AC Voltmeter should have the following characteristics:

Voltage Range: 133% of the specified or
                                  test value minimum.

Approximate Impedance
in Ohms at 60 Hz   1000 ohms/volt
Frequency Range: 50-125 Hz minimum.
Accuracy: ±5% at the specified
                                  voltage and frequency.
Temperature Stability: 0.1%/°C maximum at the
                                  specified or test voltage
                                  and frequency.

1.2.12 ■ LEAKAGE CURRENT INDICATING DEVICE An AC ammeter with an accuracy of 5% of the allowable leakage rate and covering the applicable range.

1.2.13 ■ AC VOLTAGE SOURCE An AC source, variable in voltage magnitude and frequency to accommodate the specified values and having an isolation transformer of very low capacitance at its output.

1.2.14 ■ RATIO TRANSFORMER A variable device with an accuracy and resolution of 1/10 the value to be measured.

1.2.15 ■ AC VACUUM TUBE VOLTMETER The AC Vacuum Tube Voltmeter should have the following characteristics:

Voltage Range: 133% of specified or test
                                  value minimum.
Frequency Response: Flat to within 1 db over
                                  the frequency range.
Frequency Range: 10 - 250K Hz
Accuracy: 2% over the voltage range
                                  within the frequency range.
Input Impedance: 1 megohm minimum

1.2.16 ■ AC AMMETER The AC Ammeter should have the following characteristics:

Current Range: 133% of specified or test value.
Accuracy: 1% of full scale value
Frequency Range: 25 - 125 Hz minimum.
Temperature Stability: 0.1%/°C maximum at the
                                  specified or test current
                                  and frequency.

1.2.17 ■ AC MICROAMMETER The AC Microammeter should have the following characteristics:

Current Range: 133% of the specified or
                                  test value minimum.
Accuracy: 2.5% of full scale value
Frequency Range: 25 - 125 Hz minimum
Temperature Stability: 0.5%/°C maximum at the
                                  specified or test current
                                  and frequency.

1.2.18 ■ CONFORMITY TESTER The conformity tester shown in Figure 1.2.18A is comprised of a precision master potentiometer housed in a travel measuring device that mechanically locks the master potentiometer shaft with a test potentiometer shaft for simultaneous movement. The coupling must not introduce misalignment errors between the two shafts.

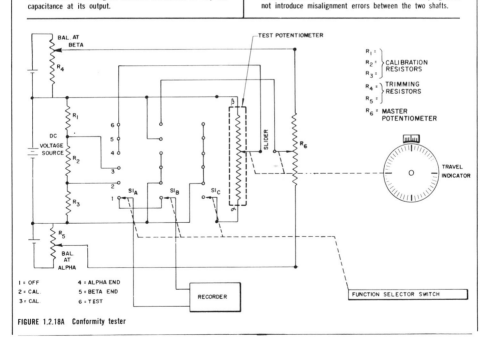

FIGURE 1.2.18A Conformity tester

The master potentiometer must be aligned with respect to the travel measuring device such that the End Points of the master potentiometer correspond as nearly as possible to the mechanical end point indications of the travel measuring device.

A constant speed drive motor should be incorporated into its design so as to cause a constant shaft velocity during a conformity test. For rotary potentiometers a nominal speed of 1 RPM in a CW direction is recommended. For translatory potentiometers a speed of 10″/min. in an extending direction is recommended.

The total equipment error from the combined effects of master potentiometer conformity error and shaft coupling errors must not exceed 1/10 of the minimum conformity tolerance to be tested.

1.2.19 ■ NULL DETECTOR The null detector should have the following characteristics:

Sensitivity:	
Current	1×10^{-9} amp/mm
Voltage	1×10^{-6} volts/mm
Stability:	1 mm drift per hour
Damping:	Critical
Response:	To final value within 1 second.

1.2.20 ■ THERMOCOUPLE BRIDGE The most common form of thermocouple bridge consists of a highly accurate potentiometer used in conjunction with an appropriate thermocouple. The accuracy of the measurement equipment should be 0.5% and temperature changes of 0.5°C should be detectable.

1.2.21 ■ INSULATION RESISTANCE TEST SET A suitable commercial megohm bridge, megohmmeter or equivalent with an accuracy of 1/10 the value to be measured and a built-in source voltage of the specified magnitude.

2   input-output measurements

2.1 □ ■ END VOLTAGE

2.1.1 □ END VOLTAGE — WIREWOUND

2.1.1.1 □ OBJECT To measure the voltage between the wiper terminal and an end terminal when the shaft is positioned at the corresponding End Point. End Voltage is expressed as a percentage of the Total Applied Voltage.

2.1.1.2 □ EQUIPMENT

Travel measuring device	1.1.3
Voltage ratio equipment	1.2.1

2.1.1.3 □ TEST PROCEDURE Mount the potentiometer to the travel measuring device. Connect the zero potential reference lead of the voltage ratio equipment with the appropriate end terminal of the potentiometer. Locate and position the shaft at the End Point (3.4). The Output Ratio measured at this position, expressed as a percentage, is the End Voltage corresponding to that end terminal.

2.1.2 ■ END VOLTAGE — NONWIREWOUND

2.1.2.1 ■ OBJECT To measure the voltage between the wiper terminal and an end terminal when the shaft is positioned at the corresponding Theoretical End Point. End Voltage is expressed as a percentage of the Total Applied Voltage.

2.1.2.2 ■ EQUIPMENT

Travel measuring device	1.1.3
Voltage ratio equipment	1.2.1

2.1.2.3 ■ TEST PROCEDURE Mount the potentiometer to the travel measuring device and phase the shaft to the Index Point (3.6). Connect the voltage ratio equipment to the appropriate potentiometer terminals such that the zero potential reference lead is connected to the end terminal concerned. Locate and position the shaft at the Theoretical End Point (3.5). The Output Ratio measured at this position, expressed as a percentage, is the End Voltage corresponding to that end terminal.

2.2 ■ MINIMUM VOLTAGE

2.2.1 ■ OBJECT To measure the voltage between the wiper terminal and an end terminal when the shaft is positioned near the corresponding end of the Electrical Continuity Travel such that a minimum voltage reading is obtained. Minimum Voltage is usually expressed as a percentage of the Total Applied Voltage.

2.2.2 ■ EQUIPMENT

Travel measuring device	1.1.3
Voltage ratio equipment	1.2.1

2.2.3 ■ TEST PROCEDURE Mount the potentiometer to the travel measuring device and connect the voltage ratio equipment to the appropriate potentiometer terminals such that the zero potential reference lead is connected to the end terminal concerned. The travel measuring device provides fine control of the shaft position. Move the shaft until a minimum reading is indicated on the voltage ratio equipment. This reading is expressed as a percentage and is the Minimum Voltage.

2.3 □ JUMP-OFF VOLTAGE (APPLICABLE TO WIREWOUND ONLY)

2.3.1 □ OBJECT To measure the magnitude of the first measurable voltage change as the wiper is displaced from the overtravel region onto the Actual Electrical Travel region. Jump-Off Voltage is expressed as a percentage of the Total Applied Voltage.

2.3.2 □ EQUIPMENT

Travel measuring device	1.1.3
Voltage ratio equipment	1.2.1

2.3.3 □ TEST PROCEDURE Mount the potentiometer to the travel measuring device. Connect the voltage ratio equipment to the appropriate terminals such that the zero potential reference lead is connected to the end terminal concerned. Locate and position the shaft at the End Point (3.4). Record the Output Ratio. Move the shaft toward the Actual Electrical Travel (3.7) to the position of the first incremental change in output ratio $\geq$ 20% of Voltage Resolution in that region and record that Output Ratio. The difference between the two readings, expressed as a percentage of Total Applied Voltage, is the Jump-Off Voltage corresponding to that End Point.

2.4 ■ SHORTED SEGMENT

2.4.1 ■ OBJECT To determine that the Output Ratio remains constant within specified limits as the wiper traverses the shorted segment with a specified load resistance. Since the Output Ratio may remain within the specified limits of the shorted segment for a segment longer than specified, it is the purpose of this procedure to assure that the Output Ratio is within limits for at least the minimum shorted segment length specified.

2.4.2 ■ EQUIPMENT

Travel measuring device	1.1.3
Voltage ratio equipment	1.2.1

See applicable sections for locating a reference point.

2.4.3 ■ TEST PROCEDURE Mount the potentiometer to the travel measuring device by normal means. Locate the reference point for the Shorted Segment in accordance with the applicable sections of this document. Connect the specified load resistance and the voltage ratio equipment to the potentiometer. If the reference point is not coincident with the beginning of the short, move the shaft in the direction of the short (be certain Backlash has been removed) the minimum specified distance from the reference point. If the Output Ratio is not within the specified limits, continue shaft movement to the first point at which the Output Ratio is within limits and note the shaft position. Continuing in the same direction, displace the shaft the minimum specified length of the Shorted Segment or until the Output Ratio exceeds the specified limits, whichever occurs first. Compare the results with the specification as to both location of the short with respect to the reference point and the length of the Shorted Segment.

3   rotation and translation

3.1 ■ TOTAL MECHANICAL TRAVEL

3.1.1 ■ OBJECT To measure the total travel of the shaft between integral stops, under a specified stop load. In potentiometers without stops, the mechanical travel is continuous.

3.1.2 ■ EQUIPMENT
Travel measuring device	1.1.3
Load device	1.1.4

3.1.3 ■ TEST PROCEDURE Mount the potentiometer to the travel measuring device and move the shaft to each stop, applying a stop load of 150% of the maximum specified actuating force or starting torque against each stop. Note reading of the travel measuring device at each stop with load applied. The difference between the two readings is the Total Mechanical Travel.

Note: In no case should the applied load exceed 75% of the Static Stop Strength.

3.2 □ ■ MECHANICAL OVERTRAVEL

3.2.1 □ MECHANICAL OVERTRAVEL — WIREWOUND

3.2.1.1 □ OBJECT To measure the shaft travel between the limits of the Actual Electrical Travel and the corresponding limits of Total Mechanical Travel. Theoretical Electrical Travel is substituted for Actual Electrical Travel when Absolute Linearity or Absolute Conformity is specified.

3.2.1.2 □ EQUIPMENT
Travel measuring device	1.1.3
Load device	1.1.4
Voltage ratio equipment	1.2.1

3.2.1.3 □ TEST PROCEDURE Mount the potentiometer to the travel measuring device. Locate the limits of Actual Electrical Travel, which are the End Points, (3.4) and the limits of Total Mechanical Travel (3.1) and note the readings on the travel measuring device at these points. The Mechanical Overtravel at each end is the difference between the readings at the End Point and the corresponding limit of Total Mechanical Travel. When Absolute Linearity or Absolute Conformity is specified, the Theoretical End Points, as determined from the Index Point (3.6), are used in place of End Points.

3.2.2 ■ MECHANICAL OVERTRAVEL — NONWIREWOUND

3.2.2.1 ■ OBJECT To measure the shaft travel between the limits of the Theoretical Electrical Travel and the corresponding limits of Total Mechanical Travel.

3.2.2.2 ■ EQUIPMENT
Travel measuring device	1.1.3
Load device	1.1.4
Voltage ratio equipment	1.2.1

3.2.2.3 ■ TEST PROCEDURE Mount the potentiometer to the travel measuring device. Locate the Theoretical End Points (3.5), and the limits of Total Mechanical Travel (3.1) and note the readings on the travel measuring device at these points. The Mechanical Overtravel at each end is the difference between the readings at the Theoretical End Points and the corresponding limits of Total Mechanical Travel.

3.3 ■ BACKLASH

3.3.1 ■ OBJECT To measure the maximum difference in shaft position that occurs when the shaft is moved to the same actual output ratio point from opposite directions.

3.3.2 ■ EQUIPMENT
Travel measuring device	1.1.3
Voltage ratio equipment	1.2.1

3.3.3 ■ TEST PROCEDURE The potentiometer is mounted in the travel measuring device, connected to the voltage ratio equipment and the operating shaft is displaced to approximately the 40 percent voltage point (not in a tap or shorted area). The voltage ratio equipment is then adjusted to obtain zero indication on the null indicator. The shaft is moved approximately 1/4 revolution (or 10%) in the direction of decreasing output voltage and then in a reverse direction until the null detector first approaches a zero reading. At this point a travel reading is taken. The shaft is then moved to approximately the 80 percent voltage point and returned in the opposite direction until the null detector passes through zero and shows its first perceptible change from zero. At this point the position of the shaft is again noted. The difference between the two shaft position readings is the Backlash.

3.4 □ END POINT (APPLICABLE TO WIREWOUND ONLY)

3.4.1 □ OBJECT To determine the location of the End Points which are the shaft positions immediately before the first and after the last measurable change(s) in Output Ratio, after wiper continuity has been established, as the shaft moves in the specified direction.

3.4.2 □ EQUIPMENT
Travel measuring device	1.1.3
Voltage ratio equipment	1.2.1
Load device (when applicable)	1.1.4

3.4.3 □ TEST PROCEDURE Mount the potentiometer to the travel measuring device and connect the output ratio equipment to the appropriate terminals. Move the potentiometer shaft in the direction opposite "specified direction" until the voltage ratio equipment indicates the wiper is in the overtravel region by an amount $>>$ Backlash. The specified direction is CW for rotary potentiometers and extending for translatories. Move the shaft in the specified direction until the voltage ratio equipment indicates the first single discrete change in actual output ratio $\geq$ 20% of Voltage Resolution in that region, after wiper continuity has been established. The shaft position immediately preceding this change is the first End Point. Continue the shaft motion in the same direction to the point where the last discrete change in output ratio $\geq$ 20% of Voltage Resolution occurs. The shaft position immediately after this change is the other End Point. In the case where normal resolution steps are observed to the limits of Total Mechanical Travel (3.1) the End Points are the limits of Total Mechanical Travel.

3.5 ■ THEORETICAL END POINT

3.5.1 ■ OBJECT To determine the location of the Theoretical End Points which are the shaft positions corresponding to the ends of the Theoretical Electrical Travel as determined from the Index Point.

3.5.2 ■ EQUIPMENT

Travel measuring device	1.1.3
Voltage ratio equipment	1.2.1
Load device (when applicable)	1.1.4

3.5.3 ■ TEST PROCEDURE Mount the potentiometer to the travel measuring device and connect the voltage ratio equipment to the appropriate terminals. Phase the potentiometer and travel measuring device at the Index Point (3.6). Move the potentiometer shaft in the direction opposite to the "specified direction" until the voltage ratio equipment indicates the wiper is in the overtravel region by an amount $>>$ Backlash. The specified direction is CW for rotary potentiometers and extending for translatories, unless otherwise indicated. Move the shaft in the specified direction until the travel measuring device indicates the beginning of the Theoretical Electrical Travel (3.8). This is one Theoretical End Point. Continuing in the same direction move the shaft until the travel measuring device indicates the opposite limit of Theoretical Electrical Travel which is the other Theoretical End Point.

3.6 ■ INDEX POINT

3.6.1 ■ OBJECT To provide a procedure for establishing a shaft position reference for the related specified Output Ratio when the shaft is moved in the specified direction.

3.6.2 ■ EQUIPMENT

Travel measuring device	1.1.3
Voltage ratio equipment	1.2.1

3.6.3 ■ TEST PROCEDURE The potentiometer is mounted in the travel measuring device and is electrically connected to the voltage ratio equipment. Move the travel measuring device in the specified direction (unless otherwise stated, this is CW for rotary and extending for translatories) to the position at which the Output Ratio of the Index Point is first reached. Holding the potentiometer shaft fixed, adjust the indicator of the travel measuring device to read precisely the specified shaft position. The travel measuring device and voltage ratio equipment now read the corresponding values of the Index Point.

3.7 ☐ ACTUAL ELECTRICAL TRAVEL (APPLICABLE TO WIREWOUND ONLY)

3.7.1 ☐ OBJECT To measure the total travel of the shaft between the End Points.

3.7.2 ☐ EQUIPMENT

Travel measuring device	1.1.3
Voltage ratio equipment	1.2.1
Load device (when applicable)	1.1.4

3.7.3 ☐ TEST PROCEDURE Mount the potentiometer to the travel measuring device and locate the End Points (3.4). Note the reading on the travel measuring device at both End Points. The difference between these two readings is the Actual Electrical Travel.

3.8 ■ THEORETICAL ELECTRICAL TRAVEL

3.8.1 ■ OBJECT To determine the limits of the Theoretical Electrical Travel (the Theoretical End Points) utilizing the Index Point.

3.8.2 ■ EQUIPMENT

Travel measuring device	1.1.3
Voltage ratio equipment	1.2.1
Load device (when applicable)	1.1.4

3.8.3 ■ TEST PROCEDURE The Theoretical Electrical Travel is a value given as part of the potentiometer specifications whose limits have been defined as the Theoretical End Points. To physically locate these limits see Paragraph 3.5.

3.9 ☐ ■ ELECTRICAL OVERTRAVEL

3.9.1 ☐ ELECTRICAL OVERTRAVEL — WIREWOUND

3.9.1.1 ☐ OBJECT To measure the shaft travel between the limits of the Actual Electrical Travel and the corresponding limits of Electrical Continuity Travel. Theoretical Electrical Travel is substituted for Actual Electrical Travel when Absolute Conformity or Linearity is specified.

3.9.1.2 ☐ EQUIPMENT

Travel measuring device	1.1.3
Resistance measuring device	1.2.2
Load device (when applicable)	1.1.4
Voltage ratio equipment	1.2.1

3.9.1.3 ☐ TEST PROCEDURE Mount the potentiometer to the travel measuring device. Locate the limits or End Points of the Actual Electrical Travel (3.4) and the limits of Electrical Continuity Travel (3.10) and note the readings on the travel measuring device at these points. The Electrical Overtravel at each end is the difference between the reading at the End Point and the corresponding limit of Electrical Continuity Travel. When Absolute Linearity or Absolute Conformity is specified, the limits of the Theoretical Electrical Travel, as determined from the Index Point (3.6), are used in place of the End Points.

3.9.2 ■ ELECTRICAL OVERTRAVEL — NONWIREWOUND

3.9.2.1 ■ OBJECT To measure the shaft travel between the Theoretical End Points and the corresponding limits of Electrical Continuity Travel.

3.9.2.2 ■ EQUIPMENT

Travel measuring device	1.1.3
Resistance measuring device	1.2.2
Load device (when applicable)	1.1.4
Voltage ratio equipment	1.2.1

3.9.2.3 ■ TEST PROCEDURE Mount the potentiometer to the travel measuring device. Locate the Theoretical End Points (3.5) and the limits of Electrical Continuity Travel (3.10) and note the readings on the travel measuring device at these points. The Electrical Overtravel at each end is the difference between the reading at the Theoretical Point and the corresponding limit of Electrical Continuity Travel.

3.10 ■ ELECTRICAL CONTINUITY TRAVEL

3.10.1 ■ OBJECT To measure that portion of shaft travel over which electrical continuity is maintained between the wiper and the resistance element.

3.10.2 ■ EQUIPMENT

Travel measuring device	1.1.3
Resistance measuring device	
(1.0 ma max through specimen)	1.2.2
Load device (when applicable)	1.1.4

3.10.3 ■ TEST PROCEDURE Mount the potentiometer to the travel measuring device. Ascertain that the current in the potentiometer from the resistance measuring device will not exceed 1.0 ma, then connect the resistance measuring device between the wiper and the interconnected terminals. Move the shaft in the specified direction (CW for rotary potentiometers and extending for translatories) until the resistance measuring device first indicates loss of continuity; note this position. Move the shaft in the reverse direction until loss of continuity is again observed on the resistance measuring device and continue beyond this point by an amount $>>$ Backlash. Move the shaft in the initial direction until continuity is first observed; note this position. The difference between the two noted positions is the Electrical Continuity Travel.

Note: In some potentiometers there may not be a position at which a discontinuity is indicated. In these cases, the limits of Electrical Continuity Travel are the end(s) of continuity or the end(s) of Total Mechanical Travel, whichever occurs first.

3.11 ■ TAP LOCATION

3.11.1 ■ OBJECT To measure the location of a tap from some reference point. This is commonly expressed in terms of Output Ratio or shaft position. When a shaft position is specified, the tap position is the center of the Effective Tap Width, exclusive of Backlash.

3.11.2 ■ EQUIPMENT

Travel measuring device	1.1.3
Voltage ratio equipment	1.2.1
Null indicator	1.2.19
Power supply	1.2.3

3.11.3 ■ TEST PROCEDURE

3.11.3.1 ■ TRAVEL LOCATED TAP Mount the potentiometer to the travel measuring device. Locate the End Point (3.4 or 3.5), or other specified reference point and note the reading on the travel measuring device. Connect the power supply to the potentiometer input terminals and connect the null indicator between the wiper terminal and the tap terminal to be measured. Using the same direction of motion as when locating the End Point, or other reference point, move the shaft to the center of the Effective Tap Width (3.12). The displacement from the specified reference point to the center of the Effective Tap Width is the Tap Location.

3.11.3.2 ■ VOLTAGE RATIO LOCATED TAP Position the wiper on the Electrical Overtravel or beyond the Electrical Continuity Travel, or if neither of these is possible, in the region having the least effect on the measurement. Measure the voltage ratio between the tap and the specified reference terminal.

3.12 ■ EFFECTIVE TAP WIDTH

3.12.1 ■ OBJECT To measure the travel of the shaft during which the voltage at the wiper terminal and the tap terminal are the same as the wiper is moved past the tap in one direction.

3.12.2 ■ EQUIPMENT

Travel measuring device	1.1.3
Null indicator	1.2.19
Power supply	1.2.3

3.12.3 ■ TEST PROCEDURE Mount the potentiometer to the travel measuring device. Connect the power supply to the potentiometer input terminals and connect the null detector between the wiper terminal and the tap terminal to be measured. Move the shaft to obtain a null voltage between the wiper and the tap. Continue to move the shaft in that same direction to the point immediately preceding the first measurable change in voltage; note the reading of the travel measuring device. Move the shaft in the opposite direction past the region of null by an amount much greater than the Backlash. Move the shaft in the original direction to the point at which the last measurable change in voltage occurs as the null is reached; note the reading on the travel measuring device.

The difference in shaft displacement between the two positions is the Effective Tap Width.

3.13 ■ PHASING POINT

3.13.1 ■ OBJECT To provide a method of locating a point of reference on each electrical element used to describe the relative alignment of the electrical elements of a gang.

3.13.2 ■ EQUIPMENT Refer to equipment used for measurement of the specific characteristic by which the phasing point is described.

3.13.3 ■ TEST PROCEDURE Since phasing points are commonly specified in terms of Index Points (3.6), Tap Locations (3.11), Output Voltage Ratios, etc., refer to the applicable section of this document for the appropriate procedure.

3.14 ■ PHASING

3.14.1 ■ OBJECT To measure the relative alignment of the Phasing Points of each cup of a gang potentiometer.

3.14.2 ■ EQUIPMENT Refer to the equipment used for measurement of the specific characteristic by which the Phasing Point is described.

3.14.3 ■ TEST PROCEDURE When shaft displacement is required, set the potentiometer to the Phasing Point (3.13) of the first cup utilizing the applicable section of this document. Moving in the same direction (to eliminate Backlash) displace the shaft to the respective Phasing Point or Points of the subsequent cups of the gang. The measurements of the electrical or mechanical relationship of the Phasing Points is compared with the specified Phasing allowance.

4 *resistance*

4.1 ■ TOTAL RESISTANCE (DC INPUT IMPEDANCE)

4.1.1 ■ OBJECT To measure the DC resistance between the input terminals with the shaft positioned so as to give a maximum value.

4.1.2 ■ EQUIPMENT

Resistance measuring device	1.2.2

4.1.3 ■ TEST PROCEDURE With the aid of a resistance measuring device (10 ma maximum current) connected between the wiper and one input terminal, position the wiper on the Electrical Overtravel. If this is not possible due to the limitations of Total Mechanical Travel or the existence of continuous Electrical Travel and the Total Resistance measurement is critically close to the tolerance limits, the shaft should be moved to a region which maximizes the resistance reading during the Total Resistance measurement. Connect the resistance measuring device to the input terminals of the potentiometer. The maximum reading observed is the Total Resistance. The voltage applied during measurement should be minimized to avoid errors in the measurement due to heating.

4.2 ■ DC OUTPUT IMPEDANCE

4.2.1 ■ OBJECT To determine the maximum DC resistance between the wiper and either end terminal with the input shorted.

4.2.2 ■ EQUIPMENT

Wheatstone bridge	1.2.2.2

4.2.3 ■ TEST PROCEDURE The resistance measuring bridge is connected to the potentiometer between the wiper and either end terminal with the input terminals shorted. The resistance is monitored as the pot shaft is rotated or translated over the Total Variable Output. The maximum resistance reading is the DC Output Impedance.

4.3 □ ■ MINIMUM RESISTANCE

4.3.1 □ MINIMUM RESISTANCE — WIREWOUND

LEGEND:

☐ WIREWOUND
■ WIREWOUND AND NONWIREWOUND
■ NONWIREWOUND

4.3.1.1 ☐ OBJECT To determine the resistance value between the wiper terminal and any other specified terminal with the shaft positioned to give a minimum value.

4.3.1.2 ☐ EQUIPMENT

Shaft positioning device	1.1.2
Resistance measuring device	1.2.2
(10 ma maximum output into potentiometer or the wiper/tap current rating whichever is less)	

4.3.1.3 ☐ TEST PROCEDURE Mount the potentiometer to the shaft positioning device and connect the resistance measuring device between the wiper terminal and the specified terminal. The shaft positioning device provides fine control of the shaft position. Move the shaft until a minimum reading is indicated on the resistance measuring device. This reading is the Minimum Resistance.

4.3.2 ■ MINIMUM RESISTANCE — NONWIREWOUND Refer to Tap Resistance (4.5) for applicable test procedure.

4.4 ☐ ■ END RESISTANCE

4.4.1 ☐ END RESISTANCE — WIREWOUND

4.4.1.1 ☐ OBJECT To determine the resistance value between the wiper terminal and an end terminal with the shaft positioned at the corresponding End Point.

4.4.1.2 ☐ EQUIPMENT

Shaft positioning device	1.1.2
Voltage ratio equipment	1.2.1
Resistance measuring device	1.2.2
(10 ma maximum output into potentiometer of the wiper/tap current rating whichever is less)	

4.4.1.3 ☐ TEST PROCEDURE Mount the potentiometer to the shaft positioning device which provides fine control of the shaft position. With the voltage ratio equipment, position the shaft at the applicable End Point (3.4).

The End Resistance is determined with the resistance measuring device connected between the wiper and the end terminal.

4.4.2 ■ END RESISTANCE — NONWIREWOUND Refer to End Voltage (2.1) for applicable test procedure.

4.5 ■ TAP RESISTANCE (APPLICABLE TO NONWIREWOUND ONLY)

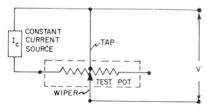

FIGURE 4.5A Tap resistance

4.5.1 ■ OBJECT To measure the resistance between a tap terminal (other than end terminations) and the wiper terminal, with the wiper positioned to give a minimum value, without drawing wiper current.

4.5.2 ■ EQUIPMENT

Shaft positioning device	1.1.2
Constant current source	1.2.5

4.5.3 ■ TEST PROCEDURE Connect the constant current source to the test pot as indicated in Figure 4.5A and adjust the source to 1.0 ma or the maximum current rating of the tap and/or resistance element whichever is less. Using the shaft positioning device move the shaft to the position on the resistance element that minimizes the Voltage V. The Tap Resistance is determined from V and I_c.

4.6 ■ APPARENT CONTACT RESISTANCE (APPLICABLE TO NONWIREWOUND ONLY) Refer to Output Smoothness (6.2) for applicable test procedure.

4.7 ☐ ■ EQUIVALENT NOISE RESISTANCE (ENR)

4.7.1 ☐ EQUIVALENT NOISE RESISTANCE — WIREWOUND Refer to Noise (6.1) for applicable test procedure.

4.7.2 ■ EQUIVALENT NOISE RESISTANCE — NONWIREWOUND Refer to Output Smoothness (6.2) for applicable test procedure.

4.8 ☐ TEMPERATURE COEFFICIENT OF RESISTANCE (APPLICABLE TO WIREWOUND ONLY)

4.8.1 ☐ OBJECT To determine the unit change in resistance per degree celsius change for a reference temperature, expressed in parts per million per degree celsius.

4.8.2 ☐ EQUIPMENT

Resistance measuring device	1.2.2
Temperature test chamber	1.1.16
Thermocouple bridge	1.2.20

4.8.3 ☐ TEST PROCEDURE Position the shaft of the potentiometer with the wiper off the Actual Electrical Travel or at a point that minimizes the Total Resistance (4.1) if no overtravel exists. Subject the pot to two standard series of test temperatures in the sequence described. The first series is room temperature (defined at 25°C) down to −55°C or the lowest rated operating temperature with two intermediate temperature steps at 0°C and −25°C; the second series is room temperature to +125°C or the highest rated operating temperature (whichever is less) with two intermediate temperature steps at +50°C and +85°C. The Total Resistance is measured after the temperature chamber has been stabilized for each temperature (a minimum of 30 minutes, but avoid overaging) with 25°C as the reference temperature for both series. The Temperature Coefficient of Resistance at each test temperature is computed with the following formula:

$$\text{Tempco} = \frac{R_2 - R_1}{R_1(T_2 - T_1)} \times 10^6$$

Where:

R_1 = Resistance at reference temperature in ohms.
R_2 = Resistance at test temperature in ohms.
T_1 = Reference temperature in degrees celsius.
T_2 = Test temperature in degrees celsius.

The Temperature Coefficient of Resistance of the potentiometer is the maximum value calculated.

4.9 ■ RESISTANCE-TEMPERATURE CHARACTERISTIC (APPLICABLE TO NONWIREWOUND ONLY)

4.9.1 ■ OBJECT To determine the maximum change in total resistance over a specified temperature range expressed as a percent of the total resistance at a specified reference temperature.

4.9.2 ■ EQUIPMENT

Resistance measuring device	1.2.2
Temperature test chamber	1.1.16
Thermocouple bridge	1.2.20

4.9.3 ■ TEST PROCEDURE Position the shaft of the potentiometer with the wiper off the Theoretical Electrical Travel

234

or at a point that maximizes the Total Resistance (4.1) if no overtravel exists. Subject the pot to two standard series of test temperatures in the sequence described. The first series is room temperature (defined at 25°C) down to −55°C or the lowest rated operating temperature with two intermediate temperature steps at 0°C and −25°C; the second series is room temperature to +125°C or the highest rated operating temperature (whichever is less) with two intermediate temperature steps at +50°C and +85°C. The Total Resistance is measured after the temperature chamber has been stabilized for each temperature (a minimum of 30 minutes, but avoid overaging) with 25°C as the reference temperature for both series. Compute the Resistance-Temperature Characteristic for each temperature interval with the following formula:

$$RTC = \frac{R_2 - R_1}{R_1} \times 100$$

Where:

R_1 = Resistance at reference temperature in ohms.
R_2 = Maximum or minimum resistance at any of the test temperatures, in ohms.

The Resistance-Temperature Characteristic of the potentiometer is the maximum value calculated.

5   conformity and linearity

5.1 ■ ABSOLUTE CONFORMITY

5.1.1 ■ OBJECT To measure the maximum deviation expressed as a percent of Total Applied Voltage of the actual function characteristic from theoretical function characteristic extending between the specified Output Ratios which are separated by the Theoretical Electrical Travel. An Index Point on the actual output is required.

5.1.2 ■ EQUIPMENT

Travel measuring device   1.1.3
Voltage ratio equipment   1.2.1

5.1.3 ■ TEST PROCEDURE Mount the potentiometer in the travel measuring device and connect electrically to the voltage ratio equipment. Set the pot shaft to the Index Point (3.6). Then move the shaft to the beginning of the Theoretical Electrical Travel. The Index Point and beginning of Theoretical Electrical Travel must be approached from the same direction to eliminate effects of Backlash. Continuing in the same direction compare the actual output ratio of the test potentiometer to the theoretical function output ratio and note deviations for each shaft position. When an empirical function is specified the deviations are noted at the given data points only. When the function is described by a mathematical equation, deviations are noted at every 3-1/2% of Theoretical Electrical Travel or 45°, whichever is less. The maximum deviation from the theoretical, expressed as a percentage of the Total Applied Voltage, is the Absolute Conformity.

Note: In no case should the applied voltage exceed the voltage or power rating of the unit being tested.

5.1.4 ■ ALTERNATE PROCEDURE If a master of sufficient accuracy is available, test by the same procedure used for Absolute Linearity (5.2), substituting the proper non-linear master for the linear master.

5.2 ■ ABSOLUTE LINEARITY

5.2.1 ■ OBJECT To measure the maximum deviation (expressed as a percent of the Total Applied Voltage) of the actual function characteristic from a fully defined straight reference line over the Theoretical Electrical Travel as determined from the Index Point.

5.2.2 ■ EQUIPMENT

Conformity tester   1.2.18
Recorder   1.2.4
Power Supply   1.2.3
Voltage ratio equipment   1.2.1

5.2.3 ■ TEST PROCEDURE — CONTINUOUS METHOD Locate the Index Point (3.6) using the conformity tester as a travel measuring device. Disconnect the voltage ratio equipment and electrically connect the test potentiometer to the conformity tester as shown in Figure 1.2.18A and proceed as follows:

A. With switch in position 1, adjust recorder to null at center of chart.

B. With switch in position 6
   1. Move the travel indicator to the Beta limit of Theoretical Electrical Travel.
   2. By external means, short the slider terminal to the Beta terminal of the potentiometer under test.
   3. Adjust "BAL AT BETA" control to produce a null on the recorder.
   4. Remove the jumper (step 2) and move the travel indicator to the Alpha limit of Theoretical Electrical Travel.
   5. By external means, short the slider terminal to the Alpha terminal of the potentiometer under test.
   6. Adjust "BAL AT ALPHA" control to produce a null on the recorder.
   7. Remove the jumper (step 5).

Note: Since "BAL" controls interact to some extent, it may be necessary to repeat step B until no further adjustment is needed.

C. With switch in position 2 or 3 as desired, adjust recorder gain to produce desired deflection.

D. After completing balancing and calibrating operations, set switch to position 6 and move the travel indicator over the full extent of the Theoretical Electrical Travel in the specified direction at a uniform speed noting linearity deviations on the recorder. The maximum deviation from the reference line, expressed in percent, is the Absolute Linearity.

5.2.4 ■ TEST PROCEDURE — POINT-BY-POINT METHOD The point-by-point method for testing Absolute Linearity is the same as the procedure for Absolute Conformity with the function described by a linear equation.

5.3 ☐ TERMINAL BASED LINEARITY (APPLICABLE TO WIREWOUND ONLY)

5.3.1 ☐ OBJECT To measure the maximum deviation (expressed as a percent of the Total Applied Voltage) of the actual function characteristic from a straight reference line drawn through the specified minimum and maximum output voltage ratios, which are separated by the Actual Electrical Travel. Unless otherwise specified minimum and maximum output ratios are respectively zero and 100% of Total Applied Voltage.

5.3.2 ☐ EQUIPMENT

Conformity tester   1.2.18
Recorder   1.2.4
Power supply   1.2.3

5.3.3 ☐ TEST PROCEDURE

A. Connect the potentiometer Electrically to the conformity tester as shown in Figure 1.2.18A.

LEGEND:

□ WIREWOUND
■ WIREWOUND AND NONWIREWOUND
■ NONWIREWOUND

B. Set function selector switch at position 1 and adjust recorder to null at center line of chart paper.

C. Set function selector switch to position 2 or 3 as desired and adjust recorder gain control to produce desired deflection.

D. Set function selector switch to position 4 and locate the Alpha End Point (3.4) approximately, moving the shaft by hand.

E. Set the travel indicator of the conformity tester at or near zero and mount the potentiometer in the conformity tester by normal means.

F. Locate the exact position of the Alpha End Point (3.4) and note the shaft position.

G. Set the function selector switch to position 6, short the slider terminal to the Alpha terminal by external means and adjust the "BAL AT ALPHA" control to obtain a null on the recorder.

H. Remove the short, set function selector switch to position 5, locate the Beta End Point exactly and note the shaft position.

J. Set the function selector switch to position 6, short the slider terminal to the Beta terminal by external means and adjust the "BAL AT BETA" control to obtain a null on the recorder.

Note: It may be necessary to repeat steps F through J several times because of interaction of the balance controls.

K. After completing balancing and calibrating operations, set selector switch to position 6 and move the travel indicator over the full extent of the Actual Electrical Travel (3.7), in the specified direction at uniform speed, noting the linearity deviations on the recorder. The maximum deviation from the reference line, expressed as a percent of Total Applied Voltage, is the Terminal Based Linearity.

5.4 □ ZERO BASED LINEARITY (APPLICABLE TO WIREWOUND ONLY)

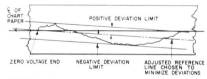

FIGURE 5.4A Determination of zero based linearity

5.4.1 □ OBJECT To determine the maximum deviation, expressed as a percent of Total Applied Voltage, of the Actual function characteristic from a straight reference line drawn through the specified minimum Output Ratio, extended over the Actual Electrical Travel, and rotated to minimize the maximum deviations. Unless otherwise specified, the specified minimum Output Ratio will be zero.

5.4.2. □ EQUIPMENT

Conformity tester	1.2.18
Recorder	1.2.4
Power supply	1.2.3

5.4.3 □ TEST PROCEDURE To employ the specified conformity tester in this procedure, connect the specified minimum (or zero) output ratio end of the potentiometer to the "ALPHA" end of the tester.

A. Connect the potentiometer electrically to the conformity tester as shown in Figure 1.2.18A.

B. Set function selector switch at position 1 and adjust recorder to null at centerline of chart paper.

C. Set function selector switch to position 2 or 3 as desired and adjust recorder gain control to produce desired deflection.

D. Set function selector switch to position 4 and locate the Alpha End Point (3.4) approximately, moving the shaft by hand.

E. Set the travel indicator of the conformity tester at or near zero and mount the potentiometer in the conformity tester by its normal mounting means.

F. Locate the exact position of the Alpha End Point (3.4) and note the shaft position.

G. Set the function selector switch to position 6, short the slider terminal to the alpha terminal by external means and adjust the "BAL AT ALPHA" control to obtain a null on the recorder.

H. Remove the short, set function selector switch to position 5, locate the Beta End Point exactly and note the shaft position.

J. Set the function selector switch to position 6, and adjust the "BAL AT BETA" control to obtain a null on the recorder.

Note: It may be necessary to repeat steps F through J several times due to the interaction of the balance controls.

K. After completing balancing and calibrating operations, set selector switch to position 6 and move the travel indicator over the full extent of the Actual Electrical Travel (3.7), in the specified direction at a uniform speed, noting the linearity deviations on the recorder.

Referring to Figure 5.4A, determine Zero Based Linearity by drawing a reference line through the recording so that it intersects the centerline of the recording at the "zero voltage" end and is rotated about this point until the positive and negative deviations from it are minimized. Draw lines parallel to the reference line through the maximum deviations from it to establish the limits of the Zero Based Linearity. Express as a percent of the Total Applied Voltage.

5.5 □ ■ INDEPENDENT LINEARITY

5.5.1 □ INDEPENDENT LINEARITY — WIREWOUND

5.5.1.1 □ OBJECT To measure the maximum deviation, expressed as a percent of the Total Applied Voltage, of the actual function characteristic from a straight reference line whose slope and position minimize the maximum deviations over the Actual Electrical Travel, or any specified portion thereof.

5.5.1.2 □ EQUIPMENT

Conformity tester	1.2.18
Recorder	1.2.4
Power supply	1.2.3

5.5.1.3 □ TEST PROCEDURE

A. Connect the potentiometer electrically to the conformity tester as shown in Figure 1.2.18A.

B. Set function selector switch at position 1 and adjust recorder to null at centerline of chart paper.

C. Set function selector switch to position 2 or 3 as desired and adjust recorder gain control to produce desired deflection.

D. Set function selector switch to position 4 and locate the Alpha End Point (3.4) approximately, moving the shaft by hand.

E. Set the travel indicator of the conformity tester at or near zero and mount the potentiometer in the linearity tester by its normal mounting means.

F. Locate the exact position of the Alpha End Point (3.4) and note the shaft position.

□	WIREWOUND
■	WIREWOUND AND NONWIREWOUND
■	NONWIREWOUND

G. Set the function selector switch to position 6, and adjust the "BAL AT ALPHA" control to obtain a null on the recorder.

H. Set the function selector switch to position 5, locate the Beta End Point exactly, and note the shaft position.

J. Set the function selector switch to position 6, and adjust the "BAL AT BETA" control to obtain a null on the recorder.

Note: It may be necessary to repeat steps F through J several times because of the interaction of the balance controls.

K. After completing balancing and calibrating operations, set selector switch to position 6, and move the travel indicator over the full extent of the Actual Electrical Travel (3.7), in the specified direction at a uniform speed, noting the linearity deviations on the recorder.

Referring to Figure 5.5A, determine Independent Linearity by drawing the best straight line through the recording so as to minimize the maximum positive and negative deviations from the line irrespective of position or slope. Draw lines parallel to this reference line, through the maximum deviations from it to establish the limits of the Independent Linearity. Express as a percent of the Total Applied Voltage.

Note: If point-by-point method is used, the best straight line must be determined graphically by plotting.

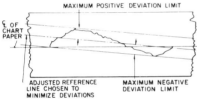

MAXIMUM POSITIVE DEVIATION LIMIT

₵ OF CHART PAPER

ADJUSTED REFERENCE LINE CHOSEN TO MINIMIZE DEVIATIONS

MAXIMUM NEGATIVE DEVIATION LIMIT

FIGURE 5.5A Determination of independent linearity

5.5.2 ■ INDEPENDENT LINEARITY — NONWIREWOUND

5.5.2.1 ■ OBJECT To measure the maximum deviation, expressed as, a percent of the Total Applied Voltage, of the actual function characteristic from a straight reference line whose slope and position minimize the maximum deviations over the Theoretical Electrical Travel, or any specified portion thereof. An Index Point on the actual output is required.

5.5.2.2 ■ EQUIPMENT

Conformity tester	1.2.18
Recorder	1.2.4
Power supply	1.2.3
Voltage ratio equipment	1.2.1

5.5.2.3 ■ TEST PROCEDURE Locate the Index Point (3.6) using the conformity tester as the travel measuring device. Disconnect the voltage ratio equipment, electrically connect the test pot to the conformity tester as shown in Figure 1.2.18A and proceed, as follows:

A. With switch in position 1, adjust recorder to null at center of chart.

B. With switch in position 6
  1. Move the travel indicator to the Beta limit of Theoretical Electrical Travel.
  2. Adjust "BAL AT BETA" control to produce a null on the recorder.

3. Move the travel indicator to the Alpha limit of Theoretical Electrical Travel.

4. Adjust "BAL AT ALPHA" control to produce a null on the recorder.

5. Because of the possible interaction of the Bal controls, repeat steps 1 through 4 until no further adjustment is necessary.

C. With switch in position 2 or 3 as desired, adjust recorder gain to produce desired deflection.

D. After completing balancing and calibration, set selector switch to position 6, and move the travel indicator over the full Theoretical Electrical Travel, in the specified direction at a uniform speed, noting the linearity deviations on the recorder.

Referring to Figure 5.5A, determine Independent Linearity by drawing the best straight line through the recording so as to minimize the maximum positive and negative deviations from the line irrespective of position or slope. Draw lines parallel to this reference line through the maximum deviations from it to establish the limits of the Independent Linearity. Express as a percent of the Total Applied Voltage.

Note: If point-by-point method is used, the best straight line must be determined graphically by plotting.

5.6 ■ SIMULTANEOUS CONFORMITY PHASING

5.6.1 ■ OBJECT To determine that the electrical elements of a gang potentiometer are so aligned that each electrical element falls within its conformity or linearity limits when a common Index Point is used.

5.6.2 ■ EQUIPMENT See Absolute Conformity (5.1), Absolute Linearity (5.2) or Independent Linearity (Nonwirewound — 5.5.2).

5.6.3 ■ TEST PROCEDURE Determine Simultaneous Conformity Phasing in the same manner as Absolute Conformity (5.1), Absolute Linearity (5.2), or Independent Linearity — Nonwirewound (5.5.2), for each electrical element, using a common Index Point (3.6). The Index Point is on the first cup unless otherwise specified.

When running point-by-point tests described in the Absolute Conformity test procedure, take Output Voltage Ratio readings for each shaft setting simultaneously on all cups in the gang.

5.7 ■ VOLTAGE TRACKING ERROR

5.7.1 ■ OBJECT To measure the difference at any shaft position between the Output Ratios of any two commonly actuated similar electrical elements; expressed as a percentage of the single Total Voltage Applied to them.

5.7.2 ■ EQUIPMENT

Travel measuring device	1.1.3
Recorder	1.2.4
Power supply	1.2.3

5.7.3 ■ TEST PROCEDURE Connect the power supply across the corresponding ends of all similar electrical elements in the gang which are to be tracked. Connect the recorder between the wiper terminal of the reference element and the wiper terminal of the element to be tracked. Move the shaft throughout the travel range over which tracking is specified and directly observe the Voltage Tracking Error on the recorder.

Compare all cups/electrical elements to the first element, unless otherwise specified.

6 *general electrical characteristics*

6.1 □ NOISE (EQUIVALENT NOISE RESISTANCE)

6.1.1 □ INTRODUCTION Perhaps the most persistent problem affecting potentiometer manufacturers and users is the so-called "Noise" problem. Countless man-hours have been expended over the last two decades, and longer, investigating the sources of potentiometer Noise and means for eliminating or controlling it. Considerable headway has already been made in this area and the future portends even greater achievements.

The need exists right now, however, for putting the problem in its proper perspective. For twenty years, or more, practically all potentiometer noise measurements have been made using the 1.0 ma constant current source and oscilloscope detection technique currently defined in NAS710 and MIL-R-12934.

This technique is arbitrary and suffers from the following disabilities:

A. Virtually no potentiometer is ever used in such a circuit. Most precision potentiometer applications involve negligible slider current.

B. The bandwidth of the measuring instrument ($>$50 KHz) is at least three orders of magnitude greater than the bandwidth of most servo systems, which constitute a large percentage of precision potentiometer applications.

The advantage of the technique is its simplicity.

The Variable Resistive Components Institute has chosen to establish a new test procedure (6.1.4) for measuring equivalent noise resistance in wirewound potentiometers. This new procedure is not unlike earlier versions. It retains their simplicity and yet provides a basis for Noise measurement technique more realistically related to actual system usage.

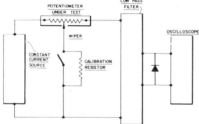

FIGURE 6.1A Noise test circuit — with low pass filter

Very simply, the new procedure recommends the insertion of a 1000 Hz low pass filter in the circuit (see Figure 6.1A).

In view of the wide disparity between bandwidths used previously, and this procedure, it would appear at first glance that the new procedure simply disguises potentiometer Noise. It is clear, in view of the "spikey" nature of pot Noise, that with the low pass filter, less Noise will be indicated. A bandwidth of 1000 Hz, however, is still well in excess of the bandwidth of most servo systems.

Unfortunately, the new procedure is no less arbitrary than its predecessors since it still measures Noise in a non-typical circuit. The ultimate in Noise tests would be to dynamically extract the desired signal from the potentiometer output and record the residual Noise under conditions of voltage and loading corresponding to its actual usage. Such a measurement requires special pot test equipment and detailed knowledge of potentiometer application. In the usual situation, the user doesn't have the proper equipment and the manufacturer is unable to obtain sufficient data on the potentiometer's

application. For purposes of standardization the two must generally meet, therefore, on the common ground of the arbitrary but simple constant current rheostat approach.

It is not anticipated that this simple modification will eliminate the Noise problem, nor that it will be a suitable procedure for all applications. For this reason the larger bandwidth test (0-50KHz) has been included as an alternate procedure (6.1.5) to be specified where the application dictates its use.

It is expected that by standardizing on the limited bandwidth test procedure a large number of unnecessary rejections will be eliminated and focus users' attention on the need to more clearly define their requirements in situations where the standard procedure is inadequate.

6.1.2 □ OBJECT To measure the spurious variations in the electrical output not present in the input, defined quantitatively in terms of the equivalent parasitic transient resistance in ohms, appearing between the contact and the resistance element when the shaft is rotated or translated. The equivalent noise resistance is defined independently of resolution, the functional characteristics, and the total travel. The magnitude of the equivalent noise resistance is the maximum departure from a specified reference line. The wiper of the potentiometer is required to be excited by a specified current and moved at a specified speed.

6.1.3 □ EQUIPMENT

Constant speed drive	1.1.12
Constant current source	1.2.5
Oscilloscope	1.2.6
Low pass filter	1.2.7
Zener diode (6v)	

6.1.4 □ TEST PROCEDURE (SEE FIGURE 6.1A) The potentiometer shaft is cycled not less than ten times over a minimum of 95% of the Electrical Continuity Travel (3.10) within the rated travel speed of the potentiometer just prior to making Noise measurements. The potentiometer shaft is then connected mechanically to the constant speed drive and electrically connected in the Noise test circuit as shown in the Figure 6.1A. With the constant speed drive engaged, the potentiometer Noise characteristic may then be noted on the oscilloscope as the wiper traverses one complete cycle over the full Electrical Continuity Travel and the maximum values are compared to the specified limit.

If only random spikes of Noise are noted, the potentiometer should be cycled again. If the random spikes are repetitive, the maximum values should be noted. Otherwise, do not consider the initial measurements Noise.

Take care to discount apparent contact resistance change due to secondary current paths present in potentiometers with shunting or padding resistors, continuous windings or windings with resistance overtravels or shorts.

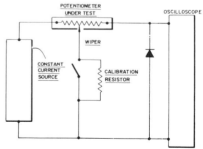

FIGURE 6.1B Noise test circuit — without low pass filter

It is recommended that a zener diode be placed across the scope input to protect the terminations of the potentiometer from excessive arcing occurring at discontinuities in the output, such as at the dead space in single turn pots.

6.1.5 □ ALTERNATE TEST PROCEDURE
Use this procedure only when specified on the procurement documents or individual control drawings.

The procedure is identical to the standard Noise test procedure (6.1.4) except that the low pass filter is removed from the test circuit as indicated in Figure 6.1B.

Test conducted with this circuit will show an off set from the baseline corresponding to the apparent contact resistance in the wiper circuit and/or the value of a wiper protecting resistor. Apparent contact resistance will be found especially in non-wirewound potentiometers with relatively large ratios of track width or cross section to wiper contact area.

6.2 □ OUTPUT SMOOTHNESS

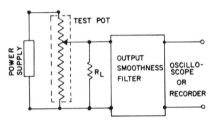

FIGURE 6.2A Output smoothness

6.2.1 ■ INTRODUCTION
The "noise" test in paragraph 6.1 measures ENR (equivalent nosie resistance), which is a variation in real or apparent contact resistance in the wiper circuit. This measurement is significant for units used in rheostat applications, but not for voltage dividers, in which very little, if any, current flows in the wiper circuit. For these applications the output smoothness test is used, which reflects sudden variations in output of voltage dividers. If the units are used with finite loads, the effect of contact resistance variation is included in the output smoothness test along with all other causes of noise in the output signal.

While different use circuits may well be sensitive to different types or frequencies of output variations, the VRCI has chosen one output smoothness filter for a standardized test which has proven adequate for a wide range of uses. It should be used, unless all conditions are sufficiently known and understood to permit design of another filter tailored to that application.

6.2.2 ■ OBJECT
To measure the spurious variations in the electrical output not present in the input. They are measured for specified travel increments over the Theoretical Electrical

Travel and expressed as a percentage of the Total Applied Voltage.

6.2.3 ■ EQUIPMENT
Power supply	1.2.3
Output smoothness filter	1.2.8
Oscilloscope	1.2.6
Constant speed drive	1.1.12

6.2.4 ■ TEST PROCEDURE
Mount the potentiometer in the 4 RPM constant speed drive and excite it with the power supply. Connect the wiper and power common lead to the input of the filter and the output of the filter to the oscilloscope as shown in Figure 6.2A. When a load is specified for a Conformity test, use that load for the Output Smoothness test. When no load is specified for the Conformity test, apply a load $R_L = 100 \times R_T$ between the wiper and the CCW end for the Output Smoothness test, unless otherwise specified. The Output Smoothness is the largest excursion voltage occurring over one specified travel increment, divided by the Total Applied Voltage. Unless otherwise specified, the travel increment is 1% of the Theoretical Electrical Travel. Excursions occurring at the point of abrupt changes in output slope (start, end, and reversal) are not considered Output Smoothness faults.

Proper trace evaluation requires examining the trace over one specified travel increment (θ_i) at a time. The travel increment is placed on the trace to include the maximum excursion within the increment, using the oscilloscope. Determine the increment by adjusting the sweep of the oscilloscope, (for example, Theoretical Electrical Travel 300°; $\theta_i = 3°$; at 4 RPM 3° = 0.12 sec.; at 1 cm/sec. sweep $\theta_i = 1.2$ cm). In most cases you can obtain equivalent results on a standard recorder on which the travel increment is easily calibrated by measuring the total trace length and evaluating the trace as shown in Figure 6.2B.

Exercise care in evaluating the magnitude of the excursion voltages because of the substantial attenuation of pot output voltage through the Output Smoothness filter.

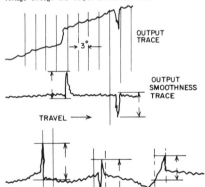

FIGURE 6.2B Evaluation of output smoothness trace

6.3 □ VOLTAGE RESOLUTION (APPLICABLE TO WIREWOUND ONLY)

6.3.1 □ OBJECT
To measure the maximum incremental change in Output Ratio, in any specified portion of the resistance element, with shaft travel in one direction.

6.3.2 ☐ EQUIPMENT

Power supply	1.2.3
Recorder	1.2.4
Variable speed drive	1.1.17

6.3.3 ☐ TEST PROCEDURE (SEE FIGURE 6.3A)

To properly measure Voltage Resolution the values of the high pass filter must first be determined. Referring to Figure 6.3A, C_F and R_A are dependent upon the desired time constant, test pot

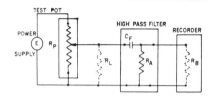

WHERE: E = INPUT VOLTAGE

R_P = TOTAL RESISTANCE OF TEST POT

R_L = LOAD RESISTANCE SIMULATING USER APPLICATION
(USED ONLY WHEN SPECIFIED.)

C_F = FILTER CAPACITANCE

R_A = FILTER SHUNT RESISTANCE

R_B = INPUT IMPEDANCE TO RECORDER

FIGURE 6.3A   Measurement of voltage resolution

resistance and input impedance to the recorder. The desired time constant is further dependent upon the rate of "cutting turns" determined from the velocity of the drive and the Theoretical Resolution. The following guides are given for the determination of the numerical values of the components of the high pass filter and shaft travel velocity:

A. R_F is a value chosen to minimize the loading effect on the test pot. It should be at least ten (10) times the Total Resistance of the test pot. Should the ratio become much less than 10:1, the apparent magnitude of the Voltage Resolution will be distorted.

B. R_A is determined from the total effective shunt resistance (R_F) and is calculated from $R_A = R_B R_F/(R_B - R_F)$.

C. The filter capacitance is calculated from R_F and the desired time constant ($R_F C_F$) for the filter network. The time constant should be approximately 1/10th the rate of "cutting turns" such that each resolution spike decays sufficiently before another is generated. Obviously, this depends upon the travel speed of the test pot and its inherent Theoretical Resolution. It should be further noted that the allowable rate of "cutting turns" must be commensurate with the frequency response of the recorder.

Example:

If: $R_P = 50 \text{ K}\Omega$
$R_B = 1\text{meg}\Omega$

Theoretical Resolution = 3000 turns or 0.033%
Frequency Response of Recorder =100 Hz

Then:

1. R_F should be at least 500KΩ
2. $R_A = R_B R_F (R_B - R_F) = 1 \text{ meg}\Omega$
3. Assuming a test pot drive speed of 1 RPM, the rate of cutting turns = 0.02 sec/turn
4. The desired time constant (T) should then be approximately 0.002 sec
Therefore, $C_F = T/R_F = 0.004 \ \mu f$
5. In this case the rate of "cutting turns" (20 milliseconds/pulse) is significantly large when compared with the time constant of the filter network and the approximate response time (0.35/100) of a 100 Hz recorder, thus permitting proper display of the resolution of the test pot.

D. The resistor R_L shown in Figure 6.3A is employed only when specified by the user. It is intended to simulate the load in the actual application such that the recorder output of the Voltage Resolution test will depict as closely as possible the output of the pot in use.

Having determined the high pass filter constants, the calibration circuit of Figure 6.3B is established. The constant K is chosen as some convenient increment of the magnitude of the Voltage Resolution to be displayed on the recorder. For example, if it is desired to calibrate the recorder to the limit of a desired resolution of 0.1% then K = 0.001.

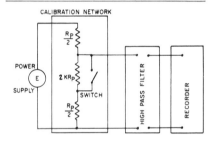

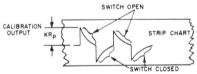

WHERE: E = INPUT VOLTAGE

R_P = TOTAL RESISTANCE OF TEST POT

K = CONSTANT EQUAL TO CALIBRATION SIGNAL LIMIT FOR EVALUATION OF VOLTAGE RESOLUTION RECORDING

FIGURE 6.3B   Calibration circuit

Set the power supply to a voltage providing sufficient sensitivity, but small enough to prevent overheating of the pot under test. With the switch closed the recorder is nulled to the center of the recording. Then, while opening and closing the switch, adjust the recorder gain until the zero to peak output being recorded is some convenient distance on the chart recording.

Then replace calibration network with the test pot as shown in Figure 6.3A. With the recorder chart speed set to a convenient value for displaying the resolution, the pot is driven in one direction through the full Actual Electrical Travel (3.7) or specified portion thereof with the variable speed drive set to a predetermined value. Compare the magnitude of the maximum incremental pulse change shown on the recording with the zero to peak calibration- signal to determine the Voltage Resolution.

Note 1: Toward the center of the test pot the Voltage Resolution appears as half the magnitude and double the number of steps unevenly spaced. This is due to the leading and trailing edge effects of the wiper contact. This effect should be considered when determining the proper time constant for the high pass filter and the drive speed for the test pot, if true representation of this region is deemed important.

Note 2: If a dropping resistor is used in series with the power supply to establish the desired input voltage, it will have the effect of shifting the aforementioned half steps to one end of the test pot and not give true indication of the Voltage Resolution. Should the dropping resistor be made sufficiently large, the half steps may be eliminated completely.

6.4 ■ DIELECTRIC WITHSTANDING VOLTAGE

6.4.1 ■ OBJECT To measure the ability of a potentiometer to withstand a specified potential of a given characteristic between the terminals of each cup and the exposed conducting surfaces of the potentiometer, or between the terminals of each cup and the terminals of every other cup in the gang under prescribed conditions without exceeding a specified leakage current value.

6.4.2 ■ EQUIPMENT

High voltage source	1.2.10
AC voltmeter	1.2.11
Leakage current indicating device	1.2.12

6.4.3 ■ TEST PROCEDURE The magnitude of the test voltage should be as specified. The voltage is taken from an alternating current supply of a commercial 60 cycle line frequency and waveform. Connect the equipment by applying the high voltage source between the potentiometer terminals (interconnected) and the shaft or case. Raise the test voltage gradually from zero to the proper maximum value at a rate of 500 volts per second maximum. Maintain the test voltage at this value while the shaft is moved through one full sweep of Total Mechanical Travel (3.1) in a time interval not less than 5 seconds nor more than 60 seconds. Monitor the leakage current indicating device throughout this test for evidence of damage, arcing, breakdown, or leakage currents in excess of 1 milliampere. Upon completion of the test, the voltage should be reduced gradually to zero prior to disconnecting the test leads, for operator safety.

For ganged potentiometers, repeat the foregoing applying the high voltage between the terminals of each cup and the terminals of every other cup.

6.5 ■ INSULATION RESISTANCE

6.5.1 ■ OBJECT To measure the resistance of a potentiometer to a specified impressed DC voltage between the terminals of each cup and the exposed conducting surfaces of the potentiometer, or between the terminals of each cup and the terminals of every other cup in the gang under prescribed conditions.

6.5.2 ■ EQUIPMENT

Insulation resistance test set	1.2.21
Shaft positioning device	1.1.2

6.5.3 ■ TEST PROCEDURE Interconnect all electrically insulated terminals of each cup of the potentiometer. Connect the insulation resistance test set to the terminal of the first cup and to some exposed conducting surface of the potentiometer (as shaft or case) and apply the specified test voltage. Unless otherwise specified the test voltage is 500V DC. Maintain the test voltage at this value for 5 to 10 seconds before initiating movement of the shaft through one full sweep of the Total Mechanical Travel (3.1) in a time interval of not less than 5 seconds nor more than 60 seconds. Monitor the indicated Insulation Resistance during this voltage application. The Insulation Resistance is the minimum value observed during the movement of the shaft.

Note: Any region where there is sharp decrease in the measured value during shaft travel should be examined at reduced speed to ascertain that damping of the indicating meter has not passed the region of lower values than that indicated.

For gang potentiometers, repeat the procedure for each cup applying the high voltage between the terminals of each cup and the exposed conducting surface of the potentiometer. The procedure is continued by applying the high voltage between the terminals of each adjacent cup in the gang.

7   *ac characteristics*

7.1 ■ TOTAL INPUT IMPEDANCE

7.1.1 ■ OBJECT To measure, at a specified voltage and frequency and with the shaft positioned to give a maximum value, the Total Input Impedance between the two input terminals with open circuit between output terminals.

7.1.2 ■ EQUIPMENT

AC voltage source	1.2.13
AC vacuum tube voltmeter	1.2.15
AC ammeter	1.2.16

7.1.3 ■ TEST PROCEDURE Connect the potentiometer to the source at the input terminals with the wiper open circuited. Move the pot shaft to a position on the Electrical Overtravel or against the stop at minimum voltage if no Electrical Overtravel or discontinuity exists. Measure the current through the resistance element and the voltage across the element at the specified voltage and frequency. The Total Input Impedance is calculated from the formula:

$$\text{Total Input impedance} = \frac{\text{Voltage applied}}{\text{Current}}$$

7.2 ☐ ■ OUTPUT IMPEDANCE

7.2.1 ☐ OUTPUT IMPEDANCE — WIREWOUND

7.2.1.1 ☐ OBJECT To measure the Output impedance defined as the maximum impedance between wiper and either end terminal with the input shorted and at a specified voltage and frequency.

7.2.1.2 ☐ EQUIPMENT

AC voltage source	1.2.13
AC vacuum tube voltmeter	1.2.15
AC microammeter	1.2.17

7.2.1.3 ☐ TEST PROCEDURE The source is connected to the potentiometer between the wiper and either end terminal with the input terminals shorted. The voltage applied and the current through the wiper are measured at the specified voltage and frequency as the pot shaft is rotated or translated over the Actual Electrical Travel (3.7). The Output Impedance is the applied voltage divided by the minimum current reading. The applied voltage should not exceed that which would cause a current to flow greater than 10 milliamperes.

7.2.2 ■ OUTPUT IMPEDANCE — NONWIREWOUND

7.2.2.1 ■ OBJECT To measure the Output Impedance defined as the maximum impedance between wiper and either end terminal with the input shorted and at a specified voltage and frequency.

7.2.2.2 ■ EQUIPMENT

AC voltage source	1.2.13
AC vacuum tube voltmeter	1.2.15
AC microammeter	1.2.17

7.2.2.3 ■ TEST PROCEDURE

Connect the source to the potentiometer between the wiper and either end terminal with the input terminals shorted. The voltage applied and the current through the wiper are measured at the specified voltage and frequency as the pot shaft is rotated or translated over the Theoretical Electrical Travel (3.8). The Output Impedance is the applied voltage divided by the minimum current reading. The applied voltage should not exceed that which would cause a current flow greater than 1.0 milliamperes.

7.3 ☐ ■ QUADRATURE VOLTAGE

7.3.1 ☐ QUADRATURE VOLTAGE — WIREWOUND

7.3.1.1 ☐ OBJECT To measure the maximum value of that portion of the output voltage which is $\pm90°$ out of time phase with the input voltage, expressed as volts per volt applied, at a specified input voltage and frequency.

7.3.1.2 ☐ EQUIPMENT

AC voltage source	1.2.13
Ratio transformer	1.2.14
AC vacuum tube voltmeters	1.2.15
Potentiometer mounting fixture	1.1.1

7.3.1.3 ☐ TEST PROCEDURE Mount the potentiometer in the potentiometer mounting fixture and connect in the circuit shown in Figure 7.3A. Set the test pot to a position on the

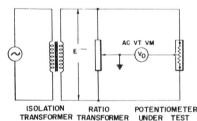

ISOLATION   RATIO   POTENTIOMETER
TRANSFORMER TRANSFORMER UNDER TEST

FIGURE 7.3A Quadrature voltage measurement

Actual Electrical Travel (3.7). Adjust the ratio transformer for a minimum null voltage (V_o). The quadrature voltage (e_q) is calculated from the formula:

$$e_q = \frac{V_o}{E}$$

This procedure is repeated over the Actual Electrical Travel until a maximum value is obtained.

Note: Wiring must be arranged to provide minimum possible stray capacitances.

7.3.2 ■ QUADRATURE VOLTAGE — NONWIREWOUND

7.3.2.1 ■ OBJECT To measure the maximum value of that portion of the output voltage which is $\pm90°$ out of time phase with the input voltage, expressed as volts per volt applied, at a specified input voltage and frequency.

7.3.2.2 ■ EQUIPMENT

AC voltage source	1.2.13
Ratio transformer	1.2.14
AC vacuum tube voltmeters	1.2.15
Potentiometer mounting fixture	1.1.1

7.3.2.3 ■ TEST PROCEDURE The potentiometer is mounted in the potentiometer mounting fixture and connected in the circuit shown in Figure 7.3A. Set the test pot to a position on the Theoretical Electrical Travel (3.8). Adjust the ratio transformer for a minimum null voltage (V_o). The quadrature voltage (e_q) is calculated from the formula:

$$e_q = \frac{V_o}{E}$$

This procedure is repeated over the Theoretical Electrical Travel until a maximum value is obtained.

Note: Wiring must be arranged to provide minimum possible stray capacitances.

7.4.1 ■ OBJECT To measure the phase difference between the sinusoidal input and output voltages at a specified input voltage and frequency with the shaft at a specified position.

7.4.2 ■ EQUIPMENT

AC voltage source	1.2.13
Ratio transformer	1.2.14
AC vacuum tube voltmeters	1.2.15
Potentiometer mounting fixture	1.1.1

7.4.3 ■ TEST PROCEDURE Phase Shift in a precision potentiometer varies in magnitude with shaft position (see Figure 7.4A). The figure shows the relationship between Phase Shift, Quadrature Voltage (7.3) and shaft position for normal conditions and indicates that Phase Shift becomes maximum at essentially zero output, a somewhat meaningless value. A more valuable measure of Phase Shift can be described at the point of maximum Quadrature Voltage, and for the purposes of this document Phase Shift will be measured in terms of Quadrature Voltage.

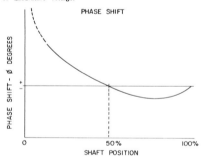

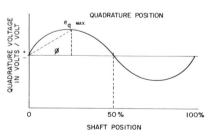

FIGURE 7.4A Phase shift determination

Since Quadrature Voltage is expressed in terms of volts/volt, Phase Shift can be calculated from:

$$\text{Phase Shift} = \tan^{-1}(e_q)$$

Therefore, Phase Shift is determined by measuring the Quadrature Voltage and calculating as above using e_q maximum unless otherwise specified.

8 *mechanical characteristics*

8.1 ■ SHAFT RUNOUT

8.1.1 ■ OBJECT To measure the eccentricity of the shaft diameter with respect to the rotational axis of the shaft and measured at a specified distance from the end of the shaft. The body of the potentiometer is held and the shaft is rotated with a specified load applied radially to the shaft.

8.1.2 ■ EQUIPMENT

Dial indicator	1.1.5
Dial indicator holding fixture	1.1.6
Potentiometer mounting fixture	1.1.1
Dead weight load	1.1.9
Cylindrical shaft adaptor	1.1.8
Surface plate or equivalent firm surface	

8.13 ■ TEST PROCEDURE (SEE FIGURE 8.1A) Mount the potentiometer firmly with the shaft axis in a horizontal position and hold rigid with respect to the dial indicator. Posi-

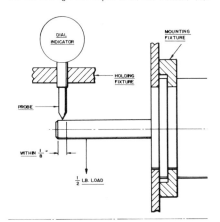

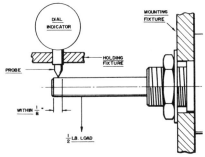

FIGURE 8.1A Measurement of shaft runout

tion the dial indicator such that its probe contacts the shaft within 1/8 inch from the end of the shaft or the edge of any interruption of the smooth cylindrical shaft surface. This measurement requires that the shaft be a smooth cylindrical surface at the point of measurement and when specified, shafts with noncylindrical surfaces such as flats, slots, or splines will require the use of the cylindrical shaft adaptor. Depress the probe sufficiently to insure a proper positive and negative indication of the dial. Apply 1/2 pound load radially to the shaft to remove Shaft Radial Play and position as close to the indicator probe as is practical. For small diameter shafts reduce the magnitude of the load applied such that it never exceeds that which would cause the shaft to permanently deform. Then rotate the shaft slowly through 360° or through its Total Mechanical Travel (3.1), whichever is less. The Shaft Runout is the total indicated reading determined by adding the maximum positive and negative readings without regard to algebraic signs.

8.2 ■ LATERAL RUNOUT

8.2.1 ■ OBJECT To measure the perpendicularity of the mounting surface of the potentiometer with respect to the rotational axis of the shaft measured on the mounting surface at a specified distance from the outside edge of the mounting surface. The shaft is held and the body of the potentiometer is rotated while specified loads are applied radially and axially to the body of the potentiometer.

8.2.2 ■ EQUIPMENT

Dial indicator	1.1.5
Dial indicator holding fixture	1.1.6
Potentiometer shaft holding fixture	1.1.7
Two dead weight loads	1.1.9
Surface plate or equivalent firm surface	

8.2.3 ■ TEST PROCEDURE (SEE FIGURE 8.2A) Make the measurement with the potentiometer mounted firmly in the shaft holding fixture and with the shaft axis in a vertical position. Clamp the shaft within 1/8 inch of the front surface of the potentiometer without interference and hold rigid with respect to the dial indicator. The potentiometer body is to remain free to rotate. Care should be taken to insure that the shaft is not distorted in any way due to the mode of clamping or the inherent weight of the potentiometer body. Position the dial indicator such that its probe contacts the smooth portion of mounting surface of the potentiometer less than 1/8 inch from the outside edge of the mounting surface. The probe should be depressed sufficiently to insure a proper positive and negative indication. A 1/2 pound load is applied normal to the centerline of the shaft axis on the potentiometer body within 1/8″ of the mounting surface. Simultaneously, a 1/2 pound load is applied axially on the centerline of the potentiometer. The loads serve to remove the Shaft Radial and End Plays. For small diameter shafts the magnitude of the load applied shall be reduced such that it never exceeds that which would cause the shaft to permanently deform. The body of the potentiometer is then slowly rotated through 360° or through the Total Mechanical Travel (3.1), whichever is less. The Lateral Runout is the total indicated reading determined by adding the maximum positive and negative readings without regard to algebraic signs.

8.3 ■ PILOT DIAMETER RUNOUT

8.3.1 ■ OBJECT To measure the eccentricity of the pilot diameter with respect to the rotational axis of the shaft indicated on the pilot diameter. The shaft is held and the body of the potentiometer is rotated while a specified load is applied radially to the body of the potentiometer.

LEGEND:

☐ WIREWOUND
■ WIREWOUND AND NONWIREWOUND
■ NONWIREWOUND

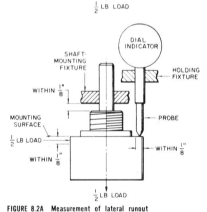

FIGURE 8.2A Measurement of lateral runout

8.3.2 ■ EQUIPMENT

Dial indicator	1.1.5
Dial indicator holding fixture	1.1.6
Potentiometer shaft holding fixture	1.1.7
Two dead weight loads	1.1.9
Surface plate or equivalent firm surface	

8.3.3 ■ TEST PROCEDURE (SEE FIGURE 8.3A) Make the measurement with the potentiometer mounted firmly in the shaft holding fixture and with the shaft axis in a vertical position. Clamp the shaft within 1/8 inch of the front surface of the potentiometer without interference and hold rigid with respect to the dial indicator. The potentiometer body is to remain free to rotate. Take care to insure that the shaft is not distorted in any way due to the mode of clamping or the inherent weight of the potentiometer body. Position the dial indicator such that its probe contacts the periphery of the pilot surface near the mid-point of the surface. Depress the probe sufficiently to insure a proper positive and negative indication. A 1/2 pound load is applied normal to the centerline of the shaft axis on the potentiometer body within 1/8

inch of the mounting surface to remove the Shaft Radial Play. For small diameter and/or long shafts reduce the magnitude of the load applied such that it never exceeds that which would cause the shaft to permanently deform. The body of the potentiometer is then slowly rotated through 360° or through the Total Mechanical Travel (3.1), whichever is less. The Pilot Diameter Runout is the total indicated reading determined by adding the maximum positive and negative readings without regard to algebraic signs.

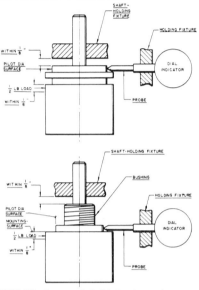

FIGURE 8.3A Measurement of pilot diameter runout

8.4 ■ SHAFT RADIAL PLAY

8.4.1 ■ OBJECT To measure the total radial excursion of the shaft with respect to the pilot diameter, indicated at a specified distance from the front surface of the potentiometer, with a specified radial load applied alternately in opposite directions at a specified point.

8.4.2 ■ EQUIPMENT

Dial indicator	1.1.5
Dial indicator holding fixture	1.1.6
Potentiometer mounting fixture	1.1.1
Dead weight load	1.1.9
Surface plate or equivalent firm surface	

8.4.3 ■ TEST PROCEDURE (SEE FIGURE 8.4A) Mount the potentiometer firmly with the shaft axis in a horizontal position and hold rigid with respect to the dial indicator. Position the dial indicator such that its probe contacts the shaft within 1/8 inch of the front surface of the potentiometer body. Depress the probe sufficiently to insure a proper positive and negative indication of the dial. A 1/2 pound load is applied normal to the shaft at a point 1/2 inch from the front surface of the potentiometer (or at the end of the shaft for shaft extensions less than 1/2 inch) in two opposite directions, one at a time, along the axis of the dial indicator probe (or perpendicular to the stylus if a pivot pointer indicator is used). Then rotate the plane of application of load 90° relative to the potentiometer body without rotating the shaft and then repeat procedure. For small diameter and/or long shafts reduce the magnitude of the load applied such that it never exceeds that which would cause the shaft to permanently deform. The Shaft Radial Play is the largest total indicated reading for either of the two readings. The total indicated reading is determined by adding the maximum positive and negative readings without regard to algebraic signs.

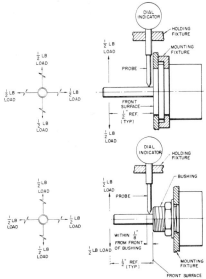

FIGURE 8.4A Measurement of shaft radial play

8.5 ■ SHAFT END PLAY

8.5.1 ■ OBJECT To measure the total axial excursion of the shaft with respect to the potentiometer body, indicated at the end of the shaft with a specified axial load applied alternately in opposite directions.

8.5.2 ■ EQUIPMENT

Dial indicator	1.1.5
Dial indicator holding fixture	1.1.6
Potentiometer mounting fixture	1.1.1
Dead weight load	1.1.9
Surface plate or equivalent firm surface	

8.5.3 ■ TEST PROCEDURE (SEE FIGURE 8.5A) The potentiometer is mounted firmly by its normal means with the shaft axis in a vertical position and held rigid with respect to the dial indicator, leaving the shaft free to rotate.

The dial indicator is positioned with its probe parallel (or normal if pivot pointer indicator is used) to the axis of the shaft and in contact with the end of the shaft on the centerline. The probe is depressed sufficiently to insure a proper positive and negative indication. A 1/2 pound load is applied alternately in opposite directions along the axis of the shaft. The Shaft End Play is the total indicated reading determined by adding the maximum positive and negative readings without regard to algebraic signs.

8.6 ■ STARTING TORQUE

8.6.1 ■ OBJECT To measure the maximum moment in the clockwise and counterclockwise directions required to initiate shaft rotation anywhere in the Total Mechanical Travel.

8.6.2 ■ EQUIPMENT

Potentiometer mounting fixture	1.1.1
Load device	1.1.4

8.6.3 ■ TEST PROCEDURE Mount the potentiometer firmly by its normal mounting means. Connect the load device to the potentiometer shaft so as to prevent relative movement between the two. A torque is applied through the load device and about the axis of the potentiometer shaft until shaft rotation is initiated. Care should be exercised to avoid applying radial or axial loads that will cause the shaft to deform or influence the measurement.

The procedure is followed for each direction of rotation at each obvious point of mechanical or electrical junction (e.g., ends of dead space, taps and shorts) and at three randomly selected points over the Total Mechanical Travel (3.1). The Starting Torque is the maximum indicated reading of the load device.

8.7 ■ RUNNING TORQUE

8.7.1 ■ OBJECT To measure the maximum moment in the clockwise and counterclockwise directions required to sustain uniform shaft rotation at a specified speed throughout the Total Mechanical Travel.

8.7.2 ■ EQUIPMENT

Potentiometer mounting fixture	1.1.1
Load device	1.1.4

8.7.3 ■ TEST PROCEDURE Mount the potentiometer firmly by its normal mounting means. Connect the load device to the potentiometer shaft so as to prevent relative movement between the two. Sufficient torque is applied through the load device and about the axis of the potentiometer shaft until a sustained uniform shaft rotation of 4 RPM is achieved. Care should be exercised to avoid applying radial or axial loads that will cause the shaft to deform or influence the measurement. This procedure is followed over the Total Mechanical Travel (3.1) in both the clockwise and counterclockwise directions. The Running Torque is considered to be the maximum reading of the load device.

8.8 ■ MOMENT OF INERTIA

8.8.1 ■ OBJECT To measure the mass Moment of Inertia of the rotating element of the potentiometer about its rotational axis.

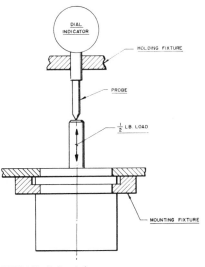

FIGURE 8.5A Shaft end play

Note: This procedure requires dismantling of the potentiometer and in this sense is considered destructive. Furthermore, it does not include the inertia of the inner races of ball bearings.

8.8.2 ■ EQUIPMENT

Steel wire (music wire)	
Moment of inertia adaptor	1.1.13
Mass of known inertia	1.1.14
Timing clock	1.1.15

8.8.3 ■ TEST PROCEDURE Suspend mass of known moment of inertia from the wire and adaptor such that the centerline of the known mass coincides with the centerline of the wire. With the wire attached to a rigid point, twist the known mass such that oscillation about the centerline will occur. Measure the period of oscillation. The system should be so located that air circulation of vibrations will not cause any swaying of the pendulum. Remove the known mass and connect the rotating elements of the potentiometer in the same manner to the wire and measure its period of oscillation. The Moment of Inertia of the rotating member of the potentiometer may then be determined from the following equation.

$$I_P = (I_g + I_A)\left(\frac{T_P}{T_g}\right)^2 - I_A$$

$I_P =$ Moment of inertia of the rotating member in gm — cm^2
$I_g =$ Moment of inertia of the known mass in gm — cm^2
$I_A =$ Moment of inertia of the adaptor in gm — cm^2
$T_P =$ Period of rotating member of the potentiometer and adaptor in seconds
$T_g =$ Period of known mass of adaptor in seconds

8.9 ■ STOP STRENGTH

8.9.1 ■ STATIC STOP STRENGTH

8.9.1.1 ■ OBJECT To determine the ability of the stop mechanism to withstand static load for a specified period of time without permanent change of the stop positions greater than specified.

This test is performed primarily for design evaluation and is not considered a routine inspection test. Therefore the measurements taken for Static Stop Strength have no relationship to the specified limits of other parameters such as Total Mechanical Travel, Mechanical Overtravel, etc.

8.9.1.2 ■ EQUIPMENT

Travel measuring device	1.1.3
Load device	1.1.4
Shaft load adaptor	1.1.10

8.9.1.3 ■ TEST PROCEDURE Mount the potentiometer in the travel measuring device and securely lock the shaft to the travel indicator. The absolute readings are recorded at each end of the Total Mechanical Travel (3.1). Then apply the specified Static Stop Strength load to the stops and hold against the stop for 10 seconds at each end of the mechanical travel. The point of application of the load to the shaft should be within 1/8 inch of the front mounting surface of the potentiometer to avoid applying unwanted moments or load to the shaft. Final readings of the absolute values of the ends of Total Mechanical Travel are then taken in the same manner as the initial values. The differences between the final readings and their corresponding initial readings are the permanent changes of the stop positions. The permanent change should be no greater than 1° or 0.005″.

Note 1: It is realized that it may be difficult (if not impossible) to apply the Static Stop Strength load while the shaft remains locked to the travel measuring indicator. However, it is necessary for the purpose of this test not to disturb this relationship so that a change in relationship between the shaft and the resistance element can be detected. This change can be caused by a twisting or elongating of the shaft

or a movement of the wiper mechanism relative to the shaft when the load is applied.

To accomplish this a coupling may be devised to connect the travel indicator to the pot shaft or modified shaft load adaptor in such a manner that the two may be disconnected and reconnected without losing the desired relationship. This assumes that the body of the potentiometer remains fixed throughout the test.

Note 2: A permanent attachment to the shaft may be required to permit large loads to be applied and prevent otherwise damaging the shaft. In this sense the test is considered destructive.

8.9.2 ■ DYNAMIC STOP STRENGTH

8.9.2.1 ■ OBJECT To determine the ability of the stop mechanism to withstand a specified inertia load, at a specified shaft velocity for a specified number of impacts, without a permanent change of the stop position greater than specified.

This test is performed primarily for design evaluation and is not considered a routine inspection test. Therefore the measurements taken for Dynamic Stop Strength have no relationship to the specified limits of other parameters such as Total Mechanical Travel, Mechanical Overtravel, etc.

8.9.2.2 ■ EQUIPMENT

Travel measuring device	1.1.3
Inertia load	1.1.11
Shaft load adaptor	1.1.10

8.9.2.3 ■ TEST PROCEDURE Mount the potentiometer in the travel measuring device and securely lock the shaft to the travel indicator. The absolute readings are recorded at each end of the Total Mechanical Travel (3.1). Without disturbing the relationship of the shaft and the travel indicator (see Note 1), the potentiometer is coupled to the specified inertia load. The point of application of the load to the shaft should be within 1/8 inch of the front mounting surface of the potentiometer to avoid applying unwanted moments or loads to the shaft. The load is then caused to move toward one stop at a specified constant velocity. The input energy is removed just prior to the pot shaft reaching the stop such that the inertia load may freely move into the stop without excessive external braking or energy input. The direction of displacement of the load is reversed and the procedure repeated. This procedure is repeated for a total of 100 impacts at each stop. At the conclusion the absolute values of the ends of the Total Mechanical Travel are again measured and recorded. The differences between the final readings and their corresponding initial readings are the permanent changes of the stop positions. The permanent change should be no greater than 1° or 0.005″.

Note 1: It is realized that it may be difficult (if not impossible) to apply the Dynamic Stop Strength load while the shaft remains locked to the travel measuring indicator. However, it is necessary for the purposes of this test not to disturb this relationship, so that a change in relationship between the shaft and the resistance element can be detected. This change can be caused by a twisting or elongating of the shaft or a movement of the wiper mechanism relative to the shaft when the load is applied.

To accomplish this a coupling may be devised to connect the travel indicator to the pot shaft or modified shaft load adaptor in such a manner that the two may be disconnected and reconnected without losing the desired relationship. This assumes that the body of the potentiometer remains fixed throughout the test.

Note 2: A permanent attachment to the shaft may be required to permit large loads to be applied and prevent otherwise damaging the shaft. In this sense, the test is considered destructive. ■ □ ■

trimming potentiometer standard

VRCI-T-215

REVISION A
INSPECTION AND TEST PROCEDURES
APPROVED BY
VARIABLE RESISTIVE COMPONENTS INSTITUTE
FEBRUARY, 1974

INTRODUCTION

The test procedures presented in this standard cover the methods of measurement of those characteristics of trimming potentiometers that are of common concern to the manufacturer and user. These procedures exclude many of the environmental exposure tests (thermal shock, moisture resistance, etc.) for which adequate standards exist elsewhere.

The purpose of this standard is to assist in obtaining better correlation of inspection results between the user's plant and the manufacturer's plant. Although the VRCI realizes that there are alternate methods to those proposed here, it believes that standardization can better achieve a common basis for evaluation. If an alternate procedure is preferred, the party using the alternate must assume the burden of proof of equivalency. Where applicable, test procedures are to be performed at, or corrected to standard conditions as follows: 25°C, 760mm of Hg, and 50% relative humidity.

Many of these procedures call for specific important test parameters with numerical values: e.g., operating speeds, applied voltages, currents, etc. Various factors governed their selection—standardization, availability of equipment, ease of test, and above all, the prevention of damage to the potentiometer. It should be noted, however, that these are only recommended typical values and are subject to modification in individual cases depending upon specific requirements of the end use.

TABLE OF CONTENTS

1 EQUIPMENT DESCRIPTION

1.1 MECHANICAL

1.1.1 SHAFT POSITIONING DEVICE: A device to provide a means for moving the shaft to any position and maintaining a stable setting during electrical measurements. The shaft positioning device must not apply any axial or radial loads on the shaft of the potentiometer.

1.1.2 TRAVEL MEASURING DEVICE: A device composed of a shaft position indicator and a potentiometer mounting fixture, which will indicate shaft position relative to the potentiometer body. The device has an overall accuracy of such value required to make total measuring errors for any test less than 1/10th the specified tolerance. The travel measuring device mounting provision should in no way distort or mar the finish of the part nor exert any radial or axial loads.

1.1.3 CONSTANT SPEED DRIVE: The drive must be able to operate at a constant velocity as specified. The device should have a slip clutch provision to prevent damage to the mechanical stops in' the unit, where applicable.

1.1.4 TEMPERATURE TEST CHAMBER: The temperature of the chamber should be adjustable within $\pm3°C$ of the test temperature. The chamber should be stable within $\pm0.5°C$ at any given point in the proximity of the test specimens. The temperature gradient in this area should not exceed $\pm1°C$. If larger gradients exist, the temperature must be monitored with a thermocouple immediately adjacent to the test specimens. Air flow around the area of the test specimens should be at least 60 CFM.

1.1.5 LIFE TEMPERATURE TEST CHAMBER: The chamber should be stable within $\pm5°C$ of the test temperature. The life fixtures will be arranged so the temperature of any one fixture will not appreciably influence the temperature of any other fixture. The rest temperature shall be monitored with a suitable indicator having an accuracy of $\pm2°$.

1.1.6 VACUUM CHAMBER: Must be capable of maintaining specified altitude within 2% for specified test time.

1.1.7 ALTITUDE MEASURING DEVICE: A manometer or equivalent device that measures the specified altitude and has an accuracy of 0.1″ Hg.

1.1.8 LOAD DEVICE: A device that provides a means for applying a force or torque of known magnitude to a pot shaft. The device, such as a spring scale, weight at a known radius, torque wrench, etc., should have an accuracy of 1/10th the specified value.

1.1.9 POTENTIOMETER MOUNTING FIXTURE: A fixture to rigidly hold the test specimen by the normal mounting means leaving the adjustment shaft free to move.

1.1.10 ROTATIONAL LIFE FIXTURE: A fixture to rigidly hold the test specimen by the normal mounting means, leaving the shaft in proper alignment with the drive. The drive shall be controlled automatically so that the test specimen wiper is cycled through not less than 90% of the adjustment travel and returned to the starting point. The drive speed shall be constant and at the necessary RPM to complete one cycle within the specified period of time.

1.1.11 PUSH-PULL DEVICE: A device that applies a push or pull force of 2 pounds to the test specimen terminal with an accuracy of $\pm2\%$.

1.1.12 BEND FIXTURE: The fixture shall firmly clamp the test specimen and each terminal pin shall be able to be bent through $90°$ at a point 1/8 inch from the body of the test specimen, with the radius of the curvature at the bend approximately 1/32 inch. The terminal pin shall be returned to the original position, bent $90°$ in the opposite direction, and again returned to the original position.

1.1.13 . HOT WATER BATH: A bath of sufficient size that no part of the test specimen is at a depth of less than 1 inch in water controlled at a temperature of $50°C$ $+5°C$ $-0°C$ or $85°C$ $+5°C$ $-0°C$ as required.

1.2 ELECTRICAL

1.2.1 OUTPUT RATIO EQUIPMENT: The Kelvin-Varley voltage divider or a modification of it is recommended for measurement of the output voltage ratio. The voltage dividers are usually of two types:

(1) Decade voltage dividers (4 or 5 decades)

(2) Digital ratiometers (4 or 5 places)

The decade voltage divider is used in conjunction with a null detector described in 1.2.4. The digital ratiometers are generally self-nulling with direct numerical readouts. The equipment accuracy, resolution and repeatability must equal or be less than 1/10th of the specified tolerance. The voltage applied should never exceed the voltage power rating of the test unit.

1.2.2 RESISTANCE MEASURING DEVICE: Care must be taken when using any resistance measuring device that the current drawn does not exceed the current carrying capacity or rating of the unit.

1.2.2.1 OHMMETER (HAND SET VOLT-OHM-METER): This type ohmmeter generally is not sufficiently accurate for quantitative measurements and, when used on its lowest scales, applies a voltage without internal impedance limits. Its use should be avoided.

1.2.2.2 WHEATSTONE BRIDGE: When resistance tolerances are less than 10 per cent and the resistance values are above 10 ohms, use a Wheatstone bridge with an accuracy of 1/10th the tolerance to be measured. For resistance values above 1 megohm, it is necessary to use a guarded Wheatstone bridge. (Commercially available Wheatstone bridges have an accuracy of 0.01 per cent to 10 megohms and 0.5 per cent to 1000 megohms.) A null detector recommended for use in the bridge circuit is described in 1.2.4.

1.2.2.3 KELVIN BRIDGE: For resistance values less than 10 ohms, a Kelvin (Thomson) bridge is used. The accuracy of the bridge required is 1/10th the specified tolerance to be measured. (Commercially available Kelvin bridges have an accuracy of 0.25 per cent from 0.0005 ohms and 0.5 per cent from 0.0001 to 0.0005 ohms.) A null detector recommended for use in the bridge circuit is described in 1.2.4.

1.2.2.4 DIGITAL OHMMETER: Digital ohmmeters of equivalent accuracy can be used in place of Wheatstone or Kelvin bridges.

1.2.3 POWER SUPPLIES: *NOTE:* A DC voltmeter should be used with all power supplies to determine actual voltage.

1.2.3.1 FOR VOLTAGE RATIO MEASUREMENT: 1-10 volts DC with no limitations on voltage stability, current regulation or line regulations. (Caution should be taken not to exceed the voltage or current limit of the unit under test.) If its capacity is sufficient for the current drawn by a potentiometer, a battery may be used.

1.2.3.2 LIFE POWER SUPPLIES: AC or DC power supplies with voltage and current range outputs within 3% of the required value. Line regulation to be within 0.5%.

1.2.4 NULL DETECTOR: The null detector should have the following characteristics:

Sensitivity:	
Current	1×10^{-9} amp/mm
Voltage	1×10^{-6} volts/mm
Stability:	1 mm drift per hour
Damping:	Critical
Response:	To final value within 1 second

1.2.5 CONSTANT CURRENT SOURCE: The source should produce any constant current ±5% between 50 microamperes and 30 milliamperes DC under load.

1.2.6 OSCILLOSCOPE: A high gain DC oscilloscope should have a high persistence screen to retain the image for a minimum of 1/2 second. It should have a minimum input impedance of one megohm and a flat frequency response from DC to a minimum of 50,000 Hz. An equivalent detector may be used in the equivalent noise resistance and contact resistance variation circuit.

1.2.7 HIGH VOLTAGE SOURCE: A source variable from zero to the maximum specified VRMS at 60 Hz of sine waveform with 5 per cent maximum distortion. A current limiter to limit the leakage current to 150 per cent of the maximum specified value should be provided.

1.2.8 AC VOLTMETER: The AC voltmeter should have the following characteristics:

Voltage range:	133% of the specified or test value minimum
Approximate impedance in ohms at 60 Hz:	1,000 ohms/volt
Frequency range:	50-125 Hz minimum
Accuracy:	±5% at the specified voltage voltage and frequency
Temperature stability:	0.1 per cent/°C maximum at the specified or test voltage and frequency

1.2.9 LEAKAGE CURRENT INDICATING DEVICE: An AC ammeter with an accuracy of 5 per cent of the allowable rate and covering the applicable range.

1.2.10 TEMPERATURE MEASURING DEVICES: The accuracy of the measurement equipment should be 0.5 per cent and temperature changes of 0.5°C should be detectable.

1.2.11 INSULATION RESISTANCE TEST SET: A suitable commercial megohm bridge, megohmmeter or equivalent with an accuracy of 1/10th the value to be measured and a built-in source voltage of the specified magnitude.

1.2.12 DC VOLTMETER: A suitable voltmeter shall be selected of sufficient voltage ranges and with an accuracy of 1/10th the value to be measured.

1.2.13 TIMING CLOCK: A timing clock with readability and repeatability to 0.1 second.

2 ROTATION AND TRANSLATION

2.1 TOTAL MECHANICAL TRAVEL-STOPS

2.1.1 EQUIPMENT: See Continuity Travel (2.4)

2.1.2 TEST PROCEDURE: Total mechanical travel-stops is synonymous with continuity travel. Refer to the procedure for continuity travel (2.4).

2.2 TOTAL MECHANICAL TRAVEL-CLUTCH

2.2.1 EQUIPMENT: See Continuity Travel (2.4)

2.2.2 TEST PROCEDURE: Total mechanical travel-clutch is synonymous with continuity travel. Refer to the procedure for continuity travel (2.4).

NOTE: Due to the fact that clutch actuation usually occurs in a region of non-changing resistance with shaft adjustment, a precise location of clutch actuation is indeterminate. Therefore, a specification or measurement of this parameter is not recommended.

2.3 ADJUSTMENT TRAVEL (ELECTRICAL)

2.3.1 EQUIPMENT

Travel Measuring Device	1.1.2
Output Ratio Equipment	1.2.1
Power Supply	1.2.3

2.3.2 TEST PROCEDURE: Mount the trimmer to the travel measuring device and connect the output ratio equipment to the appropriate terminals. Move the adjustment shaft until the output ratio equipment indicates a minimum reading (at the stop, clutch or end of continuity travel). Move the adjustment shaft in the direction of increasing output voltage until the output ratio equipment indicates a significant change in voltage output. Note the reading on the travel measuring device. Continue to move the adjustment shaft in the same direction to the point of the last significant change in output voltage and note the travel reading. The difference between the two readings is the adjustment travel.

2.4 CONTINUITY TRAVEL

2.4.1 EQUIPMENT:

Travel Measuring Device	1.1.2
Resistance Measuring Device	1.2.2

2.4.2 TEST PROCEDURE: Mount the potentiometer to the travel measuring device. Ascertain that the current in the potentiometer from the resistance measuring device will not exceed 1.0 ma or rated current whichever is less. A minimum of 40.0 μa shall be used. Then connect the resistance measuring device between the wiper and the interconnected end terminals. Move the shaft in the specified direction until the resistance measuring device first indicates loss of continuity, or stops or clutches are encountered; note this position. Move the shaft in the reverse direction until loss of continuity is again observed on the resistance measuring device, or stops or clutches are encountered; note this point. The difference between the two noted positions is the continuity travel.

NOTE: In units with clutches and stops, continuity must be maintained over the total travel and during clutching action. Several clutch actuations should be made at each end after noting travel readings to ascertain that continuity is being maintained.

3 GENERAL ELECTRICAL CHARACTERISTICS

3.1 TOTAL RESISTANCE:

3.1.1 EQUIPMENT:

Resistance Measuring Device	1.2.2

(10 milliamp max output into potentiometer, or the wiper current rating whichever is less)

3.1.2 TEST PROCEDURE: Total resistance shall be measured as specified below between the resistance-element end terminals with the contact arm positioned against a stop. The positioning of the contact arm and terminal shall be the same for all subsequent measurements of the total resistance on the same specimen.

The same measuring instrument shall be used for all resistance measurements in any one test, but not necessarily for all the tests.

Measurements of resistance shall be made by using the test voltages specified in Table I. The test voltage chosen, whether it be the maximum or a lower voltage which would still provide the sensitivity required, shall be applied across the terminals of the resistor. This same voltage shall be used whenever a subsequent resistance measurement is made.

TABLE I - DC Resistance Test Voltage

Total Resistance, Nominal	Test Voltage $+ 0\%$ -10%	
	Non-Wirewound	Wirewound
Ohms	Volts DC	Volts DC
Less than 10	0.1	0.1
10 to 100	0.3	0.3
Over 100 to 1,000 incl.	1.0	1.0
Over 1,000 to 10,000 incl.	3.0	3.0
Over 10,000 to 0.1 megohm, incl.	10	10
Over 0.1 megohm	25	—

NOTE: The test voltages should never exceed the equivalent of 10% rated power. The minimum voltage to be used is 10 MV.

3.2 ABSOLUTE MINIMUM RESISTANCE

3.2.1 EQUIPMENT:
Shaft Positioning Device   1.1.1
Resistance measuring Device   1.2.2
(10 milliamp max output into potentiometer, or the wiper current rating whichever is less)

3.2.2 TEST PROCEDURE: The wiper shall be positioned at one end of the resistance element, so that a minimum value of resistance shall be measured as specified in 3.1 between the wiper and corresponding end terminal. The same procedure shall be followed for the other end of the resistance element. Rated current through the resistance element shall not be exceeded during this measurement.

3.3 MINIMUM VOLTAGE

3.3.1 EQUIPMENT:
Shaft Positioning Device   1.1.1
Output Ratio Equipment   1.2.1

3.3.2 TEST PROCEDURE: With the output ratio equipment connected, the wiper is positioned at one end of the resistance element so that a minimum output ratio is indicated. The same procedure is followed for the other end of the resistance element.

3.4 END RESISTANCE

3.4.1 EQUIPMENT:
Shaft Positioning Device   1.1.1
Resistance Measuring Device   1.2.2
(10 milliamp max output into potentiometer, or the wiper current rating whichever is less)

3.4.2 TEST PROCEDURE: The wiper shall be positioned at the extreme counterclockwise limit of mechan-

ical travel, and the resistance shall be measured as specified in 3.1 between the wiper and the corresponding end terminal. The wiper shall then be positioned at the extreme clockwise limit of mechanical travel and the resistance shall be measured as specified in 3.1 between the wiper and the corresponding end terminal. During this test, precaution shall be taken to insure that rated current of the resistance element is not exceeded.

3.5 END VOLTAGE

3.5.1 EQUIPMENT:
Shaft Positioning Device   1.1.1
Output Ratio Equipment   1.2.1

3.5.2 TEST PROCEDURE: With the output ratio equipment properly connected, the wiper is positioned in the same manner as noted for end resistance measurements (See 3.4.1) and the appropriate output ratio readings are taken.

3.6 TEMPERATURE COEFFICIENT OF RESISTANCE

3.6.1 EQUIPMENT:
Resistance Measuring Device   1.2.2
Temperature Test Chamber   1.1.4
Temperature Measuring Device   1.2.10

3.6.2 TEST PROCEDURE: Position the wiper of the trimmer at the point required for total resistance measurement. Subject the pot to two standard series of test temperatures in the sequence described. The first series is room temperature (defined at 25°C) down to −55°C or the lowest rated operating temperature with two intermediate temperature steps at 0°C and −25°C; the second series is room temperature to +125°C or the highest rated operating temperature with at least two intermediate temperature steps at +55°C, +85 C and +125°C where applicable. The total resistance is measured after the temperature chamber has been stabilized (30 to 45 minutes) for each temperature for both series. Compute the temperature coefficient resistance for each temperature interval with the following formula:

$$TC = \frac{R_2 - R_1}{R_1(T_2 - T_1)} \times 10^6$$

Where:

R_1 — Resistance at reference temperature in ohms.
R_2 — Resistance at test temperature in ohms.
T_1 — Reference temperature in degrees celsius.
T_2 — Test temperature in degrees celsius.

The temperature coefficient of resistance of the trimmer is the maximum calculated value.

3.7 RESISTANCE-TEMPERATURE CHARACTERISTIC

3.7.1 EQUIPMENT:

Resistance Measuring Device	1.2.2
Temperature Test Chamber	1.1.4
Temperature Measuring Device	1.2.10

3.7.2 TEST PROCEDURE: Position the wiper of the trimmer at the point required for total resistance measurement. Subject the pot to two standard series of test temperatures in the sequence described. The first series is room temperature (defined at 25°C) down to −55°C or the lowest rated operating temperature with two intermediate temperature steps at 0°C and −25°C; the second series is room temperature to +125°C or the highest rated operating temperature with at least two intermediate temperature steps at +55°C, +85°C and +125°C where applicable. The total resistance is measured after the temperature chamber has been stabilized for each temperature (30 to 45 minutes) with 25°C as the reference temperature for both series. Compute the resistance-temperature characteristic for each temperature interval with the following formula:

$$RTC = \frac{R_2 - R_1}{R_1} \times 100$$

Where:

R_1 — Resistance at reference temperature in ohms.

R_2 — Resistance at any of the test temperatures in ohms.

The resistance-temperature characteristic of the potentiometer is the maximum value calculated.

3.8 CONTACT RESISTANCE VARIATION

3.8.1 EQUIPMENT:

Constant Speed Drive	1.1.3
Constant Current Source	1.2.5
Oscilloscope	1.2.6
Resistance Decade	
Filter or AC Amp	

3.8.2 TEST PROCEDURE: Contact-resistance variation shall be measured with the measuring circuit shown on Figure 1 or its equivalent. The operating shaft shall be rotated in both directions through 90 percent of the adjustment travel for a total of 6 cycles. Only the last 3 cycles shall count in determining whether or not a contact resistance variation is observed at least twice in the same area (within 5%), exclusive of the roll-on or roll-off points where the contact arm moves from the termination, on or off, the resistance element. The rate of rotation of the operating shaft shall be such that the wiper completes 1 cycle in 5 seconds, minimum, to 2 minutes, maximum. The test current used shall follow the values given in Table II below unless otherwise limited by power rating.

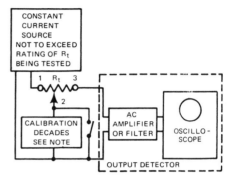

Figure 1 Contact-resistance-variation measuring circuit

R_t = Test specimen

Output detector bandwidth: 100 cycles to 50 kilocycles

Minimum input impedance

to output detector: At least 10 times the nominal resistance being tested—not exceeding 10 Ω

NOTE: At the calibration of the decade, terminals 1 and 2 must be coincident. Calibration decade is to be set for the contact-resistance variation (CRV) level of the specified nominal resistance being tested.

TABLE II

Test Current (± 20%)	Total Resistance Range
30 ma	$R_t < 50\ \Omega$
10 ma	$50\ \Omega \leq R_t < 500\ \Omega$
1 ma	$500\ \Omega \leq R_t < 100K\ \Omega$
100 ua	$100K\ \Omega \leq R_t < 2\ Meg\ \Omega$
50 ua	$R_t \geq 2\ Meg\ \Omega$

3.9 EQUIVALENT NOISE RESISTANCE

3.9.1 EQUIPMENT:

Constant Speed Drive	1.1.3
Constant Current Source	1.2.5
Oscilloscope	1.2.6
Zener Diode (6V)	

3.9.2 TEST PROCEDURE: The adjustment shaft is connected mechanically to a variable speed drive adjusted to a rate causing the wiper to make one complete cycle in 5 seconds minimum to 2 minutes maximum. A cycle is one complete traversal of at least 90% of the adjustment travel in both directions. The test trimmer is connected in the circuit shown in Figure II. During test the wiper energized with a constant current of one milliampere, shall make three complete cycles and the maximum deviations from the reference line on the oscilloscope are noted. Random spikes which are not repetitive are to be disregarded. Equivalent noise resistance is calculated from:

$$ENR \text{ (ohms)} = \frac{\text{Maximum deviation}}{.001} \text{ (volts)}$$

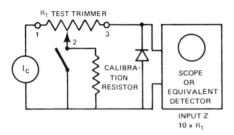

Figure II   Equivalent Noise Resistance

3.10 CONTINUITY

3.10.1 EQUIPMENT:

| See Contact Resistance Variation | 3.8 |
| or Equivalent Noise Resistance | 3.9 |

3.10.2 TEST PROCEDURE: For continuity determination, use the test procedure for contact resistance variation (3.8) or equivalent noise resistance (3.9), whichever is applicable.

3.11 SETTING STABILITY

3.11.1 EQUIPMENT:

Shaft Positioning Device	1.1.1
Output Ratio Equipment	1.2.1
Power Supply	1.2.3

3.11.2 TEST PROCEDURE: Unless otherwise specified, the wiper shall be set at approximately 40 percent of the adjustment travel. Measure the output ratio at this position. Subject unit to the specified environmental tests. Measure output ratio. The change in output ratio expressed as a percentage of input voltage is a measure of setting stability.

The difference between the initial measurement made before the environmental test and the measurement made after the test indicates the percent change, a measure of the setting stability.

3.12 DIELECTRIC WITHSTANDING VOLTAGE

3.12.1 EQUIPMENT:

High Voltage Source	1.2.7
AC Voltmeter	1.2.8
Leakage Current Indicating Device	1.2.9
Vacuum Chamber	1.1.6
Altitude Measuring Device	1.1.7

3.12.2 TEST PROCEDURE

3.12.2.1 At atmospheric pressure, resistors shall be tested in accordance with the following details:

A. Method of mounting: Resistors shall be clamped or otherwise mounted on metal plates of sufficient size to extend beyond the resistor extremities, and in such a manner that measurements can be made between the terminals tied together and any other external metal parts.

B. Magnitude of test voltage: Volts rms as specified.

C. Nature of potential: From an alternating current (ac) supply at commercial-line frequency and waveform.

D. Points of application of test voltage: Between the terminals connected together and all external metal portions of the resistors and metal-mounting plate.

E. Examinations and measurements: During the tests, the leakage current shall not exceed 1.0 ma and the resistors examined for evidence of arcing and breakdown. At the conclusion of the test, resistors shall be examined for evidence of damage.

3.12.2.2 At reduced barometric pressure, resistors shall be tested per the following details:

A. Method of mounting: As specified in 3.12.2.1 (A)

B. Test chamber equivalent to altitude of 70,000 ft.

C. Period of time at reduced pressure prior to application of potential — 1 minute.

D. Tests during subjection to reduced pressure: A specified potential from an ac supply at commercial-line frequency and waveform shall be applied for 1 minute.

E. Points of application: As specified in 3.12.1.1 (D).

F. Examinations and measurements: As specified in 3.12.1 (E).

3.13 INSULATION RESISTANCE

3.13.1 EQUIPMENT:
Insulation Resistance Test Set — 1.2.11

3.13.2 TEST PROCEDURE: Resistors shall be tested under the following conditions:

A. Test condition: 500 volts dc for 1 minute max.

B. Method of mounting: As specified in 3.12.2.1 (A).

C. Points of measurement: As specified in 3.12.2.1 (D).

3.14 LOAD LIFE

3.14.1 EQUIPMENT:

Shaft Positioning Device	1.1.1
Life Temperature Test Chamber	1.1.5
Output Ratio Equipment	1.2.1
Resistance Measuring Device	1.2.2
Power Supplies	1.2.3

3.14.2 TEST PROCEDURE: Resistors shall be tested in accordance with the following procedure:

A. Method of mounting: Resistors shall be mounted by their normal mounting means, on a 1/16 inch thick glass-base, epoxy laminate. The resistors shall be so arranged that the temperature of any one resistor shall not appreciably influence the temperature of any other resistor. There shall be no undue draft over the resistors.

B. Test temperature shall be the maximum temperature at which the maximum rated power may be dissipated. Tolerance on chamber temperature shall be ±5°C.

C. Initial measurements at room temperature: Total resistance and setting stability shall be measured as specified in 3.1 and 3.11 respectively.

D. After the resistors have been stabilized at the specified temperature (±5°C) for at least 8 hours, the resistance between the end terminals, with the wiper in the position for setting stability shall be measured.

E. Operating conditions: Rated dc working voltage or ac working voltage at commercial-line frequency and waveform shall be applied intermittently to the end terminals of the resistors, 1½ hours on and ½ hour off, for a total of 1,000 hours at the test temperature. Each resistor shall dissipate rated wattage but shall not exceed maximum voltage. Adequate precaution shall be taken to maintain a constant voltage on the resistor.

F. Test duration: 1,000 hours.

G. Measurements during test: While the resistors are still in the oven, resistance shall be measured between the end terminals, at the end of the ½ hour off periods after every 168 hours have elapsed and compared to the similar reading taken in (D). The degradations are to be compared with the specified limits allowable.

H. Measurements after test: After the resistors have been removed from the oven and returned to room temperature, setting stability and total resistance measurements are taken and compared with readings in (C). The degradations are to be compared with the specified limits allowable.

I. Examination after test: Resistors shall be examined for evidence of mechanical damage.

3.15 ROTATIONAL LIFE

3.15.1 EQUIPMENT:

Rotational Life Fixture	1.1.10
Resistance Measuring Device	1.2.2
Power Supply	1.2.3

3.15.2 TEST PROCEDURE:

3.15.2.1 Mounting: The resistors are mounted and ganged in pairs, each pair connected in series as shown in Figure III so that a nominally constant current flows through the resistors, irrespective of the wiper position during the turning of the adjustment shaft.

3.15.2.2 Procedure: Total resistance shall be measured as specified in 3.1. A dc potential, equivalent to that required to dissipate rated wattage across the entire resistive element of resistors having the same nominal total resistance as those under test, shall then be applied as shown on Figure III. The adjustment shaft shall be continuously cycled through not less than 90 percent of the adjustment travel, at the rate of 1 cycle in 5 seconds minimum to 2 minutes maximum for a total of 200 cycles. A cycle shall consist of travel through 90 percent of adjustment travel and return to the starting point. After rotation, total resistance shall be measured as specified in 3.1. Resistors shall than be examined for evidence of mechanical damage. The entire test is performed under room ambient conditions.

Figure III Rotational life test circuit

3.16 ADJUSTABILITY (OUTPUT RESISTANCE)

3.16.1 EQUIPMENT
Resistance Measuring Device utilizing
   constant current source   1.2.2
Timing Clock   1.2.13

3.16.2 TEST PROCEDURE: Connect the resistance measuring device to the CCW and wiper terminals. Position the shaft at approximately 30% of adjustment travel, but outside the specified limits of the final adjustment to be obtained. Adjust the output resistance to 30% of the nominal total resistance within the specified limits. This adjustment shall be obtained within 20 seconds maximum time, and may include shaft reversals. The final value of adjustment obtained shall be determined after removal of the means of adjustment, i.e., hand, screwdriver, etc.

Repeat at 50% and 75% of nominal total resistance.

The limits are specified as a percentage of nominal total resistance.

NOTE: Measurements of resistance shall be made using a constant current source. For most consistent results, use the current specified in Table II unless otherwise limited by power rating.

3.17 ADJUSTABILITY (OUTPUT VOLTAGE RATIO)

3.17.1 EQUIPMENT
Output Ratio Equipment   1.2.1
Timing Clock   1.2.13

3.17.2 TEST PROCEDURE: Connect the output ratio equipment to the appropriate terminals. Position the shaft at approximately 30% of adjustment travel, but outside the specified limits of the final adjustment to be obtained. Adjust the output ratio to .30 within the specified limits. This adjustment shall be obtained within 20 seconds maximum time and may include shaft reversals. The final value of adjustment obtained shall be determined after removal of the means of adjustment, i.e., hand, screwdriver, etc.

Repeat at .50 and .75 output ratio.

NOTE: The limits are specified as a percentage of total applied voltage.

4 GENERAL MECHANICAL CHARACTERISTICS

4.1 STARTING TORQUE

4.1.1 EQUIPMENT:
Load Device   1.1.8
Potentiometer Mounting Fixture   1.1.9

4.1.2 TEST PROCEDURE: Mount the trimming potentiometer firmly by its normal mounting means. Connect the load device to the adjustment shaft so as to prevent relative movement between the two. A torque is applied through the load device and about the axis of the adjustment shaft until shaft rotation is initiated. Care should be exercised to avoid applying radial or axial loads that will influence the measurement. The procedure is followed for each direction of rotation at three randomly selected points over the total mechanical travel. The starting torque is the maximum indicated reading of the load device.

4.2 STOP TORQUE

4.2.1 EQUIPMENT:
Load Device   1.1.8
Potentiometer Mounting Fixture   1.1.9
Resistance Measuring Device   1.2.2

4.2.2 TEST PROCEDURE: Mount the trimming potentiometer firmly by its normal mounting means. (Units normally mounted by the terminations alone must further have their enclosures restrained to avoid applying the full stop torque through the terminations.) Connect the resistance measuring device between the wiper and the appropriate end terminal after it has been determined that the current in the potentiometer will not exceed 1.0 ma or rated current, whichever is less. A minimum of $40.0\,\mu a$ shall be used. Applying the specified stop torque to each stop through the load device and holding that torque for 10 seconds each, note that continuity is maintained.

4.3 SOLDERABILITY

4.3.1 EQUIPMENT: The equipment specified in MIL-STD-202, Method 208, shall be used.

4.3.2 TEST PROCEDURE: The procedure of Method 208 of MIL-STD-202 is used except that the solder pot temperature shall be 288°C ±5°C. This procedure is applicable to solid wire terminations only.

4.4 WELDABILITY

4.4.1 EQUIPMENT: None required.

4.4.2 TEST PROCEDURE: The weldability of a material is dependent on the materials to be welded including size, form and composition; the degree of weld strength or conductivity required for the application and the weld process. The intent of this procedure is to control only the size, form and composition of the terminals of a trimming potentiometer. Therefore, a trimmer is said to have weldable terminals if they conform to the requirements of MIL-STD-1276.

4.5 TERMINAL STRENGTH

4.5.1 EQUIPMENT:

Potentiometer Mounting Fixture	1.1.9
Push-pull Device	1.1.11
Bend Fixture	1.1.12

4.5.2 TEST PROCEDURES:

4.5.2.1 Pull (applicable to all terminal types) resistors shall be tested in accordance with Method 211 of MIL-STD-202. The following details and exceptions shall apply.

A. Test Conditon A: Applied force—2 pounds. Resistor clamped by the resistor body, force applied to each lead individually.

B. Measurement after test: Resistors shall be examined for evidence of mechanical damage, and tested for electrical continuity.

4.5.2.2 Push (applicable to all terminals except Type L). Resistors shall be tested in accordance with Method 211 of MIL-STD-202. The following details and exceptions shall apply:

A. Test Condition A, except force shall be applied in the direction toward the resistor body. Applied force—2 pounds. Resistor clamped by the resistor body; force applied to each terminal individually.

B. Measurement After Test: Resistors shall be examined for evidence of mechanical damage and tested for electrical continuity.

4.5.2.3 Bend (applicable to all terminal types except Type L). Resistors shall be firmly clamped and each terminal shall be bent through 90° at a point 1/8 inch from the body of the resistor, with the radius of curvature at the bend approximately 1/32 inch. The pin shall be returned to the original position, bent 90° in the opposite direction, and again returned to the original position. At the conclusion of the test, the resistors shall be examined for evidence of mechanical damage and tested for electrical continuity.

4.6 IMMERSION SEAL: (Industrial/Commercial —Applicable to solid wire terminal types only)

4.6.1 EQUIPMENT:

Hot Water Bath	1.1.13

4.6.2 TEST PROCEDURE: To guard against contamination of the internal structure of trimmers by intrusion of conformal coatings or normal printed circuit board washing solvents, industry has determined that a less severe test procedure will provide verification of adequate sealing. This procedure is recommended for low cost, high production industrial components where sealing techniques do not usually produce high pressure seals.

The surface of the resistor shall be cleared of any foreign matter immediately before immersion. The bath shall consist of tap water at a temperature of 50°C +5°C, −0°C. The resistor shall be completely submerged in the bath with no part at a depth of less than 1 inch and shaken for a maximum of 5 seconds, to remove surface air, and shall remain in the bath for a period of 30 seconds, ±5 seconds. During immersion, observation shall be made for any bubbles emanating from the trimmer. No more than 3 bubbles shall be permitted.

4.7 IMMERSION SEAL: (Military — Applicable to solid wire terminal types only)

4.7.1 EQUIPMENT:

Hot Water Bath	1.1.13

4.7.2 TEST PROCEDURE: The surface of the resistor shall be cleared of any foreign matter immediately before immersion. The bath shall consist of tap water at a temperature of 85°C +5°C, −0°C. The resistor shall be completely submerged in the bath with no part at a depth of less than 1 inch and shaken for a maximum of 5 seconds, to remove surface air, and shall remain in the bath for a period of 1 minute, ±5 seconds. During immersion, observation shall be made for any bubbles emanating from the trimmer. No more than 3 bubbles shall be permitted.

Appendix II. Military Specifications

Various potentiometers are described by military specifications. In most cases, potentiometers qualified to these specifications are available from several manufacturers. These specs are used by non-military as well as military users because they are often a convenient standard or reference. Some component and standards engineers modify or use complete sections of military specifications in establishing their own potentiometer requirements.

It would be handy to have all mil-specs on potentiometers (called variable resistors by the military) here but that would require an appendix larger than this book. Also, the specifications are revised from time to time so some might soon be obsolete. Instead, a list of all relevant current military specifications and where to order them is provided in Table I.

Included for readers unfamiliar with military specifications is an overview of performance requirements and associated test methods typically prescribed for potentiometers by military users. This example is for conventional and established reliability wirewound and nonwirewound trimming potentiometers. The format parallels the manner in which the information is commonly encountered in military specifications; however, it has been condensed from narrative to tabular form in some instances (Table II) for easy comparison of requirements.

It should be noted that qualification of a product to a particular military specification is normally performed in order for the potentiometer manufacturer to obtain government approval (Defense Electronic Supply Center: DESC) and subsequent listing on the coresponding Qualified Products List (QPL) for the type of product tested. Performance criteria applicable to all styles or types of a product will be found in the basic specification for that product while specific requirements (outline dimensions, configuration, power rating, etc.) applicable to a particular style or type will be covered in a detailed specification (slash sheet) for the item. A slash sheet for RJR24 is reproduced at the end of this appendix as an example.

The narrative portion describing qualification test methods has also been condensed to the extent allowed by similarity of test methods in specifications Mil-R-22097, 27208, 39015, and 39035. Where dissimilarities in method preclude such treatment, the test method is described for each specification as applicable.

TABLE I:
GUIDE TO MILITARY SPECIFICATIONS
FOR VARIABLE RESISTIVE DEVICES

Specification	Basic Style	Description: Military Specifications for Resistors, Variable
Mil-R-19	RA	Wirewound (low operating temperature)
Mil-R-94	RV	Composition
Mil-R-12934	RR	Wirewound, precision
Mil-R-22097 1/	RJ	Nonwirewound (adjustment type)
Mil-R-23285	RVC	Nonwirewound (panel control)
Mil-R-27208 1/	RT	Wirewound (adjustment type)
Mil-R-39015 1/	RTR	Wirewound (lead screw actuated), established reliability
Mil-R-39023	RQ	Nonwirewound, precision
Mil-R-39035 1/	RJR	Nonwirewound, (adjustment), established reliability

1/ Used as example in this appendix

SUPPORTING SPECIFICATIONS:

The following specifications are used in the selection, procurement, and testing of potentiometers.

Mil-Std-202 — Detailed specification of test methods for electronic and mechanical component parts.

Mil-Std-105 — Sampling plans and procedures and tables for inspection by attributes.

Mil-Std-790 — Reliability assurance program for electronic parts and specifications.

Mil-Std-199 — Resistors, selection and use of.

Mil-Std-690 — Sampling plans and procedures for determining failure rates of established reliability devices.

Copies of these specifications are available from:

Department of the Navy
Navy Publications and Printing
Service Office
700 Robbins Avenue
Philadelphia, PA 19111

TABLE II:

COMPARISON OF REQUIREMENTS FOR QUALIFICATION INSPECTION

This table lists tests, test methods, and performance requirements in tabular form for trimming potentiometers. This is convenient for comparing the differences among these specs.

Reqmt. Para.	Test	SPECIFICATION MIL-R-				Method Para.
		39015	39035	27208	22097	
3.7	Conditioning	No Damage	No Damage	N/A	N/A	4.7.2
3.9.1	Δ TR, %, Max.	± 5	± 10	N/A	N/A	4.7.4.1
3.8	Peak Noise/CRV, Max.	100 Ω		N/A	N/A	4.7.3
	Char. C		3% or 20 Ω			
	Char. F and H		3% or 3 Ω			
	Char. J		1% or 3 Ω			
3.9.1	Total Resistance, ±%	5	10	5	10	4.7.4.1
3.10	Immersion, Bubbles, Max.	3	3	3	3	4.7.5
3.1	Visual and Mechanical	See Paragraph 3.1 Below (pg. 262)				4.7.1
3.11	Continuity	S & U*	N/A	S & U*	N/A	4.7.6
3.12	Actual Eff. Elec. Travel	As Specified				4.7.7
3.9.2	Absolute Min. Resist.	.25% or 1 Ω	N/A	.25% or 1 Ω	N/A	4.7.4.2
3.9.3	End Resistance	2% or 1 Ω	2% or 20 Ω	2% or 1 Ω	2% or 20 Ω	4.7.4.3
3.13	Dielectric	No Damage or Arcing				4.7.8
3.14	Insul. Resist. Meg Ω, Min.	1000	1000	1000	1000	4.7.9
3.15	Torque	As Specified				4.7.10
3.16	Thermal Shock	No Mechanical Damage				4.7.11
3.9.1	Δ TR%, Max.			1% + .05 Ω		4.7.4.1
	100 Ω or Greater	1				
	Below 100 Ω	1% + .05 Ω				
	Char. C		2		2	
	Char. F		1		1	
	Char. J		.05			
	Char. F & H		1			
3.16	Δ VR%, Max.	0.5% + R**		1 + R**		4.7.11.1
	Char. A				2	
	Char. C & F		1		1	
3.17	Solderability	95% Coverage				4.7.12
3.18	Res. — Temp. Char., PPM/° C	50		50		
	Char. C		250		250	
	Char. F		100		100	
	Char. H		50			
	Char. J		10			
3.19	Moisture Resistance	(20 cycles)	(20 cycles)	(10 cycles)	(10 cycles)	4.7.14
3.9.1	Δ TR, %, Max.					4.7.4.1
	100 Ω or Greater	1		1		
	Below 100 Ω	1% + .05 Ω				
	Char. C		2		2	
	Char. F		1		1% + .05 Ω	
	Char. H					
	Char. J		.1			
3.14	Ins. Resist., Meg Ω, Max.	100	100	100	100	4.7.9
3.32	Setability	N/A	As Spec.	N/A	N/A	4.7.27
3.20	Shock	No Mechanical Damage				4.7.15
3.9.1	Δ TR, %, Max.			1% + .05 Ω		4.7.4.1
	100 Ω	1				
	Below 100 Ω	1% + .05 Ω				
	Char. C, F				1% 3.05 Ω	
	Char. C, F, H		1			
	Char. J		.05			

*Smooth and Unidirectional   —   **Resolution

260

Reqmt. Para.	Test	SPECIFICATION MIL-R-				Method Para.
		39015	39035	27208	22097	
3.16	Δ VR, %, Max.	.5% + R**	1	1% + R**	1	4.7.11.1
3.21	Vibration	No Mechanical Damage				4.7.16
	Δ TR, %, Max.	(Same as for Shock)				
	Δ VR, %, Max.	(Same as for Shock)				
3.22	Salt Spray	No Appreciable Corrosion				4.7.17
3.23	Resist. to Solder Heat	No Mechanical Damage				4.7.18
	Δ TR, %, Max.	(Same as for Shock Except .1% for Char. J)				
3.24	Low Temp. Operation	No Mechanical Damage				4.7.19
3.9.1	Δ TR, %, Max.					4.7.7.1
	100Ω or Greater	1				
	Below 100Ω	1% + x05Ω				
	Char. C		2		2	
	Char. F, H		1		1	
	Char. J		.1			
3.16	Δ VR, %, Max.	.5% + R**	2		2	4.7.11.1
3.25	Low Temp. Storage	No Mechanical Damage		N/A	N/A	4.7.20
3.9.1	Δ TR, %, Max.					4.7.4.1
	100Ω or Greater	1				
	Below 100Ω	1% + .05Ω				
	Char. C		2			
	Char. F, H		1			
	Char. J		1			
3.26	High Temp. Exposure	No Mechanical Damage				4.7.21
3.9.1	Δ TR, %, Max.			1% + .05Ω		4.7.4.1
	100Ω or Greater	1				
	Below 100Ω	1% + .05Ω				
	Char. C, F, H		3			
	Char. C		.2		3	
	Char. F				2	
3.16	Δ VR, %, Max.	.5	2	1	2	4.7.11.1
3.13	Dielectric	No Damage or Arcing				4.7.8
3.14	Ins. Resist., MegΩ, Min.	1000	1000	1000	1000	4.7.9
3.27	Integrity of Shaft	No Breaks	No Breaks	N/A	N/A	4.7.22
3.28	Rotational Life					4.7.23
	Δ TR, %, Max.	2		2		
	Char. C, F. H		2		2	
	Char. J		1			
3.29	Terminal Strength	No Damage or Discontinuity				4.7.24
3.30	Life	No Mechanical Damage				4.7.25
	( )K Hrs., Δ TR, %	(2) 2% + R**	(2)	(2) 2	(1)	
	Char. C, F. H	2	3			
	Char. J		.2			
	Char. C				3	
	Char. F				2	
	Over 2K Hrs., Δ TR, %					
	Char. C, F, H	5	5			
	Char. J	1	1			
3.16	Δ VR, %, Max.			2% + R**	1	4.7.11.1
3.31	Resist. to Solvents	No Damage	No Damage			

**Resolution

COMPARISON OF
REQUIREMENTS FOR
QUALIFICATION INSPECTION

3.1   VISUAL AND MECHANICAL (See 4.7.1)

MATERIAL The material shall be suitable to meet the performance requirements of the specification. Material which is nutrient for fungus shall not be used.

> **Plastic** Plastic laminates containing a cotton-fabric base or plastic-molding compounds containing a cotton or wood-flour filler shall not be used. Plastic surfaces shall be as smooth as practicable in accordance with good manufacturing practices.

> **Ferrous Metals** The use of ferrous material, with the exception of corrosion-resistant steel and the resistance element material, is prohibited.

ENCLOSURE The resistance element shall be completely covered by a housing. The housing shall be free from holes, fissures, chips or other faults which may establish leakage paths between the parts of the units. Unplated copper alloy metals shall not be used in contact with aluminum.

MARKING The circuit diagram shall be legibly marked on a surface of the resistor. It shall be legible after all tests.

CLUTCHES Resistors shall contain clutches which permit the contact arm to idle at either end of the resistance elements without electrical or mechanical malfunction.

TABLE III:

QUALIFICATION REQUIREMENTS OF CONVENTIONAL TYPE VARIABLE RESISTORS

This table deliniates the examinations, tests, sequence of test performance, and allowable failures for qualification of trimming potentiometers to the Conventional Military Specifications, Mil-R-22097 and Mil-R-27208, covering non-wirewound and wirewound types respectively.

22097	27208	Examination or Test	Reqmt. Para.	Method Para.	Number of Test Units	Number of Failures Allowed 2/
		Group I				
X	X	Visual and Mechanical Examination 3/, 4/	3.1	4.7.1		
X	X	Total Resistance 4/	3.9.1	4.7.4.1		
	X	Continuity 4/	3.11	4.7.6		
X	X	Actual Eff.-Elec. Travel 4/	3.12	4.7.7		
	X	Absolute Minimum Resistance 4/	3.9.2	4.7.4.2	60	
X	X	End Resistance 4/	3.9.3	4.7.4.3		0
	X	Peak Noise 4/	3.8	4.7.3		
X		Contact Resist. Var.	3.8	4.7.3		
X	X	Dielectric Withstanding Voltage 4/	3.13	4.7.8		
X	X	Insulation Resistance 4/	3.14	4.7.9		
X	X	Torque 4/	3.15	4.7.10		
X	X	Thermal Shock 4/	3.16	4.7.11		
		Group II				
X	X	Res. — Temp. Char. 4/	3.18	4.7.13		
X	X	Moisture Resistance	3.19	4.7.14	12	1
	X	Peak Noise	3.8	4.7.3		
X		Contact Resist. Var.	3.8	4.7.3		
		Group III				
X	X	Shock, Medium Impact	3.20	4.7.15		
X	X	Vibration, High Frequency	3.21	4.7.16	12	
	X	Peak Noise	3.8	4.7.3		1
X		Contact Resist. Var.	3.8	4.7.3		
X	X	Salt Spray (Corrosion)	3.22	4.7.17		
		Group IV				
X	X	Resistance to Soldering Heat (Applicable to Terminal Types P, W, X, and Y only)	3.23	4.7.18		
X	X	Life	3.30	4.7.25	12	1
	X	Peak Noise	3.8	4.7.3		
X		Contact Resist. Var.	3.8	4.7.3		
		Group V				
X	X	Low-Temperature Operation	3.24	4.7.19		
X	X	High-Temperature Exposure	3.26	4.7.21	12	1
	X	Peak Noise	3.8	4.7.3		
X		Contact Resist. Var.	3.8	4.7.3		
		Group VI				
X	X	Rotational Life	3.28	4.7.23		
	X	Peak Noise	3.8	4.7.3	12	1
X		Contact Resist. Var.	3.8	4.7.3		
X	X	Terminal Strength	3.29	4.7.24		
		Group VII 5/				
X	X	Solderability (Applicable to Terminal Types P, W, and Y Only)	3.17	4.7.12	6	0
X	X	Immersion (Applicable to Terminal Types P, W, X, and Y Only)	3.10	4.7.5		

(Groups II through VI brace to a combined Number of Failures Allowed of 2)

1. See Table X.
2. Failure of a resistor in one or more tests of a group shall be charged as a single failure.
3. Marking shall be considered defective only if the marking is illegible. Marking is not applicable to unenclosed resistors.
4. Nondestructive tests.
5. Group I examination not required on Group VII resistors.

TABLE IV:
QUALIFICATION REQUIREMENTS OF ESTABLISHED RELIABILITY TYPE VARIABLE RESISTORS

This table outlines test requirements, etc. for qualification of wirewound/non-wirewound items to the Established Reliability Military Specifications 39015/39035.

39035	39015	Examination or Test	Reqmt. Para.	Method Para.	Number of Sample Units to be Insp. 2/	Number Allowed 3/
		Group I 1/				
X	X	Conditioning 4/	3.7	4.7.2	All	
	X	Peak Noise 4/	3.8	4.7.3	Sample	
X		Contact Resist. Var. 4/	3.8	4.7.3	Units	N/A
	X	Total Resist. 4/	3.9.1	4.7.4.1		
X	X	Immersion 4/	3.10	4.7.5		
		Group IA				
X	X	Visual and Mechanical Exam. 4/ 5/	3.1	4.7.1		
	X	Continuity 4/	3.11	4.7.6	All Sample	
X	X	Actual Effective Electrical Travel 4/	3.12	4.7.7	Units	
	X	Absolute Minimum Resistance 4/	3.9.2	4.7.4.2	Except	
X	X	End Resistance 4/	3.9.3	4.7.4.3	those for	
X	X	Dielectric Withstanding Voltage 4/	3.13	4.7.8	Group II	0
X	X	Insulation Resistance 4/	3.14	4.7.9		
X	X	Torque 4/	3.15	4.7.10		
X	X	Thermal Shock 4/	3.16	4.7.11		
		Group II				
X	X	Solderability	3.17	4.7.12	6 Any Value	0
		Group III				
X	X	Resistance Temp. Characeristic	3.18	4.7.13		
X	X	Moisture Resistance	3.19	4.7.14	6 Highest	
	X	Peak Noise	3.8	4.7.3	12	1
X		Contact Resist. Var.	3.8	4.7.3	6 Lowest	
		Group IV				
X		Setability	3.32	4.7.27	6 Highest	
X	X	Shock (Specified Pulse)	3.20	4.7.15	12	
X	X	Vibration, High Frequency	3.21	4.7.16	6 Lowest	1
	X	Peak Noise	3.8	4.7.3		
X		Contact Resist. Var.	3.8	4.7.3		
X	X	Salt Spray (Corrosion)	3.22	4.7.17		
		Group V				
X	X	Resistance to Soldering Heat	3.23	4.7.18		1
X	X	Low Temperature Operation	3.24	4.7.19	6 Highest	
X	X	Low Temperature Storage	3.25	4.7.20	12	1
X	X	High Temperature Exposure	3.26	4.7.21	6 Lowest	
	X	Peak Noise	3.8	4.7.3		
X		Contact Resist. Var.	3.8	4.7.3		
X	X	Integrity of Shaft	3.27	4.7.22		
		Group VI				
X	X	Rotational Life	3.28	4.7.23		
	X	Peak Noise	3.8	4.7.3	6 Highest	
X		Contact Resist. Var.	3.8	4.7.3	12	1
X	X	Terminal Strength	3.29	4.7.24	6 Lowest	
		Group VII				
X	X	Life	3.30	4.7.25	51 Highest 102 51 Lowest	1
		Group VIII				
X	X	Resistance to Solvents	3.31	4.7.26	3 Any Value	0

1. Group I tests need not to be performed if manufacturer presents certified data proving tests have been previously performed on the qualification sample units.

2. See appendix of applicable mil spec for details.

3. Failure of a resistor in one or more tests of a test group shall be charged as a single defective.

4. Nondestructive tests.

5. Marking shall be considered defective only if illegible or missing. Marking shall remain legible at the end of all tests.

QUALIFICATION TEST METHODS

4.7.1 **Visual and Mechanical Examination.** Resistors shall be examined to verify that the materials, design, construction, physical dimensions, marking, and workmanship are in accordance with the applicable requirements (see 3.1).

4.7.2 **Mil-R-39015 Conditioning (see 3.7).** Resistors shall be conditioned in accordance with method 108 of Mil-Std-202. The following details and exceptions shall apply:

(a) Method of mounting — Supported by their terminals (resistor not mounted on life test chassis). Resistors shall be so arranged that the temperature of any one resistor shall not appreciably influence the temperature of any other resistor. There shall be no undue draft on the resistors.

(b) Temperature and Tolerance — $25^\circ {}^{+10^\circ}_{-0^\circ}$ C.

(c) Initial measurements — Initial total resistance shall be measured after mounting at 25° ${}^{+10^\circ}_{-0^\circ}$ C as specified in 4.7.4.1. This initial measurement shall be used as the reference for all subsequent measurements.

(d) Operating condition — DC continuous working voltage or a continuous working voltage from an ac supply at commercial-line frequency and waveform equivalent to 1 watt power dissipation shall be applied between the end terminals intermittently 1-1/2 hours "on" and 1/2 hour "off" for a minimum of $50 {}^{+8}_{-0}$ hours at a temperature of 25° ${}^{+10^\circ}_{-0^\circ}$ C. Each resistor shall dissipate 1 watt.

(e) Measurement after conditioning — Total resistance shall be measured at the end of the $50 {}^{+8}_{-0}$ hours as specified in 4.7.4.1 after load has been removed and the resistors stabilized.

(f) Examination after conditioning — Resistors shall be examined for evidence of mechanical damage.

(g) Test duration — $50 {}^{+8}_{-0}$ hours

4.7.2 **Mil-R-39035 Conditioning (see 3.7).** Resistors shall be conditioned in accordance with method 108 of Mil-Std-202. The following details and exceptions shall apply:

(a) Method of mounting — Supported by their terminals (resistor not mounted on life test chassis). Resistors shall be so arranged that the temperature of any one resistor shall not appreciably influence the temperature of any other resistor. There shall be no undue draft on the resistors.

(b) Temperature and tolerance — Standard test condition.

(c) Initial measurements — Initial total resistance shall be measured after mounting as specified in 4.7.4.1. This initial measurement shall be used as the reference for all subsequent measurements.

(d) Operating condition — DC continuous working voltage or a continuous working voltage from an ac supply at commercial-line frequency and waveform equivalent to 1-1/2 times the specified wattage (see 3.1), shall be applied between the end terminals intermittently 1-1/2 hours "on" and 1/2 hour "off" for a minimum of $50 {}^{+8}_{-0}$ hours at a temperature of 25° $^{+10°}_{-0}$ C. Each resistor shall dissipate 1-1/2 times the rated wattage, but not to exceed the maximum rated voltage.

(e) Measurement after conditioning — Total resistance shall be measured at the end of 50 $^{+8}_{-0}$ hours as specified in 4.7.4.1 after load has been removed and the resistors stabilized.

(f) Examination after conditioning — Resistors shall be examined for evidence of mechanical damage.

(g) Test duration — 50 $^{+8}_{-0}$ hours.

4.7.3   **Contact-resistance variation (see 3.8).** Contact-resistance variation shall be measured with the measuring circuit shown on figure 1 or its equivalent. The operating shaft shall be rotated in both directions through 90 percent of the actual effective-electrical travel for a total of 6 cycles. Only the last 3 cycles shall count in determining whether or not a contact resistance variation is observed at least twice in the same location, exclusive of the roll-on or roll-off points where the contact arm moves from the termination, on or off, the resistance element. The rate of rotation of the operating shaft shall be such that the wiper completes 1 cycle in 5 seconds, minimum, to 2 minutes, maximum.

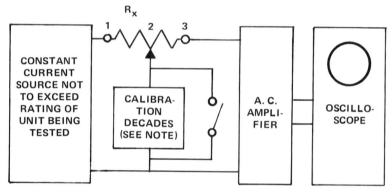

Rx — Test specimen.
Oscilloscope bandwidth: 100 Hz to 50 kHz.
Minimum input impedance: At least 10 times the nominal resistance being tested.

NOTE: At the calibration of the decade, terminals 1 and 2 must be coincident. Calibration decade is to be set for the contact-resistance (CRV) level of the specified nominal resistance being tested.

FIGURE 1   Contact-resistance-variation measuring circuit

4.7.3   **Peak noise (see 3.8).** Peak-noise resistance shall be measured with the measuring circuit shown on Figure 2, or its equivalent. The lead screw shall be rotated in both directions through 90 percent of the actual effective electrical travel for a total of 6 cycles. Only the last 3 cycles shall count in determining whether or not a noise is observed at least twice in the same location. The rate of rotation of the lead screw shall be such that the wiper completes 1 cycle in 5 seconds, minimum, to 2 minutes, maximum. The equivalent resistance shall be calculated using the following formula:

$$\text{Noise} = \frac{\text{Epn}}{0.001} \text{ ohms}$$

Where: Epn = the peak-noise signal voltage presented on the oscilloscope screen.

266

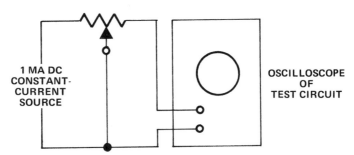

1 MA DC
CONSTANT-
CURRENT
SOURCE

OSCILLOSCOPE
OF
TEST CIRCUIT

R_x Test Specimen

Oscilloscope or test circuit bandwith; DC to 500 kilohertz minimum.

Minimum input impedance: 1.0 megohm at 400 hertz.

Figure 2   Peak-noise measuring circuit

4.7.4   **DC Resistance (see 3.9.1).** Resistors shall be tested in accordance with method 303 of Mil-Std-202. The following details shall apply:

(a) Measuring apparatus — The same measuring instrument shall be used for all resistance measurements in any one test, but not necessarily for all tests.

(b) Test voltage — Measurements of resistance shall be made by using the test voltages specified in Table V. The test voltage chosen, whether it be the maximum or a lower voltage which would still provide the sensitivity required, shall be applied across the terminals of the resistor. This same voltage shall be used whenever a subsequent resistance measurement is made.

4.7.4.1   **Total Resistance.** Total resistance shall be measured as specified in 4.7.4, between the resistance-element end terminals (terminals 1 and 3 of Figure 1), with the contact arm positioned against a stop. The positioning of the contact arm and terminal shall be the same for all subsequent measurements of the total resistance on the same specimen (see 3.9.1).

Table V   DC Resistance Test Voltage

TOTAL RESISTANCE, NOMINAL	MAXIMUM TEST VOLTAGE
Ohms	Volts
10 to 100	1.0
> 100 to 1,000	3.0
> 1,000 to 10,000	10
> 10,000 to 100,000	30
> 100,000 to 1 meg	100

4.7.4.2   **Absolute Minimun Resistance.** The contact arm shall be positioned at one end of the resistance element, so that a minimum value of resistance shall be measured as specified in 4.7.4 between the contact arm and the corresponding end terminal. The same procedure shall be followed for the other end of the resistance element. Rated current through the resistance element shall not be exceeded during this measurement (see 3.9.2).

4.7.4.3   **End Resistance.** The contact arm shall be positioned at the extreme counterclockwise limit of mechanical travel, and the resistance shall be measured as specified in 4.7.4 between the contact arm and the corresponding end terminal. The contact arm shall then be positioned at the extreme clockwise limit of mechanical travel, and the resistance shall be measured as specified in 4.7.4 between the contact arm and corresponding end terminal. During this test, precaution shall be taken to insure that rated current of the resistance element is not exceeded. Clockwise signifies the direction of rotation of the lead screw when the resistor is viewed from the screwhead. Counterclockwise signifies the direction of rotation of the lead screw when the resistor is viewed from the screwhead (see 3.9.3).

4.7.5 **Immersion (not applicable to L type terminal).** The surface of the resistor shall be cleaned of any foreign matter immediately before immersion. The bath shall consist of tap water at a temperature of $85°{}^{+5°}_{-0°}$ C. The resistor shall be completely submerged in the bath with no part at a depth of less than 1 inch and shaken for a maximum of 5 seconds to remove surface air, and shall remain in the bath for a period of 1 minute ± 5 seconds. During immersion, observation shall be made for any bubbles emanating from the resistor (see 3.10).

4.7.6 **Continuity.** The lead-screw actuator shall be rotated at a uniform rate such that the wiper traverses the effective electrical travel in both directions within 1-1/4 minutes. During rotation, a suitable electrical device shall be connected between the wiper and either end terminal and monitored for smooth and unidirectional change in voltage or resistance. Precaution shall be exercised to prevent excessive current flow in the resistor during the test. There shall be no ohmmeter discontinuity upon reversal of direction of lead screw (see 3.11).

4.7.7 **Actual effective electrical travel.** The actual effective electrical travel, which is the number of turns of the lead screw in which a change in contact arm position gives a measurable change in voltage output, shall be measured by placing the resistor in a suitable device and circuit which will indicate both mechanical position of the lead screw and voltage output (see 3.12).

4.7.8 **Dielectric withstanding voltage (see 3.13).**

4.7.8.1 **At atmospheric pressure.** Resistors shall be tested in accordance with method 301 of Mil-Std-202. The following details shall apply:

(a) Special preparation – Resistors shall be mounted on metal plates of sufficient size to extend beyond the resistor extremities, and in such a manner that measurements can be made between the terminals tied together and any other external metal parts.

(b) Magnitude of test voltage – 900 volts rms.

(c) Nature of potential – From an alternating-current (ac) supply at commercial-line frequency and waveform.

(d) Points of application of test voltage – Between the terminals connected together and all external metal portions of the resistors examined for evidence of arcing and breakdown. At the conclusion of the test, resistors shall be examined for evidence of damage.

4.7.8.2 **At reduced barometric pressure.** Resistors shall be tested in accordance with method 105 of Mil-Std–202. The following details and exceptions shall apply:

(a) Method of mounting – As specified in 4.7.8.1 (a)

(b) Test condition – C.

(c) Period of time at reduced pressure prior to application of potential – 1 minute.

(d) Tests during subjection to reduced pressure — A potential of 350 volts rms from an ac supply at commercial-line frequency and waveform shall be applied for 1 minute.

(e) Points of application — As specified in 4.7.8.1 (d).

(f) Examinations and measurements — As specified in 4.7.8.1 (d).

4.7.9 **Insulation resistance (see 3.14).** Resistors shall be tested in accordance with method 302 of Mil-Std-202. The following details shall apply:

(a) Test condition — A or B, whichever is more practicable.

(b) Special preparation — As specified in 4.7.8.1 (a).

(c) Points of measurement — As specified in 4.7.8.1 (d).

4.7.10 **Torque.**

4.7.10.1 **Operating.** The torque required to move the contact arm on the resistance element shall be determined at approximately 10, 50 and 90 percent of actual effective-electrical travel by the torque-wrench method or by any other method satisfactory to the Government (see 3.15).

4.7.10.2 **Clutch.** The contact arm shall be adjusted to each extreme limit of mechanical travel, and sufficient torque shall be applied to the lead-screw actuator to permit the contact arm to idle for 25 complete mechanical turns of the lead-screw actuator. During idle, a suitable electrical indicating device connected between the contact-arm terminal and the adjacent end terminal shall be observed for electrical continuity. After idle, the lead-screw actuator shall be rotated in the opposite direction and the indicating device observed to determine if the contact arm reversed direction without ohmmeter discontinuity (see 3.15).

4.7.11 **Thermal shock (see 3.16).** Resistors shall be tested in accordance with method 107 of Mil-Std-202. The following details and exceptions shall apply:

(a) Test condition — B (F for Mil-R-39035).

(b) Measurements before cycling — Total resistance and setting stability shall be measured as specified in 4.7.4.1 and 4.7.11.1, respectively.

(c) Measurements after cycling — Setting stability, total resistance, and continuity shall be measured as specified in 4.7.11.1, 4.7.4.1, and 4.7.6, respectively.

(d) Examination after test — Resistors shall be examined for evidence of mechanical damage.

4.7.11.1 **Setting stability.** The contact arm shall be set at approximately 40 percent of the actual effective-electrical travel. An adequate dc test potential shall be applied between the end terminals. The voltage between the end terminals, and the voltage between one end terminal and the contact arm, shall be measured and applied to the following formula:

$$\text{Setting stability in percent} = \frac{E_1 \times 100}{E_2}$$

Where: E_1 = Voltage across one end terminal and the contact-arm terminal.

E_2 = Voltage across the end terminals.

4.7.12 **Solderability (applicable to terminals P, W, X, and Y only) (see 3.17).** Resistors shall be

269

tested in accordance with method 208 of Mil-Std-202. The following detail shall apply:

The 3 pin terminals of each resistor shall be tested.

4.7.13 **Resistance-temperature characteristic (see 3.18).** Resistors shall be tested in accordance with method 304 of Mil-Std-202. The following details shall apply:

(a) Test temperatures — As specified in table VI.

(b) Measurements at end of each period — Total resistance shall be measured as specified in 4.7.4.1 (wiper against stop, measured through end terminals) at the temperature maintained during the period.

Table VI   Resistance—Temperature Characteristics

SEQUENCE	TEMPERATURE
	°C
1	25*
2**	−15
3	−55
4	25*
5**	65
6	150

*This temperature shall be considered the reference temperature for each of the succeeding temperatures.

**Not applicable in quality conformance inspection.

4.7.14 **Moisture resistance (see 3.19).** Resistors shall be tested in accordance with method 106 of Mil-Std-202. The following details and exceptions shall apply:

(a) Mounting — Units shall be mounted or clamped on a stainless-steel panel of sufficient size to extend beyond the resistor extremities and in such a manner as to allow electrical connections to be made to the terminals. Mounting means shall also provide for the insulations resistance test (see 4.7.8.1 (a)).

(b) Initial measurement — Immediately following the initial drying period, total resistance shall be measured as specified in 4.7.4.1.

(c) Polarization and loading voltage — The resistors shall be divided into two equal groups; one group shall be subjected to polarization and the other group to load.
   (1) Polarization — During steps 1 to 6 inclusive, a 100 volt dc potential shall be applied with the positive lead connected to the resistor terminals tied together, and the negative lead connected to the mounting plate.
   (2) Loading voltage — During the first 2 hours of steps 1 and 4, a dc test potential equivalent to 100 percent rated wattage shall be applied to the resistors.

(d) Test procedure — The moisture resistance cycling requirements shall be as follows:

For qualification inspection — 20 cycles for Mil-R-39015 and 39035 items; 10 cycles for Mil-R-22097 and 27208 items.

(e) Final measurements — Upon completion of step 6 of the final cycle, the resistors shall be removed from the chamber and air dried for one-half hour at room ambient conditions. Samples shall not be subjected to forced air drying. The total resistance and insulation resistance shall then be measured (30 to 60 minutes after removal from the humidity chamber), as specified in 4.7.4.1 and 4.7.9, respectively. The subsequent 24-hour conditioning period and measurements do not apply.

(f) Examination after test — Resistors shall be examined for evidence of mechanical damage.

4.7.15 **Shock (specified pulse) (see 3.20).** Resistors shall be tested in accordance with method 213 of Mil-Std-202. The following details and exceptions shall apply:

(a) Mounting — Resistors shall be mounted by their normal mounting means, with their bodies restrained from movement on an appropriate mounting fixture. The mounting fixture shall be constructed in such a manner as to insure that the mounting supports remain in a static condition with reference to the shock-test table. Resistors shall be mounted in relation to the test equipment in such a manner that the stress applied is in the direction which would be considered most detrimental.

(b) Test leads — Test leads used during this test shall be on larger than AWG size 22 stranded wire, so that the influence of the test lead on the resistor will be held to a minimum. The test-lead length shall be no longer than necessary.

(c) Measurements before shock — Total resistance and setting stability shall be measured as specified in 4.7.4.1 and 4.7.11.1, respectively.

(d) Test condition — I.

(e) Measurements during shock — Each resistor shall be monitored to determine electrical discontinuity of the resistance element, and between the contact arm and element, by a method that shall at least be sensitive enough to monitor or register, automatically, any electrical discontinuity of 0.1 millisecond or greater duration.

(f) Measurements after shock — Setting stability and total resistance shall be measured as specified in 4.7.11.1 and 4.7.4.1, respectively.

(g) Examination after shock — Resistors shall be examined for evidence of mechanical damage.

4.7.16 **Vibration, high frequency (see 3.21).** Resistors shall be tested in accordance with method 204 of Mil-Std-202. The following details and exceptions shall apply:

(a) Mounting — As specified in 4.7.15 (a).

(b) Test leads — As specified in 4.7.15 (b).

(c) Measurements before vibration — As specified in 4.7.15 (c).

(d) Test condition — D.

(e) Measurements during vibration — As specified in 4.7.15 (e).

(f) Measurements after vibration — As specified in 4.7.15 (f).

(g) Examination after vibration — Resistors shall be examined for evidence of mechanical damage.

4.7.17 **Salt spray (corrosion) (see 3.22).** Resistors shall be tested in accordance with method 101 of Mil-Std-202. The following details shall apply:

(a) Special mounting — As specified in 4.7.14 (a).

(b) Test condition — A.

(c) Examination after exposure — Resistors shall be examined for corrosion and mechanical operation.

4.7.18 **Resistance to soldering heat (applicable to terminals P, W, X, and Y) (see 3.23).** Resistors shall be tested in accordance with method 210 of Mil-Std-202. The following details shall apply:

(a) Measurement before test — Total resistance shall be measured as specified in 4.7.4.1.

(b) Test condition — A.

(c) Depth of the immersion in the molten solder — To a point within 1/8 inch to 3/16 inch from the resistor body.

(d) Measurement after test — One hour after completion of test, the total resistance shall be measured as specified in 4.7.4.1. Resistors shall be examined for evidence of mechanical damage.

4.7.19 **Low-temperature operation (see 3.24).**

4.7.19.1 **Mounting.** Resistors shall be mounted in such a manner as to allow electrical connections to be made to the terminals.

4.7.19.2 **Procedure.** Total resistance shall be measured as specified in 4.7.4.1. The resistors shall be placed in a chamber at room temperature. The temperature shall be gradually decreased to $-55°$ $^{+0°}_{-5°}$ C within a period of not less than 1-1/2 hours. For quality conformance inspection only, and at the option of the manufacturer, the resistors may be placed in the chamber when the chamber is already at the extreme low temperature. After 1 hour of stabilization at this temperature, setting stability shall be measured as specified in 4.7.11.1. Full rated continuous working voltage shall be applied for 45 minutes. The resistors may be loaded individually or in parallel. Fifteen $^{+5}_{-0}$ minutes after the removal of voltage, setting stability shall be measured as specified in 4.7.11.1. The temperature in the chamber shall be gradually increased to room temperature within a period of not more than 8 hours. The resistors shall be removed from the chamber, and maintained at a temperature of $25°$ $\pm5°$ C for a period of approximately 24 hours. Total resistance shall be measured as specified in 4.7.4.1. Resistors shall then be examined for evidence of mechanical damage.

4.7.20 **Low-temperature storage (see 3.25) (for qualification only).**

4.7.20.1 **Mounting.** Resistors shall be mounted by their normal mounting means and in such a position with respect to the air stream that the mounting offers substantially no obstruction to the flow of air across and around the resistors.

4.7.20.2 **Procedure.** Total resistance shall be measured as specified in 4.7.4.1. Within 1 hour after this measurement, the resistors shall be placed in a cold chamber at a temperature of $-65°$ $\pm2°$ C for a period of 72 $^{+8}_{-0}$ hours. The resistors shall then be removed from the chamber and maintained at a temperature of $25°$ $\pm5°$ C until thermal stabilization is achieved. Total resistance shall then be measured as specified in 4.7.4.1. Resistors shall then be examined for evidence of mechanical damage.

4.7.21 **Mil-R-39015 and 39035 High-temperature exposure (see 3.26).**

4.7.21.1 **Mounting.** Resistors shall be mounted in such a manner as to allow electrical connections to be made to the terminals.

4.7.21.2 **Procedure.** Total resistance and setting stability shall be measured as specified in 4.7.4.1 and 4.7.11.1, respectively. The resistors shall then be exposed to an ambient temperature of $150°$ $^{+5°}_{-0°}$ C for a period of 1,000 ±8 hours. Not less than 2 hours after the end of the exposure period, setting stability and total resistance shall be measured as specified in 4.7.11.1 and 4.7.4.1, respectively. Dielectric withstanding voltage (at atmospheric pressure), and insulation resistance shall be measured as specified in 4.7.8.1 and 4.7.9, respectively. Resistors shall then be examined for evidence of mechanical damage.

4.7.21a **Mil-R-22097 and 27208 High-temperature exposure (see 3.26).**

4.7.21.1a **Mounting.** Resistors shall be mounted in such a manner as to allow electrical connections to be made to the terminals.

4.7.21.2a **Procedure.** Total resistance and setting stability shall be measured as specified in 4.7.4.1 and 4.7.11.1, respectively. The resistors shall then be exposed to an ambient temperature of $125°$ $^{+5°}_{-0°}$ C for characteristic A, and $150°$ $^{+5°}_{-0°}$ C for characteristics C and F for a period of 250 ±8 hours. Not less than 2 hours after the end of the exposure period, setting stability and total resistance shall be measured as specified in 4.7.11.1 and 4.7.4.1, respectively. Torque shall be measured as specified in 4.7.10 except that it shall be determined during the movement of the contact arm from the position for setting stability to the position for total resistance. Dielectric withstanding voltage (at atmospheric pressure), and insulation resistance shall be measured as specified in 4.7.8 and 4.7.9, respectively. Resistors shall then be examined for evidence of mechanical damage.

4.7.22 **Integrity of shaft (see 3.27).**

4.7.22.1 **Mounting.** Resistors shall be mounted on an appropriate mounting fixture with the bodies restrained from movement.

4.7.22.2 **Pull force.** A force of 5 pounds shall be applied along the axis of the operating shaft away from the body of the resistor. The force shall be maintained for a minimum of 1 minute.

4.7.22.3 **Perpendicular force.** A force of 2 pounds shall be applied in a direction perpendicular to the axis of the operating shaft for a minimum of 1 minute.

4.7.22.4 **Examination after test.** Resistors shall be examined for evidence of shaft breakage.

4.7.23 **Rotational life (see 3.28).**

4.7.23.1 **Mounting.** Resistors shall be mounted by their normal mounting means, on a 1/16 inch thick, glass-base, epoxy laminate. The resistors shall be ganged in pairs, and each pair shall be connected in a series, as shown on figure 4, so that a nominally constant current flows through the resistors, irrespective of the contact-arm position during the turning of the lead-screw actuators.

4.7.23.2 **Procedure.** Total resistance shall be measured as specified in 4.7.4.1. A dc potential, equivalent to that required to dissipate rated wattage across the entire resistive element of resistors having the same nominal total resistance as those under test, shall then be applied as shown on Figure 3. The lead-screw actuators shall be continuously cycled through 90 to 100 percent of the actual effective electrical travel, at the rate of 1 cycle for 2 ±1/2 minutes, for a total of 200 cycles. A cycle shall consist of travel through 90 to 100 percent of actual effective electrical travel and return to the starting point. At no time during this test shall the contact arm be allowed to idle at either end of the travel. After rotation, total resistance shall be measured as specified in 4.7.4.1. Resistors shall be examined for evidence of mechanical damage.

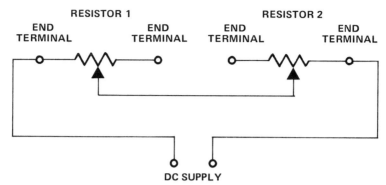

Figure 3   Rotational—Life Test Circuit

4.7.24   Terminal strength (see 3.29).

4.7.24.1 **Pull (applicable to all terminal types).** Resistors shall be tested in accordance with method 211 of Mil-Std-202. The following details and exceptions shall apply:

(a) Test condition — A. Applied force — 2 pounds. Resistor clamped by the resistor body, force applied to each lead individually.

(b) Measurement after test — Resistors shall be examined for evidence of mechanical damage, and tested for electrical continuity.

4.7.24.2 **Push (applicable to all terminals except type L).** Resistors shall be tested in accordance with method 211 of Mil-Std-202. The following details and exceptions shall apply:

(a) Test condition — A, except force shall be applied in the direction toward the resistor body. Applied force — 2 pounds. Resistor clamped by the resistor body, force applied to each terminal individually.

(b) Measurement after test — Resistors shall be examined for evidence of mechanical damage, and tested for electrical continuity.

4.7.24.3 **Bend (applicable to terminal types P, W, X, and Y only).** Resistors shall be firmly clamped and each terminal shall be bent through 90° at a point 1/8 inch from the body of the resistor, with the radius of curvature at the bend approximately 1/32 inch. The pin shall be returned to the original position, bent 90° in the opposite direction, and again returned to the original position. At the conclusion of the test, the resistors shall be examined for evidence of mechanical damage and tested for electrical continuity.

4.7.25   **Mil-R-39015 Life (see 3.30).** Resistors shall be tested in accordance with method 108 of Mil-Std-202. The following details and exceptions shall apply.:

(a) Method of mounting — Resistors shall be mounted by their normal mounting means on a 1/16 inch thick, glass-base, epoxy laminate. The resistors shall be so arranged that the temperature of any one resistor shall not appreciably influence the temperature of any other resistor. There shall be no undue draft over the resistors.

(b) Test temperature and tolerance — 85° ±5°C.

(c) Initial measurement — Resistance shall be measured between the clockwise terminal and the wiper for 50 percent of the sample units, and between the counterclockwise terminal and wiper for the remaining 50 percent of the sample units. The wiper shall be positioned at 95 percent ±2 percent of the total active element in each case. Resistance measurement shall be made after units have been stabilized at 85° ±5° C for at least 8 hours.

(d) Operating conditions — Rated dc or ac working voltage at commercial-line frequency and waveform (see 3.1) shall be applied intermittently to the wipers and active terminal of the resistors, 1-1/2 hours "on" and 1/2 hour "off" for the time duration specified in (f) at the test temperature. Each resistor shall dissipate 95 percent ±2 percent rated wattage, but shall not exceed maximum voltage. Adequate precaution shall be taken to maintain the constant voltage on the resistor.

(e) Test condition — 2,000 hours for qualification inspection with all samples continued on test to 10,000 hours.

(f) Measurements during test:

(1) Qualification inspection — Resistance (see 4.7.4.1) shall be measured at the end of the 1/2 hour "off" periods after $250 \, ^{+48}_{-0}$, $500 \, ^{+48}_{-0}$, $1,000 \, ^{+48}_{-0}$, and $2,000 \, ^{+72}_{-0}$ hours have elapsed. Units continued on test shall be measured at intervals above 2,000 hours in accordance with 4.7.25 (f)(2).

(2) Extended life testing — Resistance (see 4.7.4), shall be measured at the end of the 1/2 hour "off" periods after $250 \, ^{+48}_{-0}$, $500 \, ^{+48}_{-0}$, $1,000 \, ^{+48}_{-0}$, $2,000 \, ^{+72}_{-0}$, and every $2,000 \, ^{+96}_{-0}$ hours thereafter, until the required extended life period (10,000 hours) has elapsed. Measurements shall be made as near as possible to the specified time but may be adjusted so that measurements need not be made during other than normal working days.

(g) Examination after test — Resistors shall be examined for evidence of mechanical damage.

4.7.25 **Mil-R-39035A Life (see 3.30).** Resistors shall be tested in accordance with method 108 of Mil-Std-202. The following details and exceptions shall apply:

(a) Method of mounting — Resistors shall be mounted by their normal mounting means on a 1/16 inch thick, glass-base, epoxy laminate. The resistors shall be so arranged that the temperature of any one resistor shall not appreciably influence the temperature of any other resistor. There shall be no undue draft over the resistors.

(b) Test temperature and tolerance — 85° ±5°C.

(c) Initial measurements — Total resistance shall be measured as specified in 4.7.1.

(d) Operating conditions — Rated dc or ac continuous working voltage at commercial line frequency and waveform shall be applied intermittently to terminals 1 and 3 of the resistor, 1-1/2 hours "on" and 1/2 hour "off," for a total of 10,000 hours at the test temperature. Each resistor shall dissipate rated wattage, but shall not exceed maximum voltage. Adequate precaution shall be taken to maintain constant voltage on the resistor.

(e) Test condition — 2,000 hours for qualification inspection with all samples continued on test to 10,000 hours.

(f) Measurements during test:

(1) Qualification inspection — Resistance (see 4.7.4.1) shall be measured at the end of the 1/2 hour "off" periods after $168 ^{+48}_{-0}$, $504 ^{+48}_{-0}$, $1,008 ^{+48}_{-0}$, and $2,016 ^{+72}_{-0}$ hours have elapsed. Units continued on test shall be measured at intervals above 2,000 hours in accordance with 4.7.25 (f)(2).

(2) Extended life testing — Resistance (see 4.7.4.1) shall be measured at the end of the 1/2 hour "off" periods after $168 ^{+48}_{-0}$, $504 ^{+48}_{-0}$, $1,008 ^{+48}_{-0}$, $2,016 ^{+72}_{-0}$, and every 2,000 hours $^{+96}_{-0}$ hours thereafter, until the required extended life period (10,000 hours) has elapsed. Measurements shall be made as near as possible to the specified time but may be adjusted so that measurements need not be made during other than normal working days.

(g) Examination after test — Resistors shall be examined for evidence of mechanical damage.

4.7.25 **Mil-R-22097 and 27208 Life (see 3.30).** Resistors shall be tested in accordance with method 108 of Mil-Std-202. The following details and exceptions shall apply:

(a) Method of mounting — Resistors shall be mounted by their normal mounting means, on a 1/16 inch thick, glass-base, epoxy laminate. The resistors shall be so arranged that the temperature of any one resistor shall not appreciably influence the temperature of any other resistor. There shall be no undue draft over the resistors.

(b) Test temperatures and tolerances — $70° \pm 5°$C for characteristic A, and $85° \pm 5°$C for characteristics C and F.

(c) Initial measurements — Total resistance and setting stability shall be measured as specified in 4.7.4.1 and 4.7.11.1, respectively.

(d) After the resistors have been stabilized at their respective test temperatures for at least eight hours, the resistance between the end terminals, with the contact arm in the position for setting stability, shall be measured.

(e) Operating conditions — Rated dc or ac continuous working voltage at commercial line frequency and waveform (see 3.1) shall be applied intermittently to the end terminals of the resistors, 1-1/2 hours "on" and 1/2 hour "off," for a total of 1,000 hours at the test temperature. Each resistor shall dissipate rated wattage but shall not exceed maximum voltage. Adequate precaution shall be taken to maintain constant voltage on the resistor.

(f) Test condition — D.

(g) Measurements during test — While the resistors are still in the oven, resistance shall be measured between the end terminals at the end of the 1/2 hour off periods after 50 ±4, 100 ±8, 250 ±12, 750 ±12, and 1,000 ± 12 hours have elapsed and compared to the similar reading taken in (d).

(h) Measurements after test — After the resistors have been removed from the oven and returned to room temperature, setting stability and total resistance shall be measured as specified in 4.7.11.1 and 4.7.4.1, respectively, and compared to the similar readings taken in (c). Dielectric withstanding voltage (at atmospheric pressure) and torque shall be measured as specified in 4.7.8 and 4.7.9, respectively.

(i) Examination after test — Resistors shall be examined for evidence of mechanical damage.

4.7.26 **Resistance to Solvents (see 3.31).** Resistors shall be tested in accordance with method 215 of Mil-Std-202. The following details shall apply:

(a) The marked portion of the resistor body shall be brushed.

(b) The number of sample units shall be as specified in Tables IV and V, as applicable.

(c) Resistors shall be examined for mechanical damage and legibility of markings.

4.7.27 **Setability (see 3.32).** The resistor wiper shall be set at approximately 30 percent, 50 percent, and 75 percent of rotation. A dc voltage of up to 2.5 volts shall be applied across the end terminals, and the wiper shall then be adjusted smoothly without abrupt voltage change at each test point. The setability error shall be within the limits specified.

MIL-R-39035/2D

SUPERSEDING
MIL-R-39035/2C
5 April 1973

MILITARY SPECIFICATION

RESISTORS, VARIABLE, NONWIRE-WOUND

(ADJUSTMENT TYPE, LEAD-SCREW ACTUATED),

ESTABLISHED RELIABILITY

STYLE RJR24

This specification is approved for use by all Departments and Agencies of the Department of Defense.

1. SCOPE

1.1 This specification covers the detail requirements for style RJR24, established reliability, adjustment type, lead-screw actuated, nonwire-wound, variable resistors. This style is available in characteristics C, F, and H.

2. APPLICABLE DOCUMENTS

2.1 The following document of the issue in effect on date of invitation for bids or request for proposal forms a part of this specification to the extent specified herein.

SPECIFICATION

MILITARY

MIL-R-39035 - Resistors, Variable, Nonwire-Wound (Adjustment Type),
Established Reliability, General Specification for.

(Copies of specifications, standards, drawings, and publications required by suppliers in connection with specific procurement functions should be obtained from the procuring activity or as directed by the contracting officer.)

3 REQUIREMENTS

3.1 Requirements. Requirements shall be in accordance with MIL-R-39035, and as specified herein.

3.2 Design and construction. Resistors shall be of the design, construction, and physical dimensions specified on figure.

3.3 Terminals. Characteristics C, F, and H are available with P-, W-, X-, and L- type terminals.

3.4 Power rating. The power ratings shall be 1/2 watt for all characteristics.

3.5 Nominal resistance value and maximum rated ac or dc working voltage. Nominal resistance values and maximum rated ac or dc working voltages shall be as specified in table I.

3.6 Actual effective electrical travel. Actual effective electrical travel shall be 15 turns minimum, and 30 turns maximum.

3.7 Operating torque. Operating torque shall be a maximum of 5 ounce-inches.

3.8 Maximum voltage. Maximum rated ac or dc working voltage shall be 300 volts.

FSC 5905

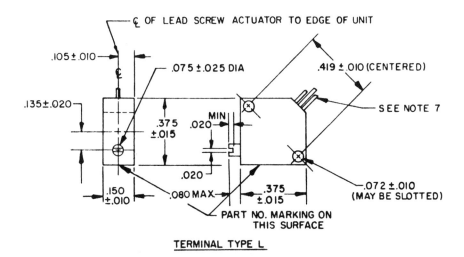

€ OF LEAD SCREW ACTUATOR TO EDGE OF UNIT

.105 ± .010

.075 ± .025 DIA

.419 ± .010 (CENTERED)

.135 ± .020

.375 ± .015

.020 MIN

SEE NOTE 7

.020

.150 ± .010

.080 MAX

.375 ± .015

.072 ± .010 (MAY BE SLOTTED)

PART NO. MARKING ON THIS SURFACE

TERMINAL TYPE L

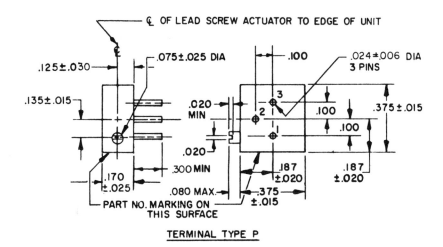

€ OF LEAD SCREW ACTUATOR TO EDGE OF UNIT

.125 ± .030

.075 ± .025 DIA

.100

.024 ± .006 DIA 3 PINS

.135 ± .015

.020 MIN

.100

.375 ± .015

.020

.100

.170 ± .025

.300 MIN

.187 ± .020

.187 ± .020

.080 MAX

.375 ± .015

PART NO. MARKING ON THIS SURFACE

TERMINAL TYPE P

FIGURE 1. Style RJR24 resistors.

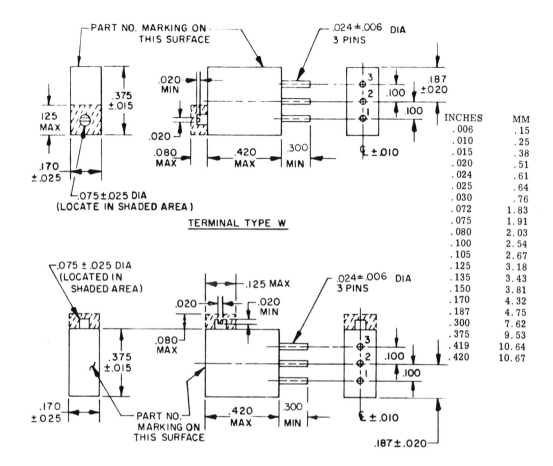

TERMINAL TYPE W

INCHES	MM
.006	.15
.010	.25
.015	.38
.020	.51
.024	.61
.025	.64
.030	.76
.072	1.83
.075	1.91
.080	2.03
.100	2.54
.105	2.67
.125	3.18
.135	3.43
.150	3.81
.170	4.32
.187	4.75
.300	7.62
.375	9.53
.419	10.64
.420	10.67

TERMINAL TYPE X

NOTES:
1. Dimensions are in inches.
2. Metric equivalents (to the nearest .01 mm) are given for general information only and are based upon 1 inch = 25.4 mm.
3. Unless otherwise specified, tolerance is ±.005(.13 mm).
4. The picturization of the styles above are given as representative of the envelope of the item. Slight deviations from the outline shown, which are contained within the envelope, and do not alter the functional aspects of the device are acceptable.
5. The entire slot of the actuating screw must be above the surface of the unit.
6. For types P, W, and X, normal mounting means is by use of pins only.
7. The three leads shall be of stranded wire, AWG size 28 to 30, having a minimum length of 6.00(152.4 mm); they shall be insulated with polytetrafluoroethylene, stripped .250± .062(6.35±1.57 mm) from the end and color coded.

FIGURE 1. Style RJR24 resistors - Continued.

TABLE 1. Nominal resistance value and maximum
rated ac or dc working voltage.

Nominal resistance value	Maximum rated ac or dc working voltage per characteristic
	C, F, and H
Ohms	
10 - - - - - - - - - - - - -	2.23
20 - - - - - - - - - - - -	3.1
50 - - - - - - - - - - - -	5.0
100 - - - - - - - - - - - -	7.0
200 - - - - - - - - - - - -	10.0
500 - - - - - - - - - - - -	15.8
1,000 - - - - - - - - - - - -	22.3
2,000 - - - - - - - - - - - -	31.6
5,000 - - - - - - - - - - - -	50.0
10,000 - - - - - - - - - - - -	70.7
20,000 - - - - - - - - - - - -	100
25,000 - - - - - - - - - - - -	111
50,000 - - - - - - - - - - - -	158
Megohms	
0.10 - - - - - - - - - - - - -	223
0.25 - - - - - - - - - - - - -	300
0.50 - - - - - - - - - - - - -	300
1.0 - - - - - - - - - - - - -	300

4. QUALITY ASSURANCE PROVISIONS

4.1 **Sampling and inspection.** Sampling and inspection shall be in accordance with MIL-R-39035, and as specified herein.

4.2 **Dielectric withstanding voltage.** The magnitude of test voltage shall be 900 volts at atmospheric pressure, and 350 volts at reduced barometric pressure.

5. PREPARATION FOR DELIVERY

5.1 Preparation for delivery shall be in accordance with MIL-R-39035.

6. NOTES

6.1 The notes specified in MIL-R-39035 are applicable to this specification.

6.2 **Weight.** The maximum weight is .00286 pound (1.3 grams).

* 6.3 **MIL-R-22097 substitution data.** Resistors of this specification regardless of their failure rate designation are substitutes for resistors of the same resistance value, tolerance, terminal characteristic and resistance temperature characteristics specified in the inactivated specifications MIL-R-22097/3, and MIL-R-22097/4.

6.4 <u>Changes from previous issue</u>. The margins of this specification are marked with an asterisk to indicate where changes (additions, modifications, corrections, deletions) from the previous issue were made. This was done as a convenience only and the Government assumes no liability whatsoever for any inaccuracies in these notations. Bidders and contractors are cautioned to evaluate the requirements of this document based on the entire content irrespective of the marginal notations and relationship to the last previous issue.

Custodians:
  Army - EL
  Navy - EC
  Air Force - 17

Review activities:
  Army - MI, MU, EL
  Navy - EC, AS, OS
  Air Force - 17, 11, 85
  DSA - ES

User activities:
  Army - ME, AT, AV
  Navy - MC
  Air Force - 19

Preparing activity:
  Air Force - 17

Agent:
  DSA - ES

(Project 5905-0847-1)

Appendix III. Bibliography of Further Reading

BIBLIOGRAPHY FOR FURTHER READING

Reference
No.

1 ADISE, H.
Precision, Film Potentiometer
IRE Wescon Convention Record, 1960

2 BARKAN, Y.
A Realistic Look at Trimming Accuracy
Electromechanical Design Mag. Sept. 1969

3 BUCHBINDER, H.
Precision Potentiometers
Electromechanical Design Mag. Jan. 1964
Potentiometer Circuits
Electromechanical Design Mag. July 1966

4 CARLSTEIN, J.
Potentiometers
Electromechanical Design Mag. July 1969
and Oct. 1969

5 DAVIS, S.
Rotating Components for Automatic Control
Product Engineering Nov. 1953
Potentiometers
Electromechanical Design Mag. April 1969
Todays Precision Potentiometers
Electromechanical Design Mag. Oct. 1970

6 DOERING, J.
Precision Potentiometer Installation
Electromechanical Design Mag. May 1971

7 DYER, S.
*Nonwirewound Pots: Subtle Tradeoffs
Yield Benefits*
Spectrol Electronics Corp.

8 ELECTROMECHANICAL DESIGN
MAGAZINE
Potentiometer Circuits July 1967 and
Jan. 1968
Optimum Potentiometers Oct. 1968
Trimming Potentiometers Sept. 1970
System Designers Handbook 1973-74

9 FIELDS, R.
*Potentiometer — How to Select and Use a
Precision One*
Instruments and Control Systems Mag.
Aug. 1974

10 FRITCHLE, F.
*Theory, Measurement and Reduction of
Precision Potentiometer Linearity Errors*
Helipot Division, Beckman Instruments,
Inc.

11 GILBERT, J.
*Use Taps to Compensate Potentiometer
Loading Errors*
Control Engineer Mag. Aug. 1956

12 GRANCHELLI, R.
Power Rating of Potentiometer Rheostats
Electromechanical Design Mag. Sept. 1971

13 GREEN, A. and K. SCHULZ
*Environmental Effects on Precision
Potentiometers*
IRE Wescon Convention Record 1956

14 HARDMAN, K.
Conductive Plastic Precision Potentiometers
Electromechanical Design Mag. Oct. 1963

15 HENWOOD, R.
*Adjustment Potentiometers — Which Way
to Go*
Electronic Products Mag. May 1972

16 HOGAN, I.
*Electrical Noise in Wirewound
Potentiometers*
Proceedings of Wescon 1952

17 HOUDYSHELL, H.
Precision Potentiometer Life and Reliability
Electronics Component Conference 1955

18 JOHNSTON, S.
*Selecting Trimmer Potentiometers for High
Frequency and Pulse Applications*
Amphenol Controls Div.

19 KARP, H.
Trimmers Take a Turn for the Better
Electronics Mag. Jan. 1972

20 KING, G.
Precision Potentiometers
Electromechanical Design May 1971

21 LERCH, J.
*Guidelines for the Selection of
Potentiometers*
EDN Mag. June 1973

22 LITTON
*Handbook Of High Precision
Potentiometers*
Litton Industries, Inc. 1965

23 MARKITE CORP.
Specifying Output Smoothness,
Tech. Data No. TD-114

24 McDONALD, R. and I. HOGAN
*Accuracy of Potentiometer Linearity
Measurements*
Tele-tech and Electronic Industries Mag.
Aug. 1953

25 NEY CO., J. M.
*Deflection Calculations for Multi-Fingered
Contact Members*
Ney Scope April-May-June 1967

26 ODESS, L.
*Impedance — Sensitivity Nomograph Aids
Design of Trimming Networks*
Electronics Mag. Sept. 1973

27 PUSATERA, E.
*A Designers Guide for Selecting
Adjustment Potentiometers*
Machine Design Mag. March 1967

28 RAGAN, R.
*Power Rating Calculations for Variable
Resistors*
Electronics Mag. July 1973

29 SCATURRO, J.
*Nonlinear Functions from Linear
Potentiometers*
E.E.E. Mag. Oct. 1963

30 SCHNEIDER, S.
An Appraisal of Cermet Potentiometers
Electronic Industries Mag. Dec. 1964
and D. Silverman
*Testing Nonwirewound Potentiometers for
Resolution and Noise*
Electro-Technology Mag. Oct. 1965
and F. Hiraoka and C. Gauldin
*Measurement and Correction of Phase
Shift in Copper Mandrel Precision
Potentiometers*

Helipot Division, Beckman Instruments,
Inc., Tech. Paper 552

31 STAPP, A.
   *Potentiometers — Changing to Meet
   Todays Needs*
   EDN Mag. Feb. 1974

32 TAYLOR, J.
   Nonwirewound Trimmers
   Electronics World Mag. April 1966

33 THOELE, W.
   Which Pot Linearity
   E.E.E. Mag. March 1965

34 T.I.C.
   T.I.C. Potentiometer Handbook
   Bomar/Technology Instrument Corp. 1968

35 TURNER, R.
   ABC's of Resistance and Resistors
   Howard W. Sams and Co., Inc. 1974

36 VARIABLE RESISTIVE COMPONENTS
   INSTITUTE (VRCI)
   Appendix I

37 VON VECK, G.
   Trimming Potentiometers
   Electromechanical Design Mag. Sept. 1970

38 WETZSTEIN, H.
   Precision Potentiometer Circuits
   Electromechanical Design Mag. Jan. 1966

39 WOODS, J. and H. PUGH
   *Contact Resistance and Contact Resistance
   Variation in Thickfilm Trimming
   Potentiometers*
   Proc. of 1971 E.C.C.

40 WORMSER, H.
   Potentiometer Output Smoothness
   Electronic Equipment Engineering Mag.
   Oct. 1962
   Conformity in Precision Potentiometers
   Electromechanical Design Mag. Mar. 1970
   *Recent Studies in Potentiometer Output
   Smoothness*
   Markite Corp. Tech. Data No. TD-111

Appendix IV. Metric Conversion Table

INCHES TO MILLIMETERS

Basis: 1 in. = 25.4 mm (exactly). All values in this table are exact.

inch	millimeter	inch	millimeter	inch	millimeter	inch	millimeter
0.001	0.025 4	0.010	0.254 0	0.100	2.540 0	1.000	25.400 0
0.002	0.050 8	0.020	0.508 0	0.200	5.080 0	2.000	50.800 0
0.003	0.076 2	0.030	0.762 0	0.300	7.620 0	3.000	76.200 0
0.004	0.101 6	0.040	1.016 0	0.400	10.160 0	4.000	101.600 0
0.005	0.127 0	0.050	1.270 0	0.500	12.700 0	5.000	127.000 0
0.006	0.152 4	0.060	1.524 0	0.600	15.240 0	6.000	152.400 0
0.007	0.177 8	0.070	1.778 0	0.700	17.780 0	7.000	177.800 0
0.008	0.203 2	0.080	2.032 0	0.800	20.320 0	8.000	203.200 0
0.009	0.228 6	0.090	2.286 0	0.900	22.860 0	9.000	228.600 0

MILLIMETERS TO INCHES

Basis: 1 mm = 1/25.4 in. (exactly). The inch value in tables below are rounded to the seventh decimal place.

millimeter	inch	millimeter	inch	millimeter	inch	millimeter	inch
0.001	0.000 039 4	0.010	0.000 393 7	0.100	0.003 937 0	1.000	0.039 370 1
0.002	0.000 078 7	0.020	0.000 787 4	0.200	0.007 874 0	2.000	0.078 740 2
0.003	0.000 118 1	0.030	0.001 181 1	0.300	0.011 811 0	3.000	0.118 110 2
0.004	0.000 157 5	0.040	0.001 574 8	0.400	0.015 748 0	4.000	0.157 480 3
0.005	0.000 196 9	0.050	0.001 968 5	0.500	0.019 685 0	5.000	0.196 850 4
0.006	0.000 236 2	0.060	0.002 362 2	0.600	0.023 622 0	6.000	0.236 220 5
0.007	0.000 275 6	0.070	0.002 755 9	0.700	0.027 559 1	7.000	0.275 590 6
0.008	0.000 315 0	0.080	0.003 149 6	0.800	0.031 496 1	8.000	0.314 960 6
0.009	0.000 354 3	0.090	0.003 543 3	0.900	0.035 433 1	9.000	0.354 330 7

millimeter	inch	millimeter	inch	millimeter	inch	millimeter	inch
10.000	0.393 700 8	20.000	0.787 401 6	30.000	1.181 102 4	40.000	1.574 803 1
11.000	0.433 070 9	21.000	0.826 771 7	31.000	1.220 472 4	41.000	1.614 173 2
12.000	0.472 440 9	22.000	0.866 141 7	32.000	1.259 842 5	42.000	1.653 543 3
13.000	0.511 811 0	23.000	0.905 511 8	33.000	1.299 212 6	43.000	1.692 913 4
14.000	0.551 181 1	24.000	0.944 881 9	34.000	1.338 582 7	44.000	1.732 283 5
15.000	0.590 551 2	25.000	0.984 252 0	35.000	1.377 952 8	45.000	1.771 653 5
16.000	0.629 921 3	26.000	1.023 622 0	36.000	1.417 322 8	46.000	1.811 023 6
17.000	0.669 291 3	27.000	1.062 992 1	37.000	1.456 692 9	47.000	1.850 393 7
18.000	0.708 661 4	28.000	1.102 362 2	38.000	1.496 063 0	48.000	1.889 763 8
19.000	0.748 031 5	29.000	1.141 732 3	39.000	1.535 433 1	49.000	1.929 133 9

Appendix V. Abreviations and Mathematical Symbols

Abbreviation	Symbol	Designates
A		Adjustability
	A_R	Adjustability of resistance
	A_V	Adjustability of output voltage
A	A	Attenuator; attenuation
a	a	Ampere
b	b	Straight line, axis intercept
C	C	Capacitor; capacitance
C	°C	Centigrade, degrees
cm	cm	Centimeter
CR	R_C	Contact Resistance
CRV	CRV	Contact Resistance Variation
CCW	CCW	Counter clockwise
CW	CW	Clockwise
D	D	Diode
d	d	Dimension; i.e., diameter or width
$\dfrac{d(1)}{d(2)}$	$\dfrac{d(1)}{d(2)}$	The rate of change of 1 with respect to 2
DIP		Dual in-line package
DOM		Digital ohmmeter
DVM		Digital voltmeter
E	V	Electromotive force; voltage d.c.
e	v	Electromotive force; voltage a.c.
E_I	E_I	Input voltage
ENR	ENR	Equivalent noise resistance
E_O	E_O	Output voltage
ER	R_E	End resistance
$f(\ )$	$f(\ )$	A function of ( )
G	G	Gain
Hz	Hz	Hertz; cycles per second
I	I	Current, d.c.
i	i	Current, a.c.
IC	IC	Integrated circuit
IR	IR	Insulation resistance
K, k	K, k	Kilo, 10^3
K	K	Conformity
k	k	Linearity
KUR		Kills (Underhandedly) Resistors
L	L	Inductor; inductance
1	1	Dimension; length
M	M	Mega; 10^6
M	M	Meter; measuring instrument
m	m	Milli; 10^{-3}
m	m	Slope of a straight line
mm	mm	Millimeter
MR	R_M	Minimum Resistance
N	N	Number of turns
OS	OS	Output smoothness
P	P	Power; electrical
p	p	Power derating factor

PC		Printed circuit
PPM	PPM	Parts per million
Q	Q	Quality factor
Q	Q	Transistor
R	R	Resistor; resistance
R$_L$	R$_L$	Load resistance
RTC	RTC	Resistance temperature characteristic
S	S	Cross sectional area
S	S	Switch
T	T	Temperature
T, t	T, t	Time; time interval
TC	TC	Temperature coefficient
TR	R$_T$	Total resistance
TCVR		Temperature compensated voltage reference
V	V	Volt(s) d.c.
v	v	Volt(s) a.c.
VOM		Volt-ohm-meter
VR		Voltage reference
VRCI		Variable Resistive Components Institute
w	w	Watt(s) of power
X		Reactance
	X$_C$	Reactance, capacitive
	X$_L$	Reactance, inductive
Z	Z	Impedance
$\propto$	$\propto$	Proportional to
∂	∂	Partial differentiation
$<$	$<$	Less than
$>$	$>$	Greater than

Greek alpha-symbols

Symbol	Name	Designates
α	alpha	index point output ratio
β	beta	output ratio
Δ	delta	a change in
δ	delta	output error
η	eta	ratio of compensation to load resistances
θ	theta	travel; wiper position
θ_A		actual travel
θ_I		travel distance to index point
θ_M		mechanical travel
θ_T		theoretical travel
θ_W		actual wiper position
λ	lamda	denotes photocell diode
ρ	rho	resistivity
Ω	omega	ohms; resistive or reactive
ω	omega	frequency

Atomic Symbols

Symbol	Designates
Ag	Silver
Au	Gold
Cr	Chromium
Cu	Copper
Ni	Nickel
Pt	Platinum

INDEX

INDEX